Mixed-Signal and DSP Design Techniques

DISCARDED

Mixed-Signal and DSP Design Techniques

by the technical staff of Analog Devices

edited by Walt Kester

A Volume in the Analog Devices Series

An imprint of Elsevier Science

Amsterdam   Boston   London   New York   Oxford   Paris
San Diego   San Francisco   Singapore   Sydney   Tokyo

Newnes is an imprint of Elsevier Science.

Copyright © 2003 by Analog Devices, Inc. All rights reserved.

No part of this publication may be reproduced, stored in a retrieval system, or transmitted in any form or by any means, electronic, mechanical, photocopying, recording, or otherwise, without the prior written permission of the publisher.

 Recognizing the importance of preserving what has been written, Elsevier Science prints its books on acid-free paper whenever possible.

Library of Congress Cataloging-in-Publication Data

ISBN: 0-75067-611-6

British Library Cataloguing-in-Publication Data

A catalogue record for this book is available from the British Library.

The publisher offers special discounts on bulk orders of this book.

For information, please contact:

Manager of Special Sales
Elsevier Science
200 Wheeler Road
Burlington, MA 01803
Tel: 781-313-4700
Fax: 781-313-4882

For information on all Newnes publications available, contact our World Wide Web home page at: http://www.newnespress.com

10 9 8 7 6 5 4 3 2 1

Printed in the United States of America

Contents

Section 1: Introduction 3
Origins of Real-World Signals and Their Units of Measurement 3
Reasons for Processing Real-World Signals 4
Generation of Real-World Signals 5
Methods and Technologies Available for Processing Real-World Signals 6
Analog Versus Digital Signal Processing 6
A Practical Example 7
References 11

Section 2: Sampled Data Systems 15
Introduction 15
Discrete Time Sampling of Analog Signals 16
ADC and DAC Static Transfer Functions and DC Errors 21
AC Errors in Data Converters 28
DAC Dynamic Performance 49
References 56

Section 3: ADCs for DSP Applications 61
Successive-Approximation ADCs 62
Sigma-Delta (S-D) ADCs 69
James M. Bryant 69
Flash Converters 81
Subranging (Pipelined) ADCs 83
Bit-Per-Stage (Serial or Ripple) ADCs 87
References 93

Section 4: DACs for DSP Applications 99
DAC Structures 99
Low Distortion DAC Architectures 101
DAC Logic 105
Interpolating DACs 107
Sigma-Delta DACs 109
Direct Digital Synthesis (DDS) 110
References 115

Section 5: Fast Fourier Transforms ... 119
- The Discrete Fourier Transform ... 119
- The Fast Fourier Transform ... 127
- FFT Hardware Implementation and Benchmarks ... 135
- DSP Requirements for Real-Time FFT Applications ... 136
- Spectral Leakage and Windowing ... 139
- References ... 143

Section 6: Digital Filters ... 147
- Finite Impulse Response (FIR) Filters ... 151
- FIR Filter Implementation in DSP Hardware Using Circular Buffering ... 156
- Designing FIR Filters ... 159
- Infinite Impulse Response (IIR) Filters ... 170
- IIR Filter Design Techniques ... 173
- Multirate Filters ... 177
- Adaptive Filters ... 181
- References ... 186

Section 7: DSP Hardware ... 191
- Microcontrollers, Microprocessors, and Digital Signal Processors (DSPs) ... 191
- DSP Requirements ... 193
- ADSP-21xx 16-Bit Fixed-Point DSP Core ... 196
- Fixed-Point Versus Floating-Point ... 212
- ADI SHARC Floating-Point DSPs ... 215
- ADSP-2116x Single-Instruction, Multiple-Data (SIMD) Core Architecture ... 220
- TigerSHARC: The ADSP-TS001 Static Superscalar DSP ... 225
- DSP Evaluation and CROSSCORE™ Development Tools ... 234
- References ... 244

Section 8: Interfacing to DSPs ... 247
- Introduction ... 247
- Parallel Interfacing to DSP Processors: Reading Data from Memory-Mapped Peripheral ADCS ... 247
- Parallel Interfacing to DSP Processors: Writing Data to Memory-Mapped DACS ... 253
- Serial Interfacing to DSP Processors ... 258
- Interfacing I/O Ports, Analog Front Ends, and Codecs to DSPs ... 265
- High Speed Interfacing ... 268
- DSP System Interface ... 269
- References ... 271

Section 9: DSP Applications 275

- High Performance Modems for Plain Old Telephone Service (POTS) 275
- Remote Access Server (RAS) Modems 281
- ADSL (Asymmetric Digital Subscriber Line) 285
- Digital Cellular Telephones 290
- GSM Handset Using SoftFone Baseband Processor and
  Othello Radio 295
- Analog Cellular Base Stations 301
- Digital Cellular Base Stations 302
- Motor Control 306
- Codecs and DSPs in Voice-Band and Audio Applications 310
- A Sigma-Delta ADC with Programmable Digital Filter 313
- Summary 315
- References 316

Section 10: Hardware Design Techniques 321

- Low Voltage Interfaces 321
- Grounding in Mixed-Signal Systems 335
- Digital Isolation Techniques 355
- Power Supply Noise Reduction and Filtering 359
- Dealing with High Speed Logic 378

Index 389

Acknowledgments

Thanks are due the many technical staff members of Analog Devices in Engineering and Marketing who provided invaluable inputs during this project. Particular credit is due the individual authors whose names appear at the beginning of their material.

Special thanks go to Wes Freeman, Ed Grokulsky, Bill Chestnut, Dan King, Greg Geerling, Ken Waurin, Steve Cox, and Colin Duggan for reviewing the material for content and accuracy.

Judith Douville compiled the index.

Section 1
Introduction

Section 1
Introduction
Walt Kester

Origins of Real-World Signals and Their Units of Measurement

In this book, we will primarily be dealing with the processing of *real-world* signals using both analog and digital techniques. Before starting, however, let's look at a few key concepts and definitions required to lay the groundwork for things to come.

Webster's *New Collegiate Dictionary* defines a *signal* as "a detectable (or measurable) physical quantity or impulse (as voltage, current, or magnetic field strength) by which messages or information can be transmitted." Key to this definition are the words: *detectable, physical quantity,* and *information.*

- Signal Characteristics
  - Signals Are Physical Quantities
  - Signals Are Measurable
  - Signals Contain Information
  - All Signals Are Analog
- Units of Measurement
  - Temperature: °C
  - Pressure: Newtons/m^2
  - Mass: kg
  - Voltage: Volts
  - Current: Amps
  - Power: Watts

Figure 1-1: Signal Characteristics

By their very nature, signals are analog, whether dc, ac, digital levels, or pulses. It is customary, however, to differentiate between *analog* and *digital* signals in the following manner: Analog (or real-world) variables in nature include all measurable physical quantities. In this book, *analog* signals are generally limited to electrical variables, their rates of change, and their associated energy or power levels. Sensors are used to convert other physical quantities such as temperature or pressure to electrical signals. The entire subject of signal conditioning deals with preparing real-world signals for processing, and includes such topics as sensors (temperature and pressure, for example), isolation amplifiers, and instrumentation amplifiers. (Reference 1.)

Some signals result in response to other signals. A good example is the returned signal from a radar or ultrasound imaging system, both of which result from a known transmitted signal.

On the other hand, there is another classification of signals, called *digital*, where the actual signal has been conditioned and formatted into a digit. These digital signals may or may not be related to real-world analog variables. Examples include the data transmitted over local area networks (LANs) or other high speed networks.

In the specific case of digital signal processing (DSP), the analog signal is converted into binary form by a device known as an analog-to-digital converter (ADC). The output of the ADC is a binary representation of the analog signal and is manipulated arithmetically by the digital signal processor. After processing, the information obtained from the signal may be converted back into analog form using a digital-to-analog converter (DAC).

Another key concept embodied in the definition of *signal* is that there is some kind of *information* contained in the signal. This leads us to the key reason for processing real-world analog signals: the *extraction of information*.

Reasons for Processing Real-World Signals

The primary reason for processing real-world signals is to extract information from them. This information normally exists in the form of signal amplitude (absolute or relative), frequency or spectral content, phase, or timing relationships with respect to other signals. Once the desired information is extracted from the signal, it may be used in a number of ways.

In some cases, it may be desirable to reformat the information contained in a signal. This would be the case in the transmission of a voice signal over a frequency division multiple access (FDMA) telephone system. In this case, analog techniques are used to "stack" voice channels in the frequency spectrum for transmission via microwave relay, coaxial cable, or fiber. In the case of a digital transmission link, the analog voice information is first converted into digital using an ADC. The digital information representing the individual voice channels is multiplexed in time (time division multiple access, or TDMA) and transmitted over a serial digital transmission link (as in the T-carrier system).

Another requirement for signal processing is to *compress* the frequency content of the signal (without losing significant information), then format and transmit the information at lower data rates, thereby achieving a reduction in required channel bandwidth. High speed modems and adaptive pulse code modulation systems (ADPCM) make extensive use of data reduction algorithms, as do digital mobile radio systems, MPEG recording and playback, and high definition television (HDTV).

Introduction

Industrial data acquisition and control systems make use of information extracted from sensors to develop appropriate feedback signals which in turn control the process itself. Note that these systems require both ADCs and DACs as well as sensors, signal conditioners, and the DSP (or microcontroller). Analog Devices offers a family of MicroConverters™ that includes precision analog conditioning circuitry, ADCs, DACs, microcontroller, and FLASH memory all on a single chip.

In some cases, the signal containing the information is buried in noise, and the primary objective is signal recovery. Techniques such as filtering, autocorrelation, and convolution are often used to accomplish this task in both the analog and digital domains.

- Extract Information about the Signal (Amplitude, Phase, Frequency, Spectral Content, Timing Relationships)
- Reformat the Signal (FDMA, TDMA, CDMA Telephony)
- Compress Data (Modems, Cellular Telephone, HDTV, MPEG)
- Generate Feedback Control Signal (Industrial Process Control)
- Extract Signal from Noise (Filtering, Autocorrelation, Convolution)
- Capture and Store Signal in Digital Format for Analysis (FFT Techniques)

Figure 1-2: Reasons for Signal Processing

Generation of Real-World Signals

In most of the above examples (the ones requiring DSP techniques), both ADCs and DACs are required. In some cases, however, only DACs are required where real-world analog signals may be generated directly using DSP and DACs. Video raster scan display systems are a good example. The digitally generated signal drives a video or RAMDAC. Another example is artificially synthesized music and speech. In reality, however, the real-world analog signals generated using purely digital techniques do rely on information previously derived from the real-world equivalent analog signals. In display systems, the data from the display must convey the appropriate information to the operator. In synthesized audio systems, the statistical properties of the sounds being generated have been previously derived using extensive DSP analysis of the entire signal chain, including sound source, microphone, preamp, and ADC.

Methods and Technologies Available for Processing Real-World Signals

Signals may be processed using analog techniques (analog signal processing, or ASP), digital techniques (digital signal processing, or DSP), or a combination of analog and digital techniques (mixed-signal processing, or MSP). In some cases, the choice of techniques is clear; in others, there is no clear-cut choice, and second-order considerations may be used to make the final decision.

With respect to DSP, the factor that distinguishes it from traditional computer analysis of data is its speed and efficiency in performing sophisticated digital processing functions such as filtering, FFT analysis, and data compression in real time.

The term *mixed-signal processing* implies that *both* analog and digital processing is done as part of the system. The system may be implemented in the form of a printed circuit board, hybrid microcircuit, or a single integrated circuit chip. In the context of this broad definition, ADCs and DACs are considered to be mixed-signal processors, since both analog and digital functions are implemented in each. Recent advances in very large scale integration (VLSI) processing technology allow complex digital processing as well as analog processing to be performed on the same chip. The very nature of DSP itself implies that these functions can be performed in *real time*.

Analog Versus Digital Signal Processing

Today's engineer faces a challenge in selecting the proper mix of analog and digital techniques to solve the signal processing task at hand. It is impossible to process real-world analog signals using purely digital techniques, since all sensors, including microphones, thermocouples, strain gages, piezoelectric crystals, and disk drive heads are analog sensors. Therefore, some sort of signal conditioning circuitry is required in order to prepare the sensor output for further signal processing, whether it be analog or digital. Signal conditioning circuits are, in reality, analog signal processors, performing such functions as multiplication (gain), isolation (instrumentation amplifiers and isolation amplifiers), detection in the presence of noise (high common-mode instrumentation amplifiers, line drivers, and line receivers), dynamic range compression (log amps, LOGDACs, and programmable gain amplifiers), and filtering (both passive and active).

Several methods of accomplishing signal processing are shown in Figure 1-3. The top portion of the figure shows the purely analog approach. The latter parts of the figure show the DSP approach. Note that once the decision has been made to use DSP techniques, the next decision must be where to place the ADC in the signal path.

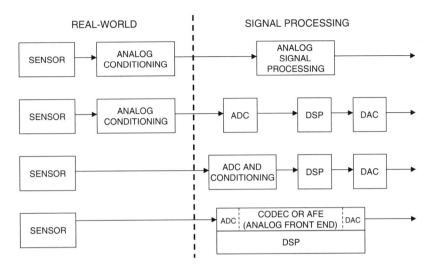

Figure 1-3: Analog and Digital Signal Processing Options

In general, as the ADC is moved closer to the actual sensor, more of the analog signal conditioning burden is now placed on the ADC. The added ADC complexity may take the form of increased sampling rate, wider dynamic range, higher resolution, input noise rejection, input filtering, programmable gain amplifiers (PGAs), and on-chip voltage references, all of which add functionality and simplify the system. With today's high resolution/high sampling rate data converter technology, significant progress has been made in integrating more and more of the conditioning circuitry within the ADC/DAC itself. In the measurement area, for instance, 24-bit ADCs are available with built-in programmable gain amplifiers (PGAs) that allow full-scale bridge signals of 10 mV to be digitized directly with no further conditioning (e.g., AD773x series). At voice-band and audio frequencies, complete coder/decoders (codecs or analog front ends) are available with sufficient on-chip analog circuitry to minimize the requirements for external conditioning components (AD1819B and AD73322). At video speeds, analog front ends are also available for such applications as CCD image processing and others (e.g., AD9814, AD9816, and the AD984x series).

A Practical Example

As a practical example of the power of DSP, consider the comparison between an analog and a digital low-pass filter, each with a cutoff frequency of 1 kHz. The digital filter is implemented in a typical sampled data system shown in Figure 1-4. Note that there are several implicit requirements in the diagram. First, it is assumed that an ADC/DAC combination is available with sufficient sampling frequency, resolution, and dynamic range to accurately process the signal. Second, the DSP must be fast enough to complete all its calculations within the sampling interval, $1/f_s$. Third, analog filters are still required at the ADC input and DAC output for antialiasing and anti-imaging, but the performance demands are not as great. Assuming these conditions have been met, the following offers a comparison between the digital and analog filters.

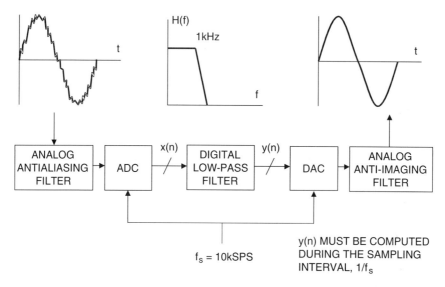

Figure 1-4: Digital Filter

The required cutoff frequency of both filters is 1 kHz. The analog filter is realized as a 6-pole Chebyshev Type 1 filter (ripple in pass band, no ripple in stop band), and the response is shown in Figure 1-5. In practice, this filter would probably be realized using three 2-pole stages, each of which requires an op amp, and several resistors and capacitors. Modern filter design CAD packages make the 6-pole design relatively straightforward, but maintaining the 0.5 dB ripple specification requires accurate component selection and matching.

On the other hand, the 129-tap digital FIR filter shown has only 0.002 dB pass band ripple, linear phase, and a much sharper roll-off. In fact, it could not be realized using analog techniques. Another obvious advantage is that the digital filter requires no component matching, and it is not sensitive to drift since the clock frequencies are crystal controlled. The 129-tap filter requires 129 multiply-accumulates (MAC) in order to compute an output sample. This processing must be completed within the sampling interval, $1/f_s$, in order to maintain real-time operation. In this example, the sampling frequency is 10 kSPS; therefore 100 µs is available for processing, assuming no significant additional overhead requirement. The ADSP-21xx family of DSPs can complete the entire multiply-accumulate process (and other functions necessary

Introduction

for the filter) in a single instruction cycle. Therefore, a 129-tap filter requires that the instruction rate be greater than 129/100 μs = 1.3 million instructions per second (MIPS). DSPs are available with instruction rates much greater than this, so the DSP certainly is not the limiting factor in this application. The ADSP-218x 16-bit fixed-point series offers instruction rates up to 75 MIPS.

The assembly language code to implement the filter on the ADSP-21xx family of DSPs is shown in Figure 1-6. Note that the actual lines of operating code have been marked with arrows; the rest are comments.

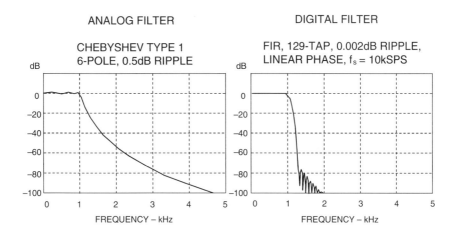

Figure 1-5: Analog Versus Digital Filter Frequency Response Comparison

In a practical application, there are certainly many other factors to consider when evaluating analog versus digital filters, or analog versus digital signal processing in general. Most modern signal processing systems use a combination of analog and digital techniques in order to accomplish the desired function and take advantage of the best of both the analog and the digital worlds.

Section One

```
            .MODULE         fir_sub;
            {               FIR Filter Subroutine
                            Calling Parameters
                                    I0 --> Oldest input data value in delay line
                                    I4 --> Beginning of filter coefficient table
                                    L0 = Filter length (N)
                                    L4 = Filter length (N)
                                    M1,M5 = 1
                                    CNTR = Filter length - 1 (N-1)
                            Return Values
                                    MR1 = Sum of products (rounded and saturated)
                                    I0 --> Oldest input data value in delay line
                                    I4 --> Beginning of filter coefficient table
                            Altered Registers
                                    MX0,MY0,MR
                            Computation Time
                                    (N - 1) + 6 cycles = N + 5 cycles
                            All coefficients are assumed to be in 1.15 format. }
            .ENTRY          fir;
  ──▶       fir:            MR=0, MX0=DM(I0,M1), MY0=PM(I4,M5)
  ──▶                       CNTR = N-1;
  ──▶                       DO convolution UNTIL CE;
  ──▶       convolution:      MR=MR+MX0*MY0(SS), MX0=DM(I0,M1), MY0=PM(I4,M5);
  ──▶                       MR=MR+MX0*MY0(RND);
  ──▶                       IF MV SAT MR;
  ──▶                       RTS;
            .ENDMOD;
```

Figure 1-6: ADSP-21xx FIR Filter Assembly Code (Single Precision)

- **Digital Signal Processing**
  - ◆ ADC/DAC Sampling Frequency Limits Signal Bandwidth
    - (Don't forget Nyquist)
  - ◆ ADC/DAC Resolution/Performance Limits Signal Dynamic Range
  - ◆ DSP Processor Speed Limits Amount of Digital Processing Available, Because:
    - All DSP Computations Must Be Completed During the Sampling Interval, $1/f_s$, for Real-Time Operation
- **Don't Forget Analog Signal Processing**
  - ◆ High Frequency/RF Filtering, Modulation, Demodulation
  - ◆ Analog Antialiasing and Reconstruction Filters with ADCs and DACs
  - ◆ Where Common Sense and Economics Dictate

Figure 1-7: Real-Time Signal Processing

References

1. **Practical Design Techniques for Sensor Signal Conditioning**, Analog Devices, 1998.
2. Daniel H. Sheingold, Editor, **Transducer Interfacing Handbook**, Analog Devices, Inc., 1972.
3. Richard J. Higgins, **Digital Signal Processing in VLSI**, Prentice-Hall, 1990.

Section 2
Sampled Data Systems

- Discrete Time Sampling of Analog Signals
- ADC and DAC Static Transfer Functions and DC Errors
- AC Errors in Data Converters
- DAC Dynamic Performance

Section 2
Sampled Data Systems
Walt Kester, James Bryant

Introduction

A block diagram of a typical sampled data DSP system is shown in Figure 2-1. Prior to the actual analog-to-digital conversion, the analog signal usually passes through some sort of signal conditioning circuitry, which performs such functions as amplification, attenuation, and filtering. The low-pass/band-pass filter is required to remove unwanted signals outside the bandwidth of interest and prevent aliasing.

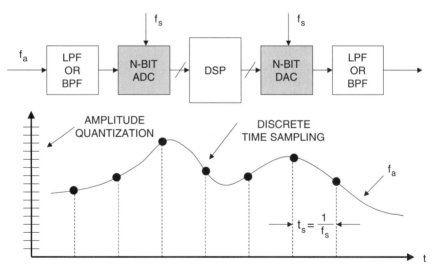

Figure 2-1: Fundamental Sampled Data System

The system shown in Figure 2-1 is a real-time system, i.e., the signal to the ADC is continuously sampled at a rate equal to f_s, and the ADC presents a new sample to the DSP at this rate. In order to maintain real-time operation, the DSP must perform all its required computation within the sampling interval, $1/f_s$, and present an output sample to the DAC before arrival of the next sample from the ADC. An example of a typical DSP function would be a digital filter.

In the case of FFT analysis, a block of data is first transferred to the DSP memory. The FFT is calculated at the same time a new block of data is transferred into the memory, in order to maintain real-time operation. The DSP must calculate the FFT during the data transfer interval so it will be ready to process the next block of data.

Section Two

Note that the DAC is required only if the DSP data must be converted back into an analog signal (as would be the case in a voice-band or audio application, for example). There are many applications where the signal remains entirely in digital format after the initial A/D conversion. Similarly, there are applications where the DSP is solely responsible for generating the signal to the DAC, such as in CD player electronics. If a DAC is used, it must be followed by an analog anti-imaging filter to remove the image frequencies.

There are two key concepts involved in the actual analog-to-digital and digital-to-analog conversion process: *discrete time sampling* and *finite amplitude resolution due to quantization*. An understanding of these concepts is vital to DSP applications.

Discrete Time Sampling of Analog Signals

The concepts of *discrete time sampling* and *quantization* of an analog signal are shown in Figure 2-1. The continuous analog data must is sampled at discrete intervals, $t_s = 1/f_s$, which must be carefully chosen to ensure an accurate representation of the original analog signal. It is clear that the more samples taken (faster sampling rates), the more accurate the digital representation, but if fewer samples are taken (lower sampling rates), a point is reached where critical information about the signal is actually lost. This leads us to the statement of Nyquist's criteria given in Figure 2-2.

- A Signal With a *Bandwidth* f_a Must Be Sampled at a Rate $f_s > 2 f_a$ or Information About the Signal Will Be Lost

- Aliasing Occurs Whenever $f_s < 2 f_a$

- The Concept of Aliasing is Widely Used in Communications Applications Such as Direct IF-to-Digital Conversion

Figure 2-2: Nyquist's Criteria

Simply stated, the Nyquist Criteria requires that the sampling frequency be at least twice the signal bandwidth, or information about the signal will be lost. If the sampling frequency is less than twice the analog signal bandwidth, a phenomenon known as aliasing will occur.

In order to understand the implications of *aliasing* in both the time and frequency domain, first consider case of a time domain representation of a single tone sine wave sampled as shown in Figure 2-3. In this example, the sampling frequency f_s is only slightly more than the analog input frequency f_a, and the Nyquist criteria is violated. Notice that the pattern of the actual samples produces an *aliased* sine wave at a lower frequency equal to $f_s - f_a$.

Sampled Data Systems

The corresponding frequency domain representation of this scenario is shown in Figure 2-4B. Now consider the case of a single frequency sine wave of frequency f_a sampled at a frequency f_s by an ideal impulse sampler (see Figure 2-4A). Also assume that $f_s > 2f_a$ as shown. The frequency-domain output of the sampler shows *aliases* or *images* of the original signal around every multiple of f_s, i.e., at frequencies equal to $|\pm Kf_s \pm f_a|$, $K = 1, 2, 3, 4, \ldots$.

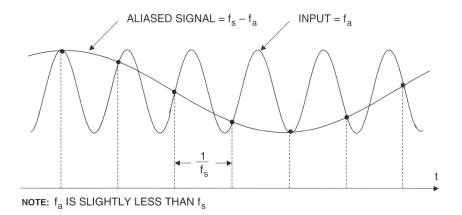

Figure 2-3: Aliasing in the Time Domain

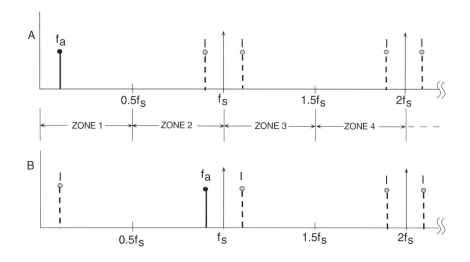

Figure 2-4: Analog Signal f_a Sampled @ f_s using Ideal Sampler Has Images (Aliases) at $|\pm Kf_s \pm f_a|$, $K = 1, 2, 3, \ldots$

The *Nyquist* bandwidth is defined to be the frequency spectrum from dc to $f_s/2$. The frequency spectrum is divided into an infinite number of *Nyquist zones*, each having a width equal to $0.5\,f_s$ as shown. In practice, the ideal sampler is replaced by an ADC followed by an FFT processor. The FFT processor only provides an output from dc to $f_s/2$, i.e., the signals or aliases that appear in the first Nyquist zone.

Now consider the case of a signal that is outside the first Nyquist zone (Figure 2-4B). The signal frequency is only slightly less than the sampling frequency, corresponding to the condition shown in the time domain representation in Figure 2-3. Notice that even though the signal is outside the first Nyquist zone, its image (or *alias*), $f_s - f_a$, falls inside. Returning to Figure 2-4A, it is clear that if an unwanted signal appears at any of the image frequencies of f_a, it will also occur at f_a, thereby producing a spurious frequency component in the first Nyquist zone.

This is similar to the analog mixing process and implies that some filtering ahead of the sampler (or ADC) is required to remove frequency components that are outside the Nyquist bandwidth, but whose aliased components fall inside it. The filter performance will depend on how close the out-of-band signal is to $f_s/2$ and the amount of attenuation required.

Baseband Antialiasing Filters

Baseband sampling implies that the signal to be sampled lies in the first Nyquist zone. It is important to note that with no input filtering at the input of the ideal sampler, *any frequency component (either signal or noise) that falls outside the Nyquist bandwidth in any Nyquist zone will be aliased back into the first Nyquist zone.* For this reason, an antialiasing filter is used in almost all sampling ADC applications to remove these unwanted signals.

Properly specifying the antialiasing filter is important. The first step is to know the characteristics of the signal being sampled. Assume that the highest frequency of interest is f_a. The antialiasing filter passes signals from dc to f_a while attenuating signals above f_a.

Assume that the corner frequency of the filter is chosen to be equal to f_a. The effect of the finite transition from minimum to maximum attenuation on system dynamic range is illustrated in Figure 2-5A.

Assume that the input signal has full-scale components well above the maximum frequency of interest, f_a. The diagram shows how full-scale frequency components above $f_s - f_a$ are aliased back into the bandwidth dc to f_a. These aliased components are indistinguishable from actual signals and therefore limit the dynamic range to the value on the diagram, which is shown as *DR*.

Some texts recommend specifying the antialiasing filter with respect to the Nyquist frequency, $f_s/2$, but this assumes that the signal bandwidth of interest extends from dc to $f_s/2$, which is rarely the case. In the example shown in Figure 2-5A, the aliased components between f_a and $f_s/2$ are not of interest and do not limit the dynamic range.

Sampled Data Systems

The antialiasing filter transition band is therefore determined by the corner frequency f_a, the stop-band frequency $f_s - f_a$, and the desired stop-band attenuation, DR. The required system dynamic range is chosen based on the requirement for signal fidelity.

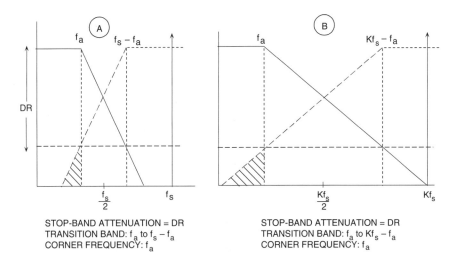

Figure 2-5: Oversampling Relaxes Requirements on Baseband Antialiasing Filter

Filters become more complex as the transition band becomes sharper, all other things being equal. For instance, a Butterworth filter gives 6 dB attenuation per octave for each filter pole. Achieving 60 dB attenuation in a transition region between 1 MHz and 2 MHz (1 octave) requires a minimum of 10 poles, not a trivial filter, and definitely a design challenge.

Therefore, other filter types are generally more suited to high speed applications where the requirement is for a sharp transition band and in-band flatness coupled with linear phase response. Elliptic filters meet these criteria and are a popular choice. There are a number of companies that specialize in supplying custom analog filters. TTE is an example of such a company (Reference 1).

From this discussion, we can see how the sharpness of the antialiasing transition band can be traded off against the ADC sampling frequency. Choosing a higher sampling rate (oversampling) reduces the requirement on transition band sharpness (hence, the filter complexity) at the expense of using a faster ADC and processing data at a faster rate. This is illustrated in Figure 2-5B, which shows the effects of increasing the sampling frequency by a factor of K, while maintaining the same analog corner frequency, f_a, and the same dynamic range, DR, requirement. The wider transition band (f_a to $Kf_s - f_a$) makes this filter easier to design than for the case of Figure 2-5A.

Section Two

The antialiasing filter design process is started by choosing an initial sampling rate of 2.5 to 4 times f_a. Determine the filter specifications based on the required dynamic range and see if such a filter is realizable within the constraints of the system cost and performance. If not, consider a higher sampling rate, which may require using a faster ADC. It should be mentioned that sigma-delta ADCs are inherently oversampling converters, and the resulting relaxation in the analog antialiasing filter requirements is therefore an added benefit of this architecture.

The antialiasing filter requirements can also be relaxed somewhat if it is certain that there will never be a full-scale signal at the stop-band frequency $f_s - f_a$. In many applications, it is improbable that full-scale signals will occur at this frequency. If the maximum signal at the frequency $f_s - f_a$ will never exceed XdB below full-scale, then the filter stop-band attenuation requirement is reduced by that same amount. The new requirement for stop-band attenuation at $f_s - f_a$ based on this knowledge of the signal is now only DR − X dB. When making this type of assumption, be careful to treat any noise signals that may occur above the maximum signal frequency f_a as unwanted signals that will also alias back into the signal bandwidth.

Undersampling (Harmonic Sampling, Band-Pass Sampling, IF Sampling, Direct IF to Digital Conversion)

Thus far we have considered the case of baseband sampling; i.e., all the signals of interest lie within the first Nyquist zone. Figure 2-6A shows such a case, where the band of sampled signals is limited to the first Nyquist zone, and images of the original band of frequencies appear in each of the other Nyquist zones.

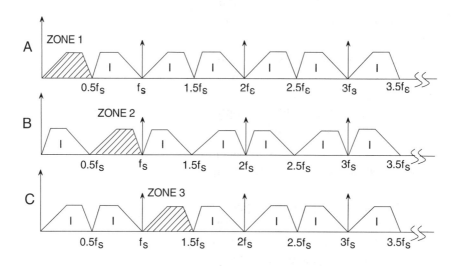

Figure 2-6: Undersampling

Sampled Data Systems

Consider the case shown in Figure 2-6B, where the sampled signal band lies entirely within the second Nyquist zone. The process of sampling a signal outside the first Nyquist zone is often referred to as *undersampling*, or *harmonic sampling*. Note that the first Nyquist zone image contains all the information in the original signal, with the exception of its original location (the order of the frequency components within the spectrum is reversed, but this is easily corrected by reordering the output of the FFT).

Figure 2-6C shows the sampled signal restricted to the third Nyquist zone. Note that the first Nyquist zone image has no frequency reversal. In fact, the sampled signal frequencies may lie in *any* unique Nyquist zone, and the first Nyquist zone image is still an accurate representation (with the exception of the frequency reversal that occurs when the signals are located in even Nyquist zones). At this point we can clearly restate the Nyquist criteria:

A signal must be sampled at a rate equal to or greater than twice its **bandwidth** *in order to preserve all the signal information.*

Notice that there is no mention of the absolute *location* of the band of sampled signals within the frequency spectrum relative to the sampling frequency. The only constraint is that the band of sampled signals be restricted to a *single* Nyquist zone, i.e., the signals must not overlap any multiple of $f_s/2$ (this, in fact, is the primary function of the antialiasing filter).

Sampling signals above the first Nyquist zone has become popular in communications because the process is equivalent to analog demodulation. It is becoming common practice to sample IF signals directly and then use digital techniques to process the signal, thereby eliminating the need for the IF demodulator. Clearly, however, as the IF frequencies become higher, the dynamic performance requirements on the ADC become more critical. The ADC input bandwidth and distortion performance must be adequate at the IF frequency, rather than only baseband. This presents a problem for most ADCs designed to process signals in the first Nyquist zone; therefore, an ADC suitable for undersampling applications must maintain dynamic performance into the higher order Nyquist zones.

ADC and DAC Static Transfer Functions and DC Errors

The most important thing to remember about both DACs and ADCs is that either the input or output is digital, and therefore the signal is quantized. That is, an N-bit word represents one of 2^N possible states, and therefore an N-bit DAC (with a fixed reference) can have only 2^N possible analog outputs, and an N-bit ADC can have only 2^N possible digital outputs. The analog signals will generally be voltages or currents.

The resolution of data converters may be expressed in several different ways, including the weight of the least significant bit (LSB), parts per million of full scale (ppm FS), millivolts (mV). Different devices (even from the same manufacturer) will be specified differently, so converter users must learn to translate between the different types of specifications if they are to successfully compare devices. The size of the least significant bit for various resolutions is shown in Figure 2-7.

RESOLUTION N	2^N	VOLTAGE (10 V FS)	ppm FS	% FS	dB FS
2-bit	4	2.5 V	250,000	25	−12
4-bit	16	625 mV	62,500	6.25	−24
6-bit	64	156 mV	15,625	1.56	−36
8-bit	256	39.1 mV	3,906	0.39	−48
10-bit	1,024	9.77 mV (10 mV)	977	0.098	−60
12-bit	4,096	2.44 mV	244	0.024	−72
14-bit	16,384	610 µV	61	0.0061	−84
16-bit	65,536	153 µV	15	0.0015	−96
18-bit	262,144	38 µV	4	0.0004	−108
20-bit	1,048,576	9.54 µV (10 µV)	1	0.0001	−120
22-bit	4,194,304	2.38 µV	0.24	0.000024	−132
24-bit	16,777,216	596 nV*	0.06	0.000006	−144

NOTES: *600 nV is the Johnson Noise in a 10 kHz BW of a 2.2 kΩ Resistor @ 25°C

10 bits and 10 V FS yields an LSB of 10 mV, 1000 ppm, or 0.1%.
All other values may be calculated by powers of 2.

Figure 2-7: Quantization—The Size of a Least Significant Bit (LSB)

Before we can consider the various architectures used in data converters, it is necessary to consider the performance to be expected, and the specifications that are important. The following sections will consider the definition of errors and specifications used for data converters. This is important in understanding the strengths and weaknesses of different ADC/DAC architectures.

The first applications of data converters were in measurement and control where the exact timing of the conversion was usually unimportant, and the data rate was slow. In such applications, the dc specifications of converters are important, but timing and ac specifications are not. Today many, if not most, converters are used in *sampling* and *reconstruction* systems where ac specifications are critical (and dc ones may not be). These will be considered in the next part of this section.

Figure 2-8 shows the ideal transfer characteristics for a 3-bit unipolar DAC, and Figure 2-9 a 3-bit unipolar ADC. In a DAC, both the input and the output are quantized, and the graph consists of eight points. While it is reasonable to discuss the line through these points, it is very important to remember that the actual transfer characteristic is *not* a line, but a number of discrete points.

Sampled Data Systems

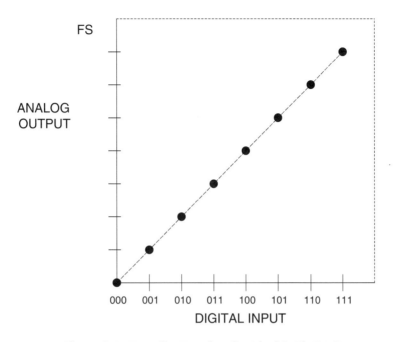

Figure 2-8: Transfer Function for Ideal 3-Bit DAC

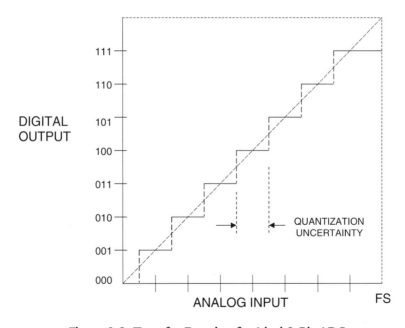

Figure 2-9: Transfer Function for Ideal 3-Bit ADC

The input to an ADC is analog and is not quantized, but its output is quantized. The transfer characteristic therefore consists of eight horizontal steps (when considering the offset, gain, and linearity of an ADC we consider the line joining the midpoints of these steps).

In both cases, digital full scale (all 1s) corresponds to 1 LSB below the analog full scale (the reference, or some multiple thereof). This is because, as mentioned above, the digital code represents the *normalized* ratio of the analog signal to the reference.

The (ideal) ADC transitions take place at 1/2 LSB above zero, and thereafter every LSB, until 1-1/2 LSB below analog full scale. Since the analog input to an ADC can take any value, but the digital output is quantized, there may be a difference of up to 1/2 LSB between the actual analog input and the exact value of the digital output. This is known as the *quantization error* or *quantization uncertainty* as shown in Figure 2-9. In ac (sampling) applications this quantization error gives rise to *quantization noise*, which will be discussed in the next section.

There are many possible digital coding schemes for data converters: *binary, offset binary, ones complement, twos complement, gray code, BCD,* and others. This section, being devoted mainly to the *analog* issues surrounding data converters, will use simple *binary* and *offset binary* in its examples and will not consider the merits and disadvantages of these, or any other forms of digital code.

The examples in Figures 2-8 and 2-9 use *unipolar* converters, whose analog port has only a single polarity. These are the simplest type, but *bipolar* converters are generally more useful in real-world applications. There are two types of bipolar converters: the simpler is merely a unipolar converter with an accurate 1 MSB of negative offset (and many converters are arranged so that this offset may be switched in and out so they can be used as either unipolar or bipolar converters at will), but the other, known as a *sign-magnitude* converter, is more complex, and has N bits of magnitude information and an additional bit that corresponds to the sign of the analog signal. Sign-magnitude DACs are quite rare, and sign-magnitude ADCs are found mostly in digital voltmeters (DVMs).

The four dc errors in a data converter are *offset error, gain error,* and two types of *linearity error*. Offset and gain errors are analogous to offset and gain errors in amplifiers, as shown in Figure 2-10 for a bipolar input range. (However, offset error and zero error, which are identical in amplifiers and unipolar data converters, are not identical in bipolar converters and should be carefully distinguished.) The transfer characteristics of both DACs and ADCs may be expressed as $D = K + GA$, where D is the digital code, A is the analog signal, and K and G are constants. In a unipolar converter, K is zero, and in an offset bipolar converter, it is -1 MSB. The offset error is the amount by which the actual value of K differs from its ideal value. The gain error is the amount by which G differs from its ideal value, and is generally expressed as the percentage difference between the two, although it may be defined as the gain error contribution (in mV or LSB) to the total error at full scale. These errors can usually be trimmed by the data converter user. Note, however, that amplifier offset is trimmed at zero input, and then the gain is trimmed near to full scale. The trim algorithm for a bipolar data converter is not so straightforward.

Sampled Data Systems

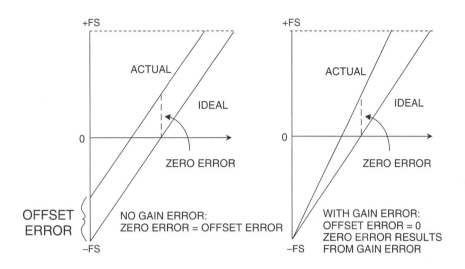

Figure 2-10: Converter Offset and Gain Error

The integral linearity error of a converter is also analogous to the linearity error of an amplifier, and is defined as the maximum deviation of the actual transfer characteristic of the converter from a straight line, and is generally expressed as a percentage of full scale (but may be given in LSBs). There are two common ways of choosing the straight line: *endpoint* and *best straight line* (see Figure 2-11).

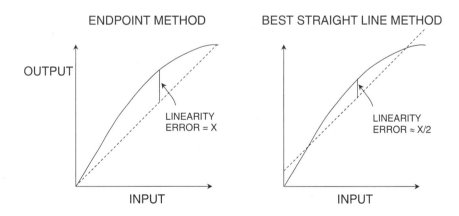

Figure 2-11: Method of Measuring Integral Linearity Errors (Same Converter on Both Graphs)

In the endpoint system, the deviation is measured from the straight line through the origin and the full-scale point (after gain adjustment). This is the most useful integral linearity measurement for measurement and control applications of data converters (since error budgets depend on deviation from the ideal transfer characteristic, not from some arbitrary "best fit"), and is the one normally adopted by Analog Devices, Inc.

The best straight line, however, does give a better prediction of distortion in ac applications, and also gives a lower value of "linearity error" on a data sheet. The best fit straight line is drawn through the transfer characteristic of the device using standard curve-fitting techniques, and the maximum deviation is measured from this line. In general, the integral linearity error measured in this way is only 50% of the value measured by endpoint methods. This makes the method good for producing impressive data sheets, but it is less useful for error budget analysis. For ac applications, it is even better to specify distortion than dc linearity, so it is rarely necessary to use the best straight line method to define converter linearity.

The other type of converter nonlinearity is *differential nonlinearity* (DNL). This relates to the linearity of the code transitions of the converter. In the ideal case, a change of 1 LSB in digital code corresponds to a change of exactly 1 LSB of analog signal. In a DAC, a change of 1 LSB in digital code produces exactly 1 LSB change of analog output, while in an ADC there should be exactly 1 LSB change of analog input to move from one digital transition to the next.

Where the change in analog signal corresponding to 1 LSB digital change is more or less than 1 LSB, there is said to be a DNL error. The DNL error of a converter is normally defined as the maximum value of DNL to be found at any transition.

If the DNL of a DAC is less than −1 LSB at any transition (see Figure 2-12), the DAC is *nonmonotonic*; i.e., its transfer characteristic contains one or more localized maxima or minima. A DNL greater than +1 LSB does not cause nonmonotonicity, but is still undesirable. In many DAC applications (especially closed-loop systems where nonmonotonicity can change negative feedback to positive feedback), it is critically important that DACs are monotonic. DAC monotonicity is often explicitly specified on data sheets, although if the DNL is guaranteed to be less than 1 LSB (i.e., |DNL| ≤ 1LSB) the device must be monotonic, even without an explicit guarantee.

ADCs can be nonmonotonic, but a more common result of excess DNL in ADCs is *missing codes* (see Figure 2-13). Missing codes (or nonmonotonicity) in an ADC are as objectionable as nonmonotonicity in a DAC. Again, they result from DNL > 1 LSB.

Sampled Data Systems

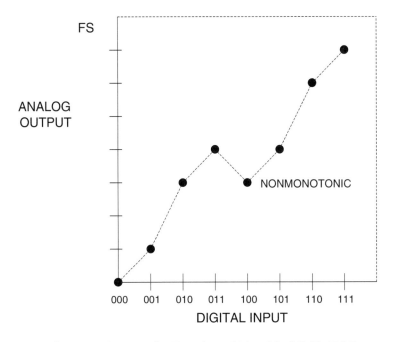

Figure 2-12: Transfer Function of Non-Ideal 3-Bit DAC

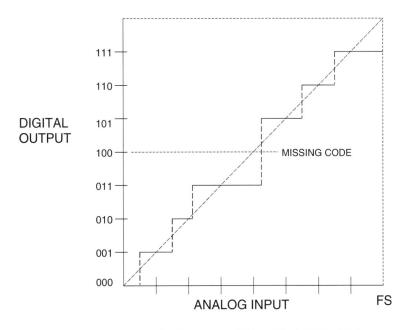

Figure 2-13: Transfer Function of Non-Ideal 3-Bit ADC

Section Two

Defining missing codes is more difficult than defining nonmonotonicity. All ADCs suffer from some transition noise as shown in Figure 2-14 (think of it as the flicker between adjacent values of the last digit of a DVM). As resolutions become higher, the range of input over which transition noise occurs may approach, or even exceed, 1 LSB. In such a case, especially if combined with a negative DNL error, it may be that there are some (or even all) codes where transition noise is present for the whole range of inputs. There are, therefore, some codes for which there is *no* input that will *guarantee* that code as an output, although there may be a range of inputs that will *sometimes* produce that code.

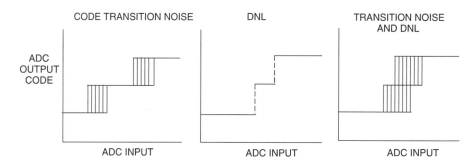

Figure 2-14: Combined Effects of ADC Code Transition Noise and DNL

For lower resolution ADCs, it may be reasonable to define *no missing codes* as a combination of transition noise and DNL that guarantees some level (perhaps 0.2 LSB) of noise-free code for all codes. However, this is impossible to achieve at the very high resolutions achieved by modern sigma-delta ADCs, or even at lower resolutions in wide bandwidth sampling ADCs. In these cases, the manufacturer must define noise levels and resolution in some other way. Which method is used is less important, but the data sheet should contain a clear definition of the method used and the performance to be expected.

AC Errors in Data Converters

Over the last decade, a major application of data converters is in ac sampling and reconstruction. In very simple terms, a *sampled data system* is a system where the instantaneous value of an ac waveform is sampled at regular intervals. The resulting digital codes may be used to store the waveform (as in CDs and DATs), or intensive computation on the samples (digital signal processing, or DSP) may be used to perform filtering, compression, and other operations. The inverse operation, reconstruction, occurs when a series of digital codes is fed to a DAC to reconstruct an ac waveform—an obvious example of this is a CD or DAT player, but the technique is very widely used in telecommunications, radio, synthesizers, and many other applications.

The data converters used in these applications must have good performance with ac signals, but may not require good dc specifications. The first high performance converters to be designed for such applications were often manufactured with good ac specifications but poor, or unspecified, dc performance. Today the design trade-offs are better understood, and most converters will have good, and guaranteed, ac and dc specifications. DACs for digital audio, however, which must be extremely competitive in price, are generally sold with comparatively poor dc specifications—not because their dc performance is poor, but because it is not tested during manufacture.

While it is easier to discuss the dc parameters of both DACs and ADCs together, their ac specifications are sufficiently different to deserve separate consideration.

Distortion and Noise in an Ideal N-Bit ADC

Thus far we have looked at the implications of the sampling process without considering the effects of ADC quantization. We will now treat the ADC as an ideal sampler, but include the effects of quantization.

The only errors (dc or ac) associated with an ideal N-bit ADC are those related to the sampling and quantization processes. The maximum error an ideal ADC makes when digitizing a dc input signal is ±1/2 LSB. Any ac signal applied to an ideal N-bit ADC will produce quantization noise whose rms value (measured over the Nyquist bandwidth, dc to $f_s/2$) is approximately equal to the weight of the least significant bit (LSB), q, divided by $\sqrt{12}$. (See Reference 2.) This assumes that the signal is at least a few LSBs in amplitude so that the ADC output always changes state. The quantization error signal from a linear ramp input is approximated as a sawtooth waveform with a peak-to-peak amplitude equal to q, and its rms value is therefore $q/\sqrt{12}$ (see Figure 2-15).

It can be shown that the ratio of the rms value of a full-scale sine wave to the rms value of the quantization noise (expressed in dB) is:

$$SNR = 6.02\,N + 1.76 \text{ dB},$$

where *N* is the number of bits in the ideal ADC. *This equation is only valid if the noise is measured over the entire Nyquist bandwidth from DC to $f_s/2$* as shown in Figure 2-16. If the signal bandwidth, BW, is less than $f_s/2$, then the SNR within the signal bandwidth BW is increased because the amount of quantization noise within the signal bandwidth is smaller. The correct expression for this condition is given by:

$$SNR = 6.02 + 1.76 \text{ dB} + 10\log\left(\frac{f_s}{2 \times BW}\right)$$

Section Two

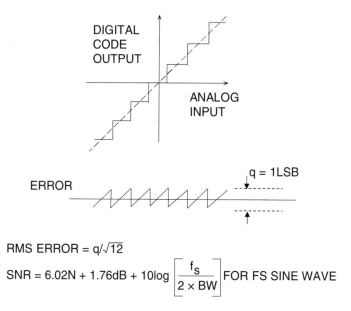

Figure 2-15: Ideal N-bit ADC Quantization Noise

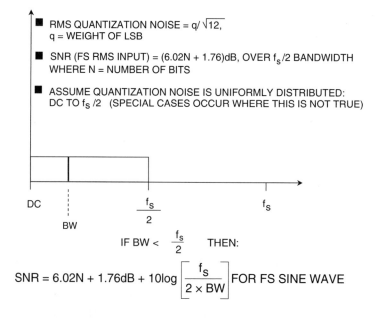

Figure 2-16: Quantization Noise Spectrum

Sampled Data Systems

The above equation reflects the condition called *oversampling*, where the sampling frequency is higher than twice the signal bandwidth. The correction term is often called *processing gain*. Notice that for a given signal bandwidth, doubling the sampling frequency increases the SNR by 3dB.

Although the rms value of the noise is accurately approximated by $q/\sqrt{12}$, its frequency domain content may be highly correlated to the AC input signal. For instance, there is greater correlation for low amplitude periodic signals than for large amplitude random signals. Quite often, the assumption is made that the theoretical quantization noise appears as white noise, spread uniformly over the Nyquist bandwidth dc to fs/2. Unfortunately, this is not true. In the case of strong correlation, the quantization noise appears concentrated at the various harmonics of the input signal, just where you don't want them.

In most applications, the input to the ADC is a band of frequencies (usually summed with some noise), so the quantization noise tends to be random. In spectral analysis applications (or in performing FFTs on ADCs using spectrally pure sine waves—see Figure 2-17), however, the correlation between the quantization noise and the signal depends upon the ratio of the sampling frequency to the input signal. This is demonstrated in Figure 2-18, where an ideal 12-bit ADC's output is analyzed using a 4096-point FFT. In the left-hand FFT plot, the ratio of the sampling frequency to the input frequency was chosen to be exactly 32, and the worst harmonic is about 76 dB below the fundamental. The right-hand diagram shows the effects of slightly offsetting the ratio, showing a relatively random noise spectrum, where the SFDR is now about 92 dBc. In both cases, the rms value of all the noise components is $q/\sqrt{12}$, but in the first case, the noise is concentrated at harmonics of the fundamental.

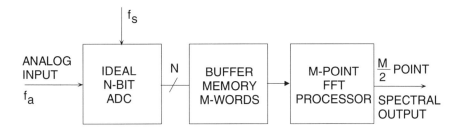

Figure 2-17: Dynamic Performance Analysis of an Ideal N-bit ADC

Section Two

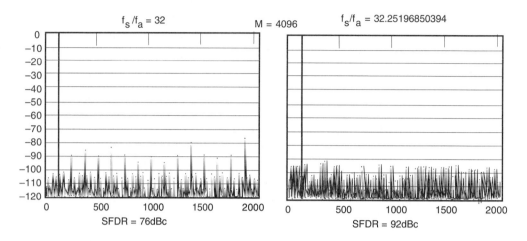

Figure 2-18: Effect of Ratio of Sampling Clock to Input Frequency on SFDR for Ideal 12-Bit ADC

Note that this variation in the apparent harmonic distortion of the ADC is an artifact of the sampling process and the correlation of the quantization error with the input frequency. In a practical ADC application, the quantization error generally appears as random noise because of the random nature of the wideband input signal and the additional fact that there is a usually a small amount of system noise that acts as a *dither* signal to further randomize the quantization error spectrum.

It is important to understand the above point, because single-tone sine wave FFT testing of ADCs is a universally accepted method of performance evaluation. In order to accurately measure the harmonic distortion of an ADC, steps must be taken to ensure that the test setup truly measures the ADC distortion, not the artifacts due to quantization noise correlation. This is done by properly choosing the frequency ratio and sometimes by injecting a small amount of noise (dither) with the input signal.

Now, return to Figure 2-18, and note that the average value of the noise floor of the FFT is approximately 100 dB below full scale, but the theoretical SNR of a 12-bit ADC is 74 dB. The FFT noise floor is *not* the SNR of the ADC, because the FFT acts as an analog spectrum analyzer with a bandwidth of f_s/M, where M is the number of points in the FFT. The theoretical FFT noise floor is therefore $10\log_{10}(M/2)$dB below the quantization noise floor due to the so-called *processing gain* of the FFT (see Figure 2-19). In the case of an ideal 12-bit ADC with an SNR of 74 dB, a 4096-point FFT would result in a processing gain of $10\log_{10}(4096/2) = 33$ dB, thereby resulting in an overall FFT noise floor of 74 + 33 = 107 dBc. In fact, the FFT noise floor can be reduced even further by going to larger and larger FFTs; just as an analog spectrum analyzer's noise floor can be reduced by narrowing the bandwidth. When testing ADCs using FFTs, it is important to ensure that the FFT size is large enough so that the distortion products can be distinguished from the FFT noise floor itself.

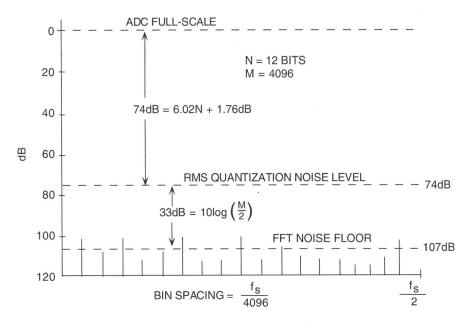

Figure 2-19: Noise Floor for an Ideal 12-Bit ADC Using 4096-Point FFT

Distortion and Noise in Practical ADCs

A practical sampling ADC (one that has an integral sample-and-hold), regardless of architecture, has a number of noise and distortion sources as shown in Figure 2-20. The wideband analog front-end buffer has wideband noise, nonlinearity, and also finite bandwidth. The SHA introduces further nonlinearity, bandlimiting, and aperture jitter. The actual quantizer portion of the ADC introduces quantization noise, and both integral and differential nonlinearity. In this discussion, assume that sequential outputs of the ADC are loaded into a buffer memory of length M and that the FFT processor provides the spectral output. Also assume that the FFT arithmetic operations themselves introduce no significant errors relative to the ADC. However, when examining the output noise floor, the FFT processing gain (dependent on M) must be considered.

Section Two

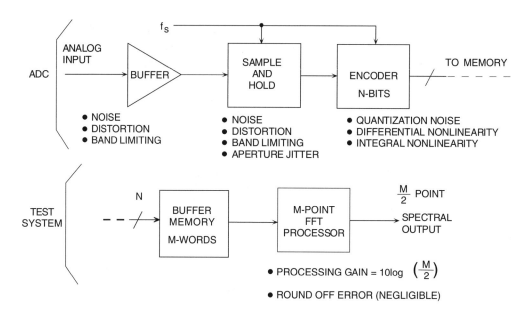

Figure 2-20: ADC Model Showing Noise and Distortion Sources

Equivalent Input Referred Noise (Thermal Noise)

The wideband ADC internal circuits produce a certain amount of wideband rms noise due to thermal and kT/C effects. This noise is present even for dc input signals, and accounts for the fact that the output of most wideband (or high resolution) ADCs is a distribution of codes, centered around the nominal value of a dc input (see Figure 2-21). To measure its value, the input of the ADC is grounded, and a large number of output samples are collected and plotted as a histogram (sometimes referred to as a *grounded-input* histogram). Since the noise is approximately Gaussian, the standard deviation of the histogram is easily calculated (see Reference 3), corresponding to the effective input rms noise. It is common practice to express this rms noise in terms of LSBs, although it can be expressed as an rms voltage.

There are various ways to characterize the ac performance of ADCs. In the early years of ADC technology (over 30 years ago) there was little standardization with respect to ac specifications, and measurement equipment and techniques were not well understood or available. Over nearly a 30-year period, manufacturers and customers have learned more about measuring the dynamic performance of converters, and the specifications shown in Figure 2-22 represent the most popular ones used today. Practically all the specifications represent the converter's performance in the frequency domain. The FFT is the heart of practically all these measurements and is discussed in detail in Section 5 of this book.

Sampled Data Systems

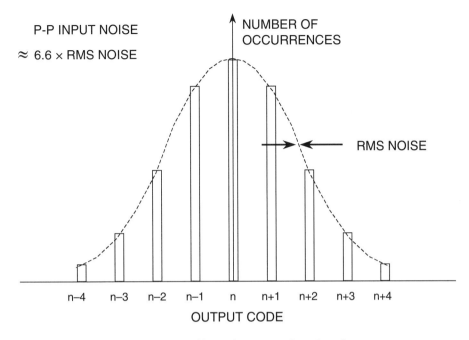

Figure 2-21: Effect of Input-Referred Noise on ADC "Grounded Input" Histogram

- Harmonic Distortion
- Worst Harmonic
- Total Harmonic Distortion (THD)
- Total Harmonic Distortion Plus Noise (THD + N)
- Signal-to-Noise-and-Distortion Ratio (SINAD, or S/N +D)
- Effective Number of Bits (ENOB)
- Signal-to-Noise Ratio (SNR)
- Analog Bandwidth (Full Power, Small Signal)
- Spurious-Free Dynamic Range (SFDR)
- Two-Tone Intermodulation Distortion
- Multitone Intermodulation Distortion

Figure 2-22: Quantifying ADC Dynamic Performance

Section Two

Integral and Differential Nonlinearity Distortion Effects

One of the first things to realize when examining the nonlinearities of data converters is that the transfer function of a data converter has artifacts that do not occur in conventional linear devices such as op amps or gain blocks. The overall integral nonlinearity of an ADC is due to the integral nonlinearity of the front end and SHA as well as the overall integral nonlinearity in the ADC transfer function. However, *differential nonlinearity is due exclusively to the encoding process* and may vary considerably, dependent on the ADC encoding architecture. Overall integral nonlinearity produces distortion products whose amplitude varies as a function of the input signal amplitude. For instance, second-order intermodulation products increase 2 dB for every 1 dB increase in signal level, and third-order products increase 3 dB for every 1 dB increase in signal level.

The differential nonlinearity in the ADC transfer function produces distortion products that not only depend on the amplitude of the signal, but on the positioning of the differential nonlinearity along the ADC transfer function. Figure 2-23 shows two ADC transfer functions having differential nonlinearity. The left-hand diagram shows an error that occurs at midscale. Therefore, for both large and small signals, the signal crosses through this point producing a distortion product that is relatively independent of the signal amplitude. The right-hand diagram shows another ADC transfer function that has differential nonlinearity errors at 1/4 and 3/4 full scale. Signals that are above 1/2 scale peak-to-peak will exercise these codes and produce distortion, while those less than 1/2 scale peak-to-peak will not.

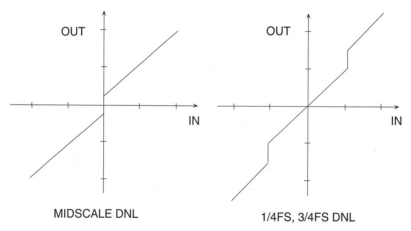

Figure 2-23: Typical ADC/DAC DNL Errors

Most high speed ADCs are designed so that differential nonlinearity is spread across the entire ADC range. Therefore, for signals that are within a few dB of full scale, the overall integral nonlinearity of the transfer function determines the distortion products. For lower level signals, however, the harmonic content becomes dominated by the differential nonlinearities and does not generally decrease proportionally with decreases in signal amplitude.

Harmonic Distortion, Worst Harmonic, Total Harmonic Distortion (THD), Total Harmonic Distortion Plus Noise (THD + N)

There are a number of ways to quantify the distortion of an ADC. An FFT analysis can be used to measure the amplitude of the various harmonics of a signal. The harmonics of the input signal can be distinguished from other distortion products by their location in the frequency spectrum. Figure 2-24 shows a 7 MHz input signal sampled at 20 MSPS and the location of the first nine harmonics. Aliased harmonics of f_a fall at frequencies equal to $|\pm Kf_s \pm nf_a|$, where n is the order of the harmonic, and K = 0, 1, 2, 3,.... The second and third harmonics are generally the only ones specified on a data sheet because they tend to be the largest, although some data sheets may specify the value of the *worst* harmonic. *Harmonic distortion* is normally specified in dBc (decibels below *carrier*), although at audio frequencies it may be specified as a percentage. Harmonic distortion is generally specified with an input signal near full scale (generally 0.5 dB to 1 dB below full scale to prevent clipping), but it can be specified at any level. For signals much lower than full scale, other distortion products due to the DNL of the converter (not direct harmonics) may limit performance.

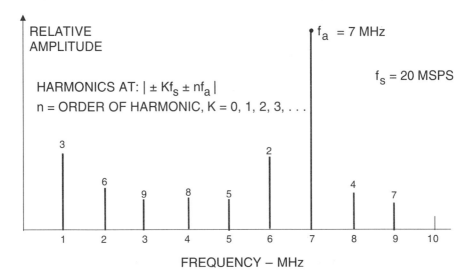

Figure 2-24: Location of Harmonic Distortion Products: Input Signal = 7 MHz, Sampling Rate = 20 MSPS

Total harmonic distortion (THD) is the ratio of the rms value of the fundamental signal to the mean value of the root-sum-square of its harmonics (generally, only the first five are significant). THD of an ADC is also generally specified with the input signal close to full scale, although it can be specified at any level.

Total harmonic distortion plus noise (THD + N) is the ratio of the rms value of the fundamental signal to the mean value of the root-sum-square of its harmonics plus all noise components (excluding dc). The bandwidth over which the noise is measured must be specified. In the case of an FFT, the bandwidth is dc to $f_s/2$. (If the bandwidth of the measurement is dc to $f_s/2$, THD+N is equal to SINAD, see below.)

Signal-to-Noise-and-Distortion Ratio (SINAD), Signal-to-Noise Ratio (SNR), and Effective Number of Bits (ENOB)

SINAD and SNR deserve careful attention, because there is still some variation between ADC manufacturers as to their precise meaning. Signal-to-noise-and-distortion (SINAD, or S/N + D) is the ratio of the rms signal amplitude to the mean value of the root-sum-square (RSS) of all other spectral components, *including harmonics* but excluding dc. SINAD is a good indication of the overall dynamic performance of an ADC as a function of input frequency because it includes all components that make up noise (including thermal noise) and distortion. It is often plotted for various input amplitudes. SINAD is equal to THD + N if the bandwidth for the noise measurement is the same. A typical plot for the AD9220 12-bit, 10 MSPS ADC is shown in Figure 2-26.

- SINAD (Signal-to-Noise-and-Distortion Ratio):
  - The Ratio of the RMS Signal Amplitude to the Mean Value of the Root-Sum-Squares (RSS) of all other Spectral Components, including Harmonics, But Excluding DC

- ENOB (Effective Number of Bits):

$$ENOB = \frac{SINAD - 1.76dB}{6.02}$$

- SNR (Signal-to-Noise Ratio, or Signal-to-Noise Ratio Without Harmonics:
  - The Ratio of the RMS Signal Amplitude to the Mean Value of the Root-Sum-Squares (RSS) of all Other Spectral Components, Excluding the First 5 Harmonics and DC

Figure 2-25: SINAD, ENOB, and SNR

Sampled Data Systems

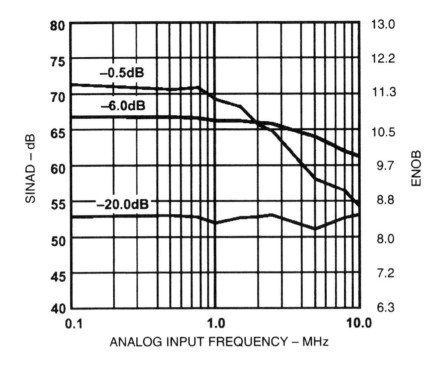

Figure 2-26: AD9220 12-Bit, 10 MSPS ADC SINAD and ENOB for Various Input Signal Levels

The SINAD plot shows where the ac performance of the ADC degrades due to high-frequency distortion and is usually plotted for frequencies well above the Nyquist frequency so that performance in undersampling applications can be evaluated. SINAD is often converted to *effective-number-of-bits* (ENOB) using the relationship for the theoretical SNR of an ideal N-bit ADC: SNR = 6.02 N + 1.76 dB. The equation is solved for N, and the value of SINAD is substituted for SNR:

Signal-to-noise ratio (SNR, or *SNR-without-harmonics*) is calculated the same as SINAD except that the signal harmonics are excluded from the calculation, leaving only the noise terms. In practice, it is only necessary to exclude the first five harmonics since they dominate. The SNR plot will degrade at high frequencies, but not as rapidly as SINAD because of the exclusion of the harmonic terms.

Many current ADC data sheets somewhat loosely refer to SINAD as SNR, so the engineer must be careful when interpreting these specifications.

Analog Bandwidth

The analog bandwidth of an ADC is that frequency at which the spectral output of the *fundamental* swept frequency (as determined by the FFT analysis) is reduced by 3 dB. It may be specified for either a small signal (SSBW—small signal bandwidth), or a full-scale signal (FPBW—full power bandwidth), so there can be a wide variation in specifications between manufacturers.

Like an amplifier, the analog bandwidth specification of a converter does not imply that the ADC maintains good distortion performance up to its bandwidth frequency. In fact, the SINAD (or ENOB) of most ADCs will begin to degrade considerably before the input frequency approaches the actual 3 dB bandwidth frequency. Figure 2-27 shows ENOB and full-scale frequency response of an ADC with a FPBW of 1 MHz; however, the ENOB begins to drop rapidly above 100 kHz.

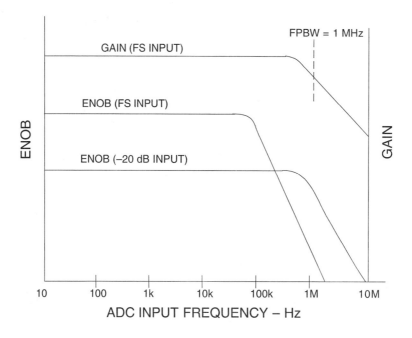

Figure 2-27: ADC Gain (Bandwidth) and ENOB vs. Frequency Shows Importance of ENOB Specification

Spurious-Free Dynamic Range (SFDR)

Probably the most significant specification for an ADC used in a communications application is its spurious-free dynamic range (SFDR). The SFDR specification is to ADCs what the third-order intercept specification is to mixers and LNAs. SFDR of an ADC is defined as the ratio of the rms signal amplitude to the rms value of the *peak spurious spectral content* (measured over the entire first Nyquist zone, dc to $f_s/2$). SFDR is generally plotted as a function of signal amplitude and may be expressed relative to the signal amplitude (dBc) or the ADC full-scale (dBFS) as shown in Figure 2-28.

Sampled Data Systems

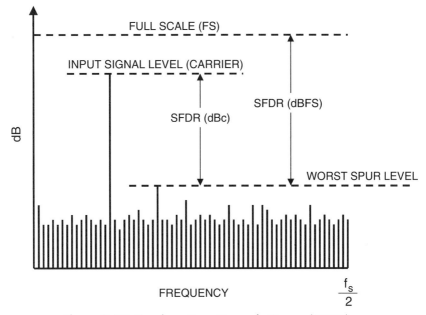

Figure 2-28: Spurious-Free Dynamic Range (SFDR)

For a signal near full scale, the peak spectral spur is generally determined by one of the first few harmonics of the fundamental. However, as the signal falls several dB below full scale, other spurs generally occur that are not direct harmonics of the input signal. This is because of the differential nonlinearity of the ADC transfer function as discussed earlier. Therefore, SFDR considers *all* sources of distortion, regardless of their origin.

The AD9042 is a 12-bit, 41 MSPS wideband ADC designed for communications applications where high SFDR is important. The SFDR for a 19.5 MHz input and a sampling frequency of 41 MSPS is shown in Figure 2-29. Note that a minimum of 80 dBc SFDR is obtained over the entire first Nyquist zone (dc to 20 MHz). The plot also shows SFDR expressed as dBFS.

SFDR is generally much greater than the ADC's theoretical N-bit SNR (6.02 N + 1.76 dB). For example, the AD9042 is a 12-bit ADC with an SFDR of 80 dBc and a typical SNR of 65 dBc (theoretical SNR is 74 dB). This is because there is a fundamental distinction between noise and distortion measurements. The process gain of the FFT (33 dB for a 4096-point FFT) allows frequency spurs well below the noise floor to be observed. Adding extra resolution to an ADC may serve to increase its SNR but may or may not increase its SFDR.

Section Two

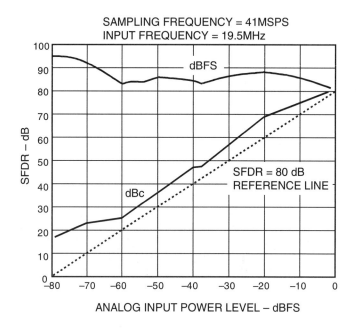

Figure 2-29: AD9042 12-Bit, 41 MSPS ADC
SFDR vs. Input Power Level

Two-Tone Intermodulation Distortion (IMD)

Two-tone IMD is measured by applying two spectrally pure sine waves to the ADC at frequencies f_1 and f_2, usually relatively close together. The amplitude of each tone is set slightly more than 6 dB below full scale so that the ADC does not clip when the two tones add in-phase. The location of the second and third-order products are shown in Figure 2-30. Notice that the second-order products fall at frequencies that can be removed by digital filters. However, the third-order products $2f_2-f_1$ and $2f_1-f_2$ are close to the original signals and are more difficult to filter. Unless otherwise specified, two-tone IMD refers to these third-order products. The value of the IMD product is expressed in dBc relative to the value of *either* of the two original tones, and not to their sum.

Note, however, that if the two tones are close to $f_s/4$, the aliased third harmonics of the fundamentals can make the identification of the actual $2f_2-f_1$ and $2f_1-f_2$ products difficult. This is because the third harmonic of $f_s/4$ is $3f_s/4$, and the alias occurs at $f_s - 3f_s/4 = f_s/4$. Similarly, if the two tones are close to $f_s/3$, the aliased second harmonics may interfere with the measurement. The same reasoning applies here; the second harmonic of $f_s/3$ is $2f_s/3$, and its alias occurs at $f_s - 2f_s/3 = f_s/3$.

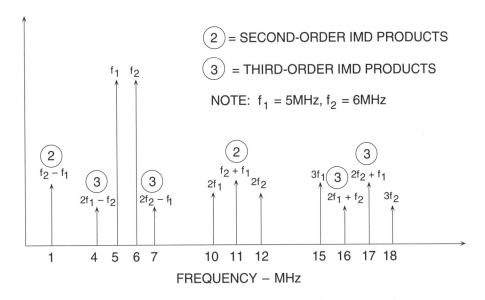

Figure 2-30: Second- and Third-Order Intermodulation Products for f_1 = 5MHz, f_2 = 6MHz

The concept of *second-and third-order intercept points* is not valid for an ADC, because the distortion products do not vary in a predictable manner (as a function of signal amplitude). The ADC does not gradually begin to compress signals approaching full scale (there is no 1 dB compression point); it acts as a *hard limiter* as soon as the signal exceeds the ADC input range, thereby suddenly producing extreme amounts of distortion because of clipping. On the other hand, for signals much below full scale, the distortion floor remains relatively constant and is independent of signal level.

Multitone SFDR is often measured in communications applications. The larger number of tones more closely simulates the wideband frequency spectrum of cellular telephone systems such as AMPS or GSM. Figure 2-31 shows the four-tone intermodulation performance of the AD6640 12-bit, 65 MSPS ADC. High SFDR increases the receiver's ability to capture small signals in the presence of large ones, and prevents the small signals from being masked by the intermodulation products of the larger ones.

Section Two

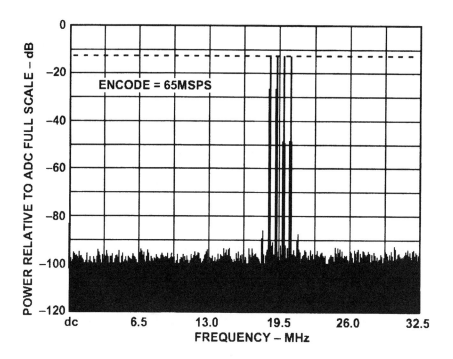

Figure 2-31: Multitone Testing: AD6640 12-Bit, 65 MSPS ADC

Noise Power Ratio (NPR)

Noise power ratio testing has been used extensively to measure the transmission characteristics of frequency division multiple access (FDMA) communications links (see Reference 4). In a typical FDMA system, 4 kHz wide voice channels are "stacked" in frequency bins for transmission over coaxial, microwave, or satellite equipment. At the receiving end, the FDMA data is demultiplexed and returned to 4 kHz individual baseband channels. In an FDMA system having more than approximately 100 channels, the FDMA signal can be approximated by Gaussian noise with the appropriate bandwidth. An individual 4 kHz channel can be measured for "quietness" using a narrow-band notch (band-stop) filter and a specially tuned receiver that measures the noise power inside the 4 kHz notch (see Figure 2-32).

Noise power ratio (NPR) measurements are straightforward. With the notch filter out, the rms noise power of the signal inside the notch is measured by the narrow-band receiver. The notch filter is then switched in, and the residual noise inside the slot is measured. The ratio of these two readings expressed in dB is the NPR. Several slot frequencies across the noise bandwidth (low, midband, and high) are tested to adequately characterize the system. NPR measurements on ADCs are made in a similar manner except the analog receiver is replaced by a buffer memory and an FFT processor.

Sampled Data Systems

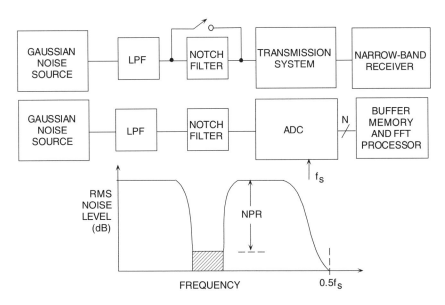

Figure 2-32: Noise Power Ratio (NPR) Measurements

NPR is usually plotted on an NPR curve. The NPR is plotted as a function of rms noise level referred to the peak range of the system. For very low noise loading level, the undesired noise (in nondigital systems) is primarily thermal noise and is independent of the input noise level. Over this region of the curve, a 1 dB increase in noise loading level causes a 1 dB increase in NPR. As the noise loading level is increased, the amplifiers in the system begin to overload, creating intermodulation products that cause the noise floor of the system to increase. As the input noise increases further, the effects of "overload" noise predominate, and the NPR is reduced dramatically. FDMA systems are usually operated at a noise loading level a few dB below the point of maximum NPR.

In a digital system containing an ADC, the noise within the slot is primarily quantization noise when low levels of noise input are applied. The NPR curve is linear in this region. As the noise level increases, there is a one-for-one correspondence between the noise level and the NPR. At some level, however, "clipping" noise caused by the hard-limiting action of the ADC begins to dominate. A theoretical curve for 10-, 11-, and 12-bit ADCs is shown in Figure 2-33 (see Reference 5).

In multichannel high frequency communication systems, NPR can also be used to simulate the distortion caused by a large number of individual channels, similar to an FDMA system. A notch filter is placed between the noise source and the ADC, and an FFT output is used in place of the analog receiver. The width of the notch filter is set for several MHz as shown in Figure 2-34 for the AD9042. NPR is the "depth" of the notch. An ideal ADC will only generate quantization noise inside the notch; however, a practical one has additional noise components due to intermodulation

Section Two

distortion caused by ADC nonlinearity. Notice that the NPR is about 60 dB compared to 62.7 dB theoretical.

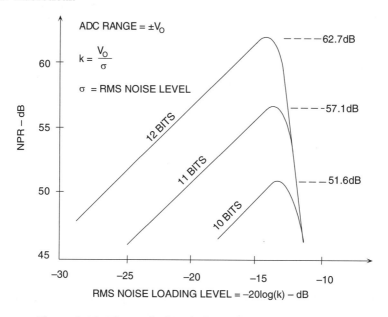

Figure 2-33: Theoretical NPR for 10-, 11-, 12 Bit ADCs

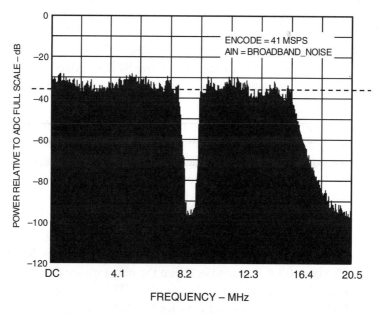

Figure 2-34: AD9042 12-Bit, 41 MSPS ADC NPR Measures 60 dB (62.7 dB Theoretical)

Aperture Jitter and Aperture Delay

Another reason that the SNR of an ADC decreases with input frequency may be deduced from Figure 2-35, which shows the effects of phase jitter (or aperture time jitter) on the sampling clock of an ADC (or internal in the sample-and-hold). The phase jitter causes a voltage error that is a function of slew rate and results in an overall degradation in SNR as shown in Figure 2-36. This is quite serious, especially at higher input/output frequencies. Therefore, extreme care must be taken to minimize phase noise in the sampling/reconstruction clock of any sampled data system. This care must extend to all aspects of the clock signal: the oscillator itself (for example, a 555 timer is absolutely inadequate, but even a quartz crystal oscillator can give problems if it uses an active device that shares a chip with noisy logic); the transmission path (these clocks are very vulnerable to interference of all sorts), and phase noise introduced in the ADC or DAC. A very common source of phase noise in converter circuitry is aperture jitter in the integral sample-and-hold (SHA) circuitry.

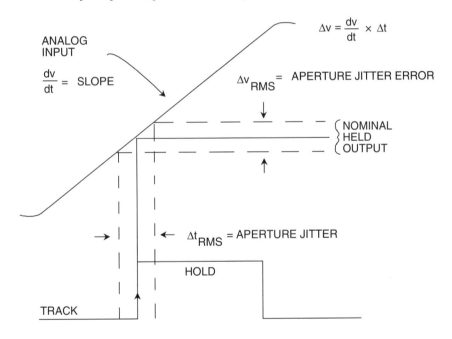

Figure 2-35: Effects of Aperture and Sampling Clock Jitter

Two decades or so ago, sampling ADCs were built up from a separate SHA and ADC. Interface design was difficult, and a key parameter was aperture jitter in the SHA. Today, most sampled data systems use *sampling* ADCs that contain an integral SHA. The aperture jitter of the SHA may not be specified as such, but this

Section Two

is not a cause of concern if the SNR or ENOB is clearly specified, since a guarantee of a specific SNR is an implicit guarantee of an adequate aperture jitter specification. However, the use of an additional high performance SHA will sometimes improve the high frequency ENOB of even the best sampling ADC by presenting "dc" to the ADC, and may be more cost-effective than replacing the ADC with a more expensive one.

It should be noted that there is also a fixed component that makes up the ADC aperture time. This component, usually called *effective aperture delay time*, does not produce an error. It simply results in a time offset between the time the ADC is asked to sample and when the actual sample takes place (see Figure 2-37), and may be positive or negative. The variation or tolerance placed on this parameter from part to part is important in simultaneous sampling applications or other applications such as I and Q demodulation where two ADCs are required to track each other.

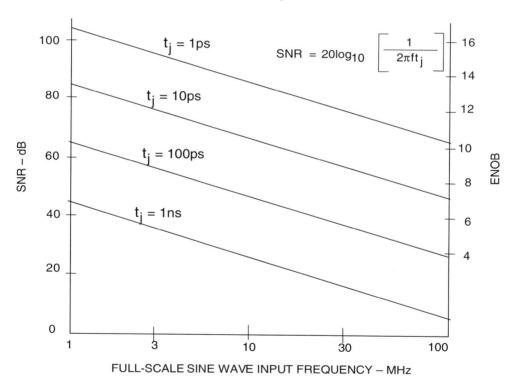

Figure 2-36: SNR Due to Aperture and Sampling Clock Jitter

Sampled Data Systems

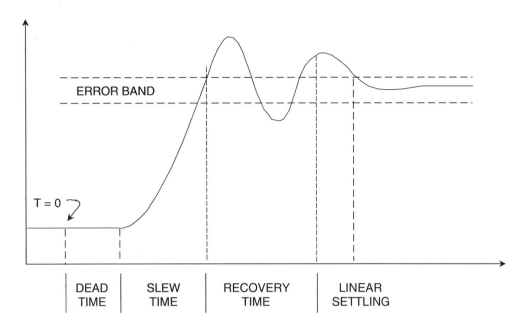

Figure 2-37: Effective Aperture Delay Time

DAC Dynamic Performance

The ac specifications that are most likely to be important with DACs are *settling time*, *glitch*, *distortion,* and *spurious-free dynamic range (SFDR)*.

The settling time of a DAC is the time from a change of digital code to when the output comes within *and remains within* some error band as shown in Figure 2-38. With amplifiers, it is hard to make comparisons of settling time, since their error bands may differ from amplifier to amplifier, but with DACs the error band will almost invariably be ± 1 or ± 1/2 LSB.

The settling time of a DAC is made up of four different periods: the *switching time* or *dead time* (during which the digital switching, but not the output, is changing), the *slewing time* (during which the rate of change of output is limited by the slew rate of the DAC output), the *recovery time* (when the DAC is recovering from its fast slew and may overshoot), and the *linear settling time* (when the DAC output approaches its final value in an exponential or near-exponential manner). If the slew time is short compared to the other three (as is usually the case with current output DACs), the settling time will be largely independent of the output step size. On the other hand, if the slew time is a significant part of the total, the larger the step, the longer the settling time.

Section Two

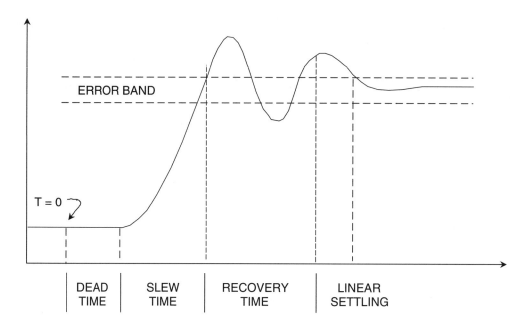

Figure 2-38: DAC Settling Time

Ideally, when a DAC output changes it should move from one value to its new one monotonically. In practice, the output is likely to overshoot, undershoot, or both (see Figure 2-39). This uncontrolled movement of the DAC output during a transition is known as *glitch*. It can arise from two mechanisms: capacitive coupling of digital transitions to the analog output, and the effects of some switches in the DAC operating more quickly than others and producing temporary spurious outputs.

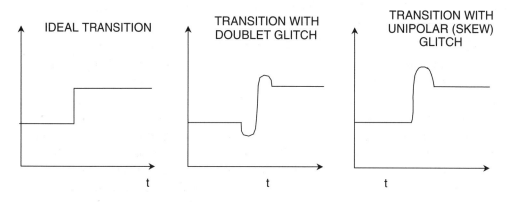

Figure 2-39: DAC Transitions (Showing Glitch)

Sampled Data Systems

Capacitive coupling frequently produces roughly equal positive and negative spikes (sometimes called a *doublet* glitch) which more or less cancel in the longer term. The glitch produced by switch timing differences is generally unipolar, much larger, and of greater concern.

Glitches can be characterized by measuring the *glitch impulse area*, sometimes inaccurately called glitch energy. The term *glitch energy* is a misnomer, since the unit for glitch impulse area is volt-seconds (or more probably μVs or pVs. The *peak glitch area* is the area of the largest of the positive or negative glitch areas. The glitch impulse area is the net area under the voltage-versus-time curve and can be estimated by approximating the waveforms by triangles, computing the areas, and subtracting the negative area from the positive area. The midscale glitch produced by the transition between the codes 0111...111 and 1000...000 is usually the worst glitch.

Glitches at other code transition points (such as 1/4 and 3/4 full scale) are generally less. Figure 2-40 shows the midscale glitch for a fast low glitch DAC. The peak and net glitch areas are estimated using triangles as described above. Settling time is measured from the time the waveform leaves the initial 1 LSB error band until it enters and remains within the final 1 LSB error band. The step size between the transition regions is also 1 LSB.

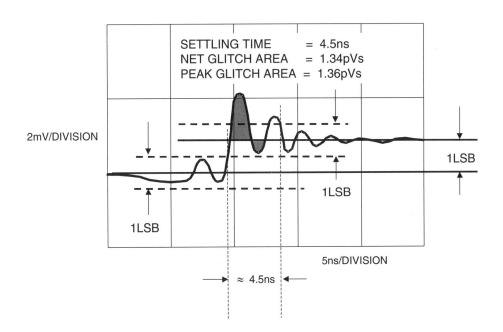

Figure 2-40: DAC Midscale Glitch Shows 1.34 pVs Net Impulse Area and Settling Time of 4.5 ns

DAC settling time is important in applications such as RGB raster scan video display drivers, but frequency-domain specifications such as SFDR are generally more important in communications.

If we consider the spectrum of a waveform reconstructed by a DAC from digital data, we find that in addition to the expected spectrum (which will contain one or more frequencies, depending on the nature of the reconstructed waveform), there will also be noise and distortion products. Distortion may be specified in terms of harmonic distortion, spurious-free dynamic range (SFDR), intermodulation distortion, or all of the above. Harmonic distortion is defined as the ratio of harmonics to fundamental when a (theoretically) pure sine wave is reconstructed, and is the most common specification. Spurious-free dynamic range is the ratio of the worst spur (usually, but not necessarily always, a harmonic of the fundamental) to the fundamental.

Code-dependent glitches will produce both out-of-band and in-band harmonics when the DAC is reconstructing a digitally generated sine wave as in a direct digital synthesis (DDS) system. The midscale glitch occurs twice during a single cycle of a reconstructed sine wave (at each midscale crossing), and will therefore produce a second harmonic of the sine wave, as shown in Figure 2-41. Note that the higher order harmonics of the sine wave, which alias back into the Nyquist bandwidth (dc to $f_s/2$), cannot be filtered.

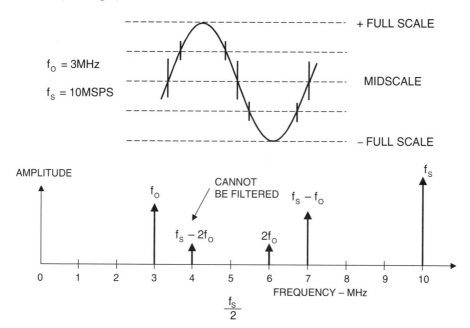

Figure 2-41: Effect of Code-Dependent Glitches on Spectral Output

It is difficult to predict the harmonic distortion or SFDR from the glitch area specification alone. Other factors, such as the overall linearity of the DAC, also contribute to distortion. It is therefore customary to test reconstruction DACs in the frequency domain (using a spectrum analyzer) at various clock rates and output frequencies as shown in Figure 2-43. Typical SFDR for the 14-bit AD9772 transmit DAC is shown in Figure 2-44. The clock rate is 65 MSPS, and the output frequency is swept to 25 MHz. As in the case of ADCs, quantization noise will appear as increased harmonic distortion if the ratio between the clock frequency and the DAC output frequency is an integer number. These ratios should be avoided when making the SFDR measurements.

- Resolution
- Integral Nonlinearity
- Differential Nonlinearity
- Code-Dependent Glitches
- Ratio of Clock Frequency to Output Frequency (Even in an Ideal DAC)
- Mathematical Analysis is Difficult

Figure 2-42: Contributors to DDS DAC Distortion

Section Two

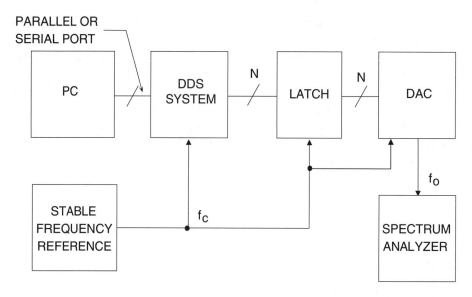

Figure 2-43: Test Setup for Measuring DAC SFDR

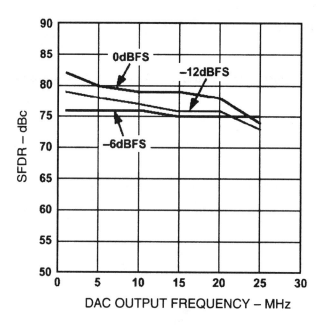

Figure 2-44: AD9772 14-Bit TxDAC™ SFDR, Data Update Rate = 65 MSPS

Sampled Data Systems

DAC SIN (x)/x Frequency Roll-Off

The output of a reconstruction DAC can be visualized as a series of rectangular pulses whose width is equal to the reciprocal of the clock rate as shown in Figure 2-45. Note that the reconstructed signal amplitude is down 3.92 dB at the Nyquist frequency, $f_c/2$. An inverse SIN (x)/x filter can be used to compensate for this effect in most cases. The images of the fundamental signal are also attenuated by the SIN (x)/x function.

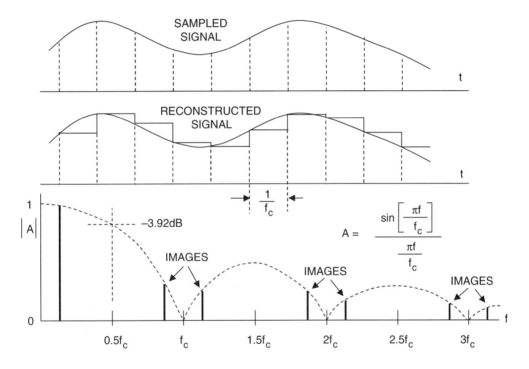

Figure 2-45: DAC SIN (x)/x Roll-off (Amplitude Normalized)

Section Two

References

1. **Active and Passive Electrical Wave Filter Catalog**, Vol. 34, TTE, Incorporated, 2251 Barry Avenue, Los Angeles, CA 90064.

2. W. R. Bennett, *"Spectra of Quantized Signals,"* **Bell System Technical Journal**, No. 27, July 1948, pp. 446–472.

3. Steve Ruscak and Larry Singer, *Using Histogram Techniques to Measure A/D Converter Noise*, **Analog Dialogue**, Vol. 29-2, 1995.

4. M.J. Tant, **The White Noise Book**, Marconi Instruments, July 1974.

5. G.A. Gray and G.W. Zeoli, *Quantization and Saturation Noise due to A/D Conversion*, **IEEE Trans. Aerospace and Electronic Systems**, Jan. 1971, pp. 222–223.

6. Chuck Lane, *A 10-bit 60MSPS Flash ADC*, **Proceedings of the 1989 Bipolar Circuits and Technology Meeting**, IEEE Catalog No. 89CH2771-4, September 1989, pp. 44–47.

7. F.D. Waldhauer, *Analog to Digital Converter*, **U.S. Patent 3-187-325**, 1965.

8. J.O. Edson and H.H. Henning, *Broadband Codecs for an Experimental 224 Mb/s PCM Terminal*, **Bell System Technical Journal**, 44, November 1965, pp. 1887–1940.

9. J.S. Mayo, *Experimental 224Mb/s PCM Terminals*, **Bell System Technical Journal**, 44, November 1965, pp. 1813–1941.

10. Hermann Schmid, **Electronic Analog/Digital Conversions**, Van Nostrand Reinhold Company, New York, 1970.

11. Carl Moreland, *An 8-Bit 150MSPS Serial ADC*, **1995 ISSCC Digest of Technical Papers**, Vol. 38, p. 272.

12. Roy Gosser and Frank Murden, *A 12-Bit 50MSPS Two-Stage A/D Converter*, **1995 ISSCC Digest of Technical Papers**, p. 278.

13. Carl Moreland, **An Analog-to-Digital Converter Using Serial-Ripple Architecture**, Masters' Thesis, Florida State University College of Engineering, Department of Electrical Engineering, 1995.

14. **Practical Analog Design Techniques**, Analog Devices, 1995, Chapters 4, 5, and 8.

15. **Linear Design Seminar**, Analog Devices, 1995, Chapters 4, 5.

16. **System Applications Guide**, Analog Devices, 1993, Chapters 12, 13, 15, 16.

17. **Amplifier Applications Guide**, Analog Devices, 1992, Chapter 7.

18. Walt Kester, *Drive Circuitry is Critical to High-Speed Sampling ADCs*, **Electronic Design Special Analog Issue**, Nov. 7, 1994, pp. 43–50.

19. Walt Kester, *Basic Characteristics Distinguish Sampling A/D Converters*, **EDN**, Sept. 3, 1992, pp. 135–144.

20. Walt Kester, *Peripheral Circuits Can Make or Break Sampling ADC Systems*, **EDN**, Oct. 1, 1992, pp. 97–105.

21. Walt Kester, *Layout, Grounding, and Filtering Complete Sampling ADC System*, **EDN**, Oct. 15, 1992, pp. 127–134.

22. Robert A. Witte, *Distortion Measurements Using a Spectrum Analyzer*, **RF Design**, September, 1992, pp. 75–84.

23. Walt Kester, *Confused About Amplifier Distortion Specs?*, **Analog Dialogue**, 27-1, 1993, pp. 27–29.

24. **System Applications Guide**, Analog Devices, 1993, Chapter 16.

25. Frederick J. Harris, *On the Use of Windows for Harmonic Analysis with the Discrete Fourier Transform*, **IEEE Proceedings**, Vol. 66, No. 1, Jan. 1978, pp. 51–83.

26. Joey Doernberg, Hae-Seung Lee, David A. Hodges, *Full Speed Testing of A/D Converters*, **IEEE Journal of Solid State Circuits**, Vol. SC-19, No. 6, Dec. 1984, pp. 820–827.

27. Brendan Coleman, Pat Meehan, John Reidy and Pat Weeks, *Coherent Sampling Helps When Specifying DSP A/D Converters*, **EDN**, October 15, 1987, pp. 145–152.

28. Robert W. Ramierez, **The FFT: Fundamentals and Concepts**, Prentice-Hall, 1985.

29. R. B. Blackman and J. W. Tukey, **The Measurement of Power Spectra**, Dover Publications, New York, 1958.

30. James J. Colotti, *Digital Dynamic Analysis of A/D Conversion Systems Through Evaluation Software Based on FFT/DFT Analysis*, **RF Expo East 1987 Proceedings**, Cardiff Publishing Co., pp. 245–272.

31. **HP Journal**, Nov. 1982, Vol. 33, No. 11.

32. **HP Product Note** 5180A-2.

33. **HP Journal**, April 1988, Vol. 39, No. 2.

34. **HP Journal**, June 1988, Vol. 39, No. 3.

35. Dan Sheingold, Editor, **Analog-to-Digital Conversion Handbook, Third Edition**, Prentice-Hall, 1986.

36. Lawrence Rabiner and Bernard Gold, **Theory and Application of Digital Signal Processing**, Prentice-Hall, 1975.

37. Matthew Mahoney, **DSP-Based Testing of Analog and Mixed-Signal Circuits**, IEEE Computer Society Press, Washington, D.C., 1987.

38. **IEEE Trial-Use Standard for Digitizing Waveform Recorders**, No. 1057-1988.

39. Richard J. Higgins, **Digital Signal Processing in VSLI**, Prentice-Hall, 1990.

40. M. S. Ghausi and K. R. Laker, **Modern Filter Design: Active RC and Switched Capacitors**, Prentice Hall, 1981.

41. Mathcad™ 4.0 software package available from MathSoft, Inc., 201 Broadway, Cambridge MA, 02139.

42. Howard E. Hilton, *A 10MHz Analog-to-Digital Converter with 110dB Linearity*, **H.P. Journal**, October 1993, pp. 105–112.

Section 3
ADCs for DSP Applications

- Successive-Approximation ADCs
- Sigma-Delta ADCs
- Flash Converters
- Subranging (Pipelined) ADCs
- Bit-Per-Stage (Serial, or Ripple) ADCs

Section 3
ADCs for DSP Applications
Walt Kester and James Bryant

The trend in ADCs and DACs is toward higher speeds and higher resolutions at reduced power levels and supply voltages. Modern data converters generally operate on ±5 V (dual supply), 5 V or 3 V (single supply). In fact, the number of 3 V devices is rapidly increasing because of many new markets such as digital cameras, camcorders, and cellular telephones. This trend has created a number of design and applications problems that were much less important in earlier data converters, where ±15 V supplies and ±10 V input ranges were the standard.

Lower supply voltages imply smaller input voltage ranges, and hence more susceptibility to noise from all potential sources: power supplies, references, digital signals, EMI/RFI, and probably most important, improper layout, grounding, and decoupling techniques. Single-supply ADCs often have an input range that is not referenced to ground. Finding compatible single-supply drive amplifiers and dealing with level shifting of the input signal in direct-coupled applications also becomes a challenge.

In spite of these issues, components are now available that allow extremely high resolutions at low supply voltages and low power. This section discusses the applications problems associated with such components and shows techniques for successfully designing them into systems.

The most popular ADCs for DSP applications are based on five fundamental architectures: *successive approximation, sigma-delta, flash, subranging (or pipelined),* and *bit-per-stage (or ripple).*

- Typical Supply Voltages: ±5 V, 5 V, 5 V/3 V, 3 V
- Lower Signal Swings Increase Sensitivity to All Types of Noise including Device, Power Supply and Logic
- Device Noise Increases at Low Currents
- Common-Mode Input Voltage Restrictions
- Input Buffer Amplifier Selection Critical
- Autocalibration Modes Desirable at High Resolutions

Figure 3-1: Low Power, Low Voltage ADC Design Issues

Section Three

- **Successive-Approximation**
  - Resolutions to 16 Bits
  - Minimal Throughput Delay Time (No Output Latency, "Single-Shot" Operation Possible)
  - Used in Multiplexed Data Acquisition Systems
- **Sigma-Delta**
  - Resolutions to 24 Bits
  - Excellent Differential Linearity
  - Internal Digital Filter (Can be Linear Phase)
  - Long Throughput Delay Time (Output Latency)
  - Difficult to Multiplex Inputs Due to Digital Filter Settling Time
- **High Speed Architectures**
  - Flash Converter
  - Subranging or Pipelined
  - Bit-Per-Stage (Ripple)

Figure 3-2: ADCs for DSP Applications

Successive-Approximation ADCs

The successive-approximation ADC has been the mainstay of signal conditioning for many years. Recent design improvements have extended the sampling frequency of these ADCs into the megahertz region. The use of internal switched capacitor techniques, along with autocalibration techniques, extends the resolution of these ADCs to 16 bits on standard CMOS processes without the need for expensive thin-film laser trimming.

The basic successive-approximation ADC is shown in Figure 3-3. It performs conversions on command. On the assertion of the CONVERT START command, the sample-and-hold (SHA) is placed in the *hold* mode, and all the bits of the successive-approximation register (SAR) are reset to "0," except the MSB, which is set to "1." The SAR output drives the internal DAC. If the DAC output is greater than the analog input, this bit in the SAR is reset; otherwise it is left set. The next most significant bit is then set to "1." If the DAC output is greater than the analog input, this bit in the SAR is reset; otherwise it is left set. The process is repeated with each bit in turn. When all the bits have been set, tested, and reset or not as appropriate, the contents of the SAR correspond to the value of the analog input, and the conversion is complete. These bit "tests" can form the basis of a serial output version SAR-based ADC.

The end of conversion is generally indicated by an end-of-convert (EOC), data-ready (DRDY), or a busy signal (actually, *not*-BUSY indicates end of conversion). The polarities and name of this signal may be different for different SAR ADCs, but the fundamental concept is the same. At the beginning of the conversion interval, the signal goes high (or low) and remains in that state until the conversion is completed, at which time it goes low (or high). The trailing edge is generally an indication of valid output data.

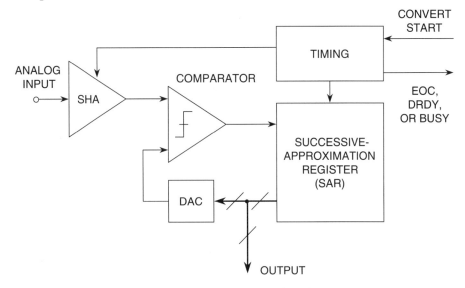

Figure 3-3: Successive-Approximation ADC

An N-bit conversion takes N steps. It would seem on superficial examination that a 16-bit converter would have twice the conversion time of an 8-bit one, but this is not the case. In an 8-bit converter, the DAC must settle to 8-bit accuracy before the bit decision is made, whereas in a 16-bit converter, it must settle to 16-bit accuracy, which takes a lot longer. In practice, 8-bit successive-approximation ADCs can convert in a few hundred nanoseconds, while 16-bit ones will generally take several microseconds.

Notice that the overall accuracy and linearity of the SAR ADC is determined primarily by the internal DAC. Until recently, most precision SAR ADCs used laser-trimmed thin-film DACs to achieve the desired accuracy and linearity. The thin-film resistor trimming process adds cost, and the thin-film resistor values may be affected when subjected to the mechanical stresses of packaging.

For these reasons, switched capacitor (or charge-redistribution) DACs have become popular in newer SAR ADCs. The advantage of the switched capacitor DAC is that the accuracy and linearity is primarily determined by photolithography, which in turn controls the capacitor plate area and the capacitance as well as matching. In addition, small capacitors can be placed in parallel with the main capacitors which, can be switched in and out under control of autocalibration routines to achieve high accuracy and linearity without the need for thin-film laser trimming. Temperature tracking between the switched capacitors can be better than 1 ppm/°C, thereby offering a high degree of temperature stability.

A simple 3-bit capacitor DAC is shown in Figure 3-4. The switches are shown in the *track*, or *sample* mode where the analog input voltage, A_{IN}, is constantly charging and discharging the parallel combination of all the capacitors. The *hold* mode is initiated by opening S_{IN}, leaving the sampled analog input voltage on the capacitor array. Switch S_C is then opened, allowing the voltage at node A to move as the bit switches are manipulated. If S1, S2, S3, and S4 are all connected to ground, a voltage equal to $-A_{IN}$ appears at node A. Connecting S1 to V_{REF} adds a voltage equal to $V_{REF}/2$ to $-A_{IN}$. The comparator then makes the MSB bit decision, and the SAR either leaves S1 connected to V_{REF} or connects it to ground, depending on the comparator output (which is high or low depending on whether the voltage at node A is negative or positive, respectively). A similar process is followed for the remaining two bits. At the end of the conversion interval, S1, S2, S3, S4, and S_{IN} are connected to A_{IN}, S_C is connected to ground, and the converter is ready for another cycle.

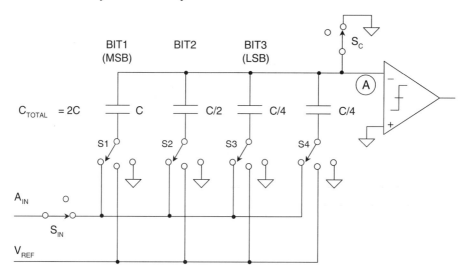

SWITCHES SHOWN IN TRACK (SAMPLE) MODE

Figure 3-4: 3-Bit Switched Capacitor DAC

Note that the extra LSB capacitor (C/4 in the case of the 3-bit DAC) is required to make the total value of the capacitor array equal to 2C so that binary division is accomplished when the individual bit capacitors are manipulated.

ADCs for DSP Applications

The operation of the capacitor DAC (cap DAC) is similar to an R/2R resistive DAC. When a particular bit capacitor is switched to V_{REF}, the voltage divider created by the bit capacitor and the total array capacitance (2C) adds a voltage to node A equal to the weight of that bit. When the bit capacitor is switched to ground, the same voltage is subtracted from node A.

Because of their popularity, successive-approximation ADCs are available in a wide variety of resolutions, sampling rates, input and output options, package styles, and costs. It would be impossible to attempt to list all types, but Figure 3-5 shows a number of recent Analog Devices' SAR ADCs that are representative. Note that many devices are complete data acquisition systems with input multiplexers that allow a single ADC core to process multiple analog channels.

	RESOLUTION	SAMPLING RATE	POWER	CHANNELS
AD7472	12 BITS	1.5 MSPS	9 mW	1
AD7891	12 BITS	500 kSPS	85 mW	8
AD7858/59	12 BITS	200 kSPS	20 mW	8
AD7887/88	12 BITS	125 kSPS	3.5 mW	8
AD7856/57	14 BITS	285 kSPS	60 mW	8
AD7660	16 BITS	100 kSPS	15 mW	1
AD974	16 BITS	200 kSPS	120 mW	4
AD7664	16 BITS	570 kSPS	150 mW	1

Figure 3-5: Resolution/Conversion Time Comparison for Representative Single-Supply SAR ADCs

While there are some variations, the fundamental timing of most SAR ADCs is similar and relatively straightforward (see Figure 3-6). The conversion process is initiated by asserting a CONVERT START signal. The $\overline{\text{CONVST}}$ signal is a negative-going pulse whose positive-going edge actually initiates the conversion. The internal sample-and-hold (SHA) amplifier is placed in the hold mode on this edge, and the various bits are determined using the SAR algorithm. The negative-going edge of the $\overline{\text{CONVST}}$ pulse causes the $\overline{\text{EOC}}$ or BUSY line to go high. When the conversion is complete, the BUSY line goes low, indicating the completion of the conversion process. In most cases the trailing edge of the BUSY line can be used as an indication that the output data is valid and can be used to strobe the output data into an external register. However, because of the many variations in terminology and design, the individual data sheet should always be consulted when using a specific ADC.

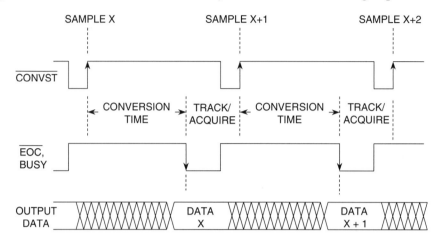

Figure 3-6: Typical SAR ADC Timing

It should also be noted that some SAR ADCs require an external high frequency clock in addition to the CONVERT START command. In most cases, there is no need to synchronize the two. The frequency of the external clock, if required, generally falls in the range of 1 MHz to 30 MHz, depending on the conversion time and resolution of the ADC. Other SAR ADCs have an internal oscillator that is used to perform the conversions and require only the CONVERT START command. Because of their architecture, SAR ADCs allow single-shot conversion at any repetition rate from dc to the converter's maximum conversion rate.

In an SAR ADC, the output data for a sampled input is valid at the end of the conversion interval for that sampled input. In other ADC architectures, such as sigma-delta or the two-stage subranging architecture shown in Figure 3-7, this is not the case. The subranging ADC shown in the figure is a two-stage *pipelined* or subranging 12-bit converter. The first conversion is done by the 6-bit ADC, which drives a 6-bit DAC. The output of the 6-bit DAC represents a 6-bit approximation to the analog input. Note that SHA2 delays the analog signal while the 6-bit ADC makes its decision and the 6-bit DAC settles. The DAC approximation is then subtracted from the analog signal from SHA2, amplified, and digitized by a 7-bit ADC. The outputs of the two conversions are combined, and the extra bit used to correct errors made in the first conversion. The typical timing associated with this type of converter is shown in Figure 3-8. Note that the output data presented immediately after sample X actually corresponds to sample X–2, i.e., there is a two-clock-cycle "pipeline" delay. The pipelined ADC architecture is generally associated with high speed ADCs, and in most cases the pipeline delay, or *latency*, is not a major system problem in most applications where this type of converter is used.

ADCs for DSP Applications

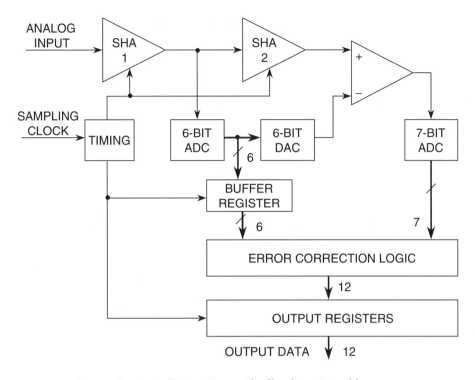

Figure 3-7: 12-Bit Two-Stage Pipelined ADC Architecture

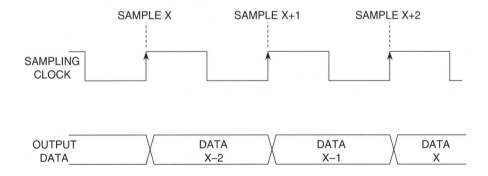

ABOVE SHOWS TWO CLOCK CYCLES PIPELINE DELAY

Figure 3-8: Typical Pipelined ADC Timing

Pipelined ADCs may have more than two clock cycles of latency depending on the particular architecture. For instance, the conversion could be done in three, or four, or perhaps even more pipelined stages causing additional latency in the output data.

Therefore, if the ADC is to be used in an event-triggered (or single-shot) mode where there must be a one-to-one time correspondence between each sample and the corresponding data, the pipeline delay can be troublesome and the SAR architecture is advantageous. Pipeline delay or latency can also be a problem in high speed servo-loop control systems or multiplexed applications. In addition, some pipelined converters have a *minimum* allowable conversion rate and must be kept running to prevent saturation of internal nodes.

Switched capacitor SAR ADCs generally have unbuffered input circuits similar to the circuit shown in Figure 3-9 for the AD7858/AD7859 ADC. During the acquisition time, the analog input must charge the 20 pF equivalent input capacitance to the correct value. If the input is a dc signal, then the source resistance, R_S, in series with the 125 Ω internal switch resistance creates a time constant. In order to settle to 12-bit accuracy, approximately nine time constants must be allowed for settling, and this defines the minimum allowable acquisition time. (Settling to 14 bits requires about 10 time constants, and 16 bits requires about 11).

$$t_{ACQ} > 9 \times (R_S + 125) \, \Omega \times 20 \, pF$$

For example, if $R_S = 50 \, \Omega$, the acquisition time per the above formula must be at least 310 ns.

For ac applications, a low impedance source should be used to prevent distortion due to the nonlinear ADC input circuit. In a single-supply application, a fast settling rail-to-rail op amp such as the AD820 should be used. Fast settling allows the op amp to settle quickly from the transient currents induced on its input by the internal ADC switches. In Figure 3-9, the AD820 drives a low-pass filter consisting of the 50 Ω series resistor and the 10 nF capacitor (cutoff frequency approximately 320 kHz). This filter removes high frequency components that could result in aliasing and decreases the noise.

Using a single-supply op amp in this application requires special consideration of signal levels. The AD820 is connected in the inverting mode and has a signal gain of –1. The noninverting input is biased at a common-mode voltage of 1.3 V with the 10.7 kΩ/10 kΩ divider, resulting in an output voltage of 2.6 V for $V_{IN} = 0$ V, and 0.1 V for $V_{IN} = 2.5$ V. This offset is provided because the AD820 output cannot go all the way to ground, but is limited to the V_{CESAT} of the output stage NPN transistor, which under these loading conditions is about 50 mV. The input range of the ADC is also offset by 100 mV by applying the 100 mV offset from the 412 Ω/10 kΩ divider to the AIN– input.

ADCs for DSP Applications

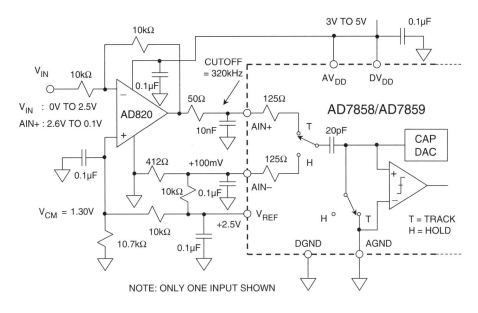

Figure 3-9: Driving Switched Capacitor Inputs of AD7858/AD7859 12-Bit, 200 kSPS ADC

Sigma-Delta (Σ-Δ) ADCs

James M. Bryant

Sigma-delta analog-digital converters (Σ-Δ ADCs) have been known for nearly 30 years, but only recently has the technology (high density digital VLSI) existed to manufacture them as inexpensive monolithic integrated circuits. They are now used in many applications where a low cost, low bandwidth, low power, high resolution ADC is required.

There have been innumerable descriptions of the architecture and theory of Σ-Δ ADCs, but most commence with a maze of integrals and deteriorate from there. In the Applications Department at Analog Devices, we frequently encounter engineers who do not understand the theory of operation of Σ-Δ ADCs and are convinced, from study of a typical published article, that it is too complex to easily comprehend.

There is nothing particularly difficult to understand about Σ-Δ ADCs, as long as one avoids the detailed mathematics, and this section has been written in an attempt to clarify the subject. A Σ-Δ ADC contains very simple analog electronics (a comparator, voltage reference, a switch, and one or more integrators and analog summing circuits), and quite complex digital computational circuitry. This circuitry consists of a digital signal processor (DSP) that acts as a filter (generally, but not invariably, a low-pass filter). It is not necessary to know precisely how the filter works to appreciate what it does. To understand how a Σ-Δ ADC works, familiarity with the concepts of *oversampling, quantization noise shaping, digital filtering,* and *decimation* is required.

- Low Cost, High Resolution (to 24 Bits)
- Excellent DNL
- Low Power, but Limited Bandwidth (Voice Band, Audio)
- Key Concepts are Simple, but Math is Complex
  - Oversampling
  - Quantization Noise Shaping
  - Digital Filtering
  - Decimation
- Ideal for Sensor Signal Conditioning
  - High Resolution
  - Self, System, and Autocalibration Modes
- Wide Applications in Voice-Band and Audio Signal Processing

Figure 3-10: Sigma-Delta ADCs

Let us consider the technique of oversampling with an analysis in the frequency domain. Where a dc conversion has a *quantization error* of up to 1/2 LSB, a sampled data system has *quantization noise.* A perfect classical N-bit sampling ADC has an RMS quantization noise of $q/\sqrt{12}$ uniformly distributed within the Nyquist band of DC to $f_s/2$ (where q is the value of an LSB and f_s is the sampling rate) as shown in Figure 3-11A. Therefore, its SNR with a full-scale sine wave input will be (6.02N + 1.76) dB. If the ADC is less than perfect, and its noise is greater than its theoretical minimum quantization noise, then its *effective* resolution will be less than N-bits. Its actual resolution (often known as its effective number of bits or ENOB) will be defined by

$$\text{ENOB} = \frac{\text{SNR} - 1.76 \text{ dB}}{6.02 \text{ dB}}$$

If we choose a much higher sampling rate, Kf_s (see Figure 3-11B), the rms quantization noise remains $q/\sqrt{12}$, but the noise is now distributed over a wider bandwidth dc to $Kf_s/2$. If we then apply a digital low-pass filter (LPF) to the output, we remove much of the quantization noise, but do not affect the wanted signal—so the ENOB is improved. We have accomplished a high resolution A/D conversion with a low resolution ADC. The factor K is generally referred to as the *oversampling ratio*. It should be noted at this point that oversampling has an added benefit in that it relaxes the requirements on the analog antialiasing filter.

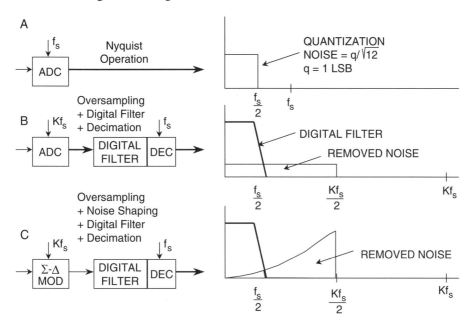

Figure 3-11: Oversampling, Digital Filtering, Noise Shaping, and Decimation

Since the bandwidth is reduced by the digital output filter, the output data rate may be lower than the original sampling rate (Kf_s) and still satisfy the Nyquist criterion. This may be achieved by passing every Mth result to the output and discarding the remainder. The process is known as "decimation" by a factor of M. Despite the origins of the term (*decem* is Latin for 10), M can have any integer value, provided that the output data rate is more than twice the signal bandwidth. Decimation does not cause any loss of information (see Figure 3-11B).

If we simply use oversampling to improve resolution, we must oversample by a factor of 2^{2N} to obtain an N-bit increase in resolution. The Σ-Δ converter does not need such a high oversampling ratio because it not only limits the signal pass band, but also shapes the quantization noise so that most of it falls outside this pass band as shown in Figure 3-11C.

If we take a 1-bit ADC (generally known as a comparator), drive it with the output of an integrator, and feed the integrator with an input signal summed with the output of a 1-bit DAC fed from the ADC output, we have a first-order Σ-Δ modulator as shown in Figure 3-12. Add a digital low-pass filter (LPF) and decimator at the digital output, and we have a Σ-Δ ADC: the Σ-Δ modulator shapes the quantization noise so that it

Section Three

lies above the pass band of the digital output filter, and the ENOB is therefore much larger than would otherwise be expected from the oversampling ratio.

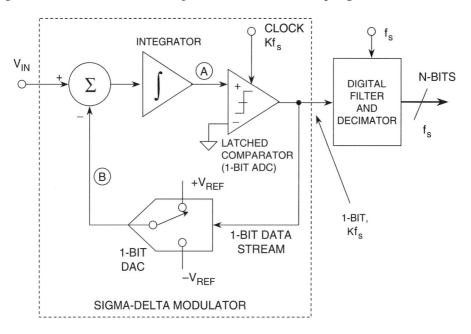

Figure 3-12: First-Order Sigma-Delta ADC

Intuitively, a Σ-Δ ADC operates as follows. Assume a dc input at V_{IN}. The integrator is constantly ramping up or down at node A. The output of the comparator is fed back through a 1-bit DAC to the summing input at node B. The negative feedback loop from the comparator output through the 1-bit DAC back to the summing point will force the average dc voltage at node B to be equal to V_{IN}. This implies that the average DAC output voltage must equal to the input voltage V_{IN}. The average DAC output voltage is controlled by the *ones-density* in the 1-bit data stream from the comparator output. As the input signal increases towards $+V_{REF}$, the number of "ones" in the serial bit stream increases, and the number of "zeros" decreases. Similarly, as the signal goes negative towards $-V_{REF}$, the number of "ones" in the serial bit stream decreases, and the number of "zeros" increases. From a very simplistic standpoint, this analysis shows that the average value of the input voltage is contained in the serial bit stream out of the comparator. The digital filter and decimator process the serial bit stream and produce the final output data.

The concept of noise shaping is best explained in the frequency domain by considering the simple Σ-Δ modulator model in Figure 3-13.

ADCs for DSP Applications

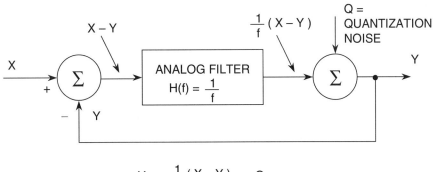

Figure 3-13: Simplified Frequency Domain Linearized Model of a Sigma-Delta Modulator

The integrator in the modulator is represented as an analog low-pass filter with a transfer function equal to H(f) = 1/f. This transfer function has an amplitude response that is inversely proportional to the input frequency. The 1-bit quantizer generates quantization noise, Q, which is injected into the output summing block. If we let the input signal be X, and the output Y, the signal coming out of the input summer must be X – Y. This is multiplied by the filter transfer function, 1/f, and the result goes to one input to the output summer. By inspection, we can then write the expression for the output voltage Y as:

$$Y = \frac{1}{f}(X - Y) + Q$$

This expression can easily be rearranged and solved for Y in terms of X, f, and Q:

$$Y = \frac{X}{f+1} + \frac{Q \times f}{f+1}$$

Note that as the frequency *f* approaches zero, the output voltage *Y* approaches *X* with no noise component. At higher frequencies, the amplitude of the signal component approaches zero, and the noise component approaches *Q*. At high frequency, the output consists primarily of quantization noise. In essence, the analog filter has a low-pass effect on the signal, and a high-pass effect on the quantization noise. Thus the analog filter performs the noise shaping function in the Σ-Δ modulator model.

For a given input frequency, higher order analog filters offer more attenuation. The same is true of Σ-Δ modulators, provided certain precautions are taken.

By using more than one integration and summing stage in the Σ-Δ modulator, we can achieve higher orders of quantization noise shaping and even better ENOB for a given oversampling ratio as is shown in Figure 3-14 for both a first- and second-order Σ-Δ modulator. The block diagram for the second-order Σ-Δ modulator is shown in Figure 3-15. Third, and higher order Σ-Δ ADCs were once thought to be potentially unstable at some values of input—recent analyses using *finite* rather than infinite gains in the comparator have shown that this is not necessarily so, but even if instability does start to occur, it is not important, since the DSP in the digital filter and decimator can be made to recognize incipient instability and react to prevent it.

Figure 3-16 shows the relationship between the order of the Σ-Δ modulator and the amount of oversampling necessary to achieve a particular SNR. For instance, if the oversampling ratio is 64, an ideal second-order system is capable of providing an SNR of about 80 dB. This implies approximately 13 effective number of bits (ENOB). Although the filtering done by the digital filter and decimator can be done to any degree of precision desirable, it would be pointless to carry more than 13 binary bits to the outside world. Additional bits would carry no useful signal information, and would be buried in the quantization noise unless postfiltering techniques were employed. Additional resolution can be obtained by increasing the oversampling ratio and/or by using a higher order modulator.

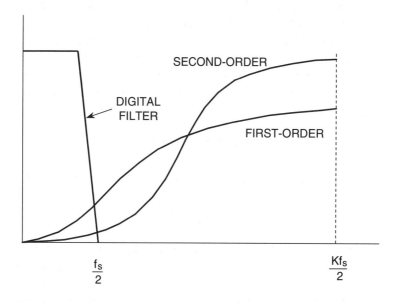

Figure 3-14: Sigma-Delta Modulators Shape Quantization Noise

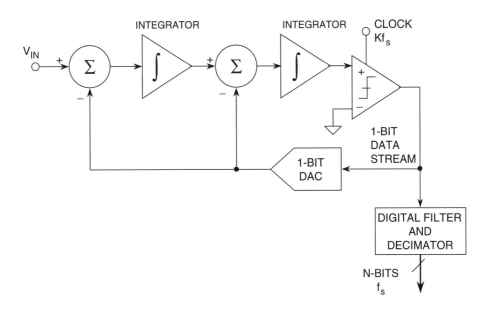

Figure 3-15: Second-Order Sigma-Delta ADC

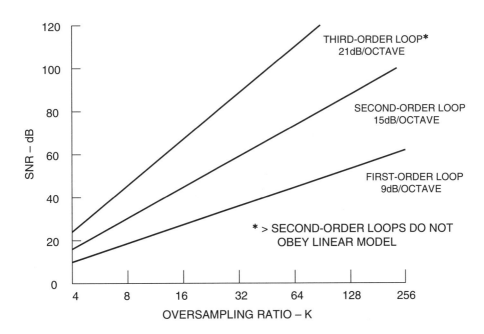

Figure 3-16: SNR vs. Oversampling Ratio for First-, Second-, and Third-Order Loops

The AD1877 is a 16-bit 48 kSPS stereo sigma-delta DAC suitable for demanding audio applications. Key specifications are summarized in Figure 3-17. This device has a 64× oversampling ratio and a fourth-order modulator. The internal digital filter is a linear phase FIR filter whose response is shown in Figure 3-18. The pass-band ripple is 0.006 dB, and the attenuation is greater than 90 dB in the stop band. The width of the transition region from pass band to stop band is only $0.1f_s$, where f_s is the effective sampling frequency of the AD1877 (maximum of 48 kSPS). Such a filter would obviously be impossible to implement in analog form.

- Single 5 V Power Supply
- Single-Ended Dual-Channel Analog Inputs
- 92 dB (typ) Dynamic Range
- 90 dB (typ) S/(THD+N)
- 0.006 dB Decimator Pass-Band Ripple
- Fourth-Order, 64× Oversampling Σ-Δ Modulator
- Three-Stage, Linear-Phase Decimator
- Less Than 100 mW (typ)
- Power-Down Mode
- Input Overrange Indication
- On-Chip Voltage Reference
- Flexible Serial Output Interface
- 28-Lead SOIC Package

Figure 3-17: AD1877 16-Bit, 48 kSPS Stereo Sigma-Delta ADC

All sigma-delta ADCs have a settling time associated with the internal digital filter, and there is no way to remove it. In multiplexed applications, the input to the ADC is a step function if there are different input voltages on adjacent channels. In fact, the multiplexer output can represent a full-scale step voltage to the sigma-delta ADC when channels are switched. Adequate filter settling time must be allowed, therefore, in such applications. This does not mean that sigma-delta ADCs shouldn't be used in multiplexed applications, just that the settling time of the digital filter must be considered.

For example, the group delay through the AD1877 FIR filter is $36/f_s$, and represents the time it takes for a step function input to propagate through one-half the number of taps in the digital filter. The total time required for settling is therefore $72f_s$, or approximately 1.5 ms when sampling at 48 kSPS with a 64× oversampling rate.

ADCs for DSP Applications

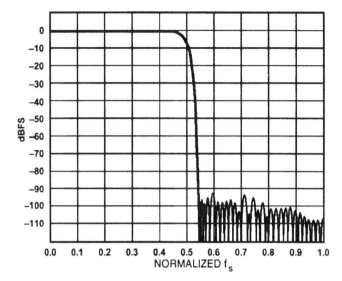

- f_s = OUTPUT WORD RATE, 32 kSPS, 44.1 kSPS, OR 48 kSPS TYPICAL
- PASS-BAND TO STOP-BAND TRANSITION REGION: $0.45f_s$ TO $0.55f_s$
- SETTLING TIME = $72/f_s$ = 1.5 ms FOR f_s = 48 kSPS
- GROUP DELAY = $36/f_s$ = 0.75 ms FOR f_s = 48 kSPS

Figure 3-18: AD1877 16-Bit, 48 kSPS Stereo Sigma-Delta ADC FIR Filter Characteristics

In other applications, such as low frequency, high resolution 24-bit measurement sigma-delta ADCs (such as the AD77xx series), other types of digital filters may be used. For instance, the SINC^3 response is popular because it has zeros at multiples of the throughput rate. For instance, a 10 Hz throughput rate produces zeros at 50 Hz and 60 Hz, which aid in ac power line rejection.

So far we have considered only sigma-delta converters that contain a single-bit ADC (comparator) and a single-bit DAC (switch). The block diagram of Figure 3-19 shows a multibit sigma-delta ADC that uses an n-bit flash ADC and an n-bit DAC. Obviously, this architecture will give a higher dynamic range for a given over-sampling ratio and order of loop filter. Stabilization is easier, since second-order and higher loops can be used. Idling patterns tend to be more random, thereby minimizing tonal effects.

The real disadvantage of this technique is that the linearity depends on the DAC linearity, and thin film laser trimming is generally required to approach 16-bit performance levels. This makes the multibit architecture extremely difficult to implement on sigma-delta ADCs. It is, however, currently used in sigma-delta audio DACs (AD1852, AD1853, AD1854) where special "bit scrambling" techniques are used to ensure linearity and eliminate idle tones.

Section Three

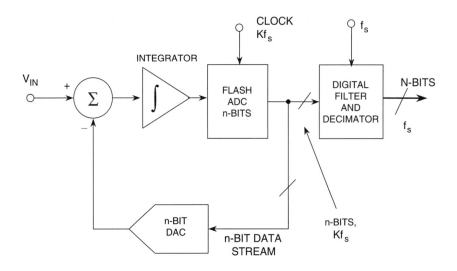

Figure 3-19: Multibit Sigma-Delta ADC

The Σ-Δ ADCs that we have described so far contain integrators, which are low-pass filters, whose pass band extends from dc. Thus, their quantization noise is pushed up in frequency. At present, most commercially available Σ-Δ ADCs are of this type (although some that are intended for use in audio or telecommunications applications contain band-pass rather than low-pass digital filters to eliminate any system dc offsets). But there is no particular reason why the filters of the Σ-Δ modulator should be LPFs, except that traditionally, ADCs have been thought of as being baseband devices, and that integrators are somewhat easier to construct than band-pass filters. If we replace the integrators in a Σ-Δ ADC with band-pass filters (BPFs) as shown in Figure 3-20, the quantization noise is moved up and down in frequency to leave a virtually noise-free region in the pass band (see Reference 1). If the digital filter is then programmed to have its pass band in this region, we have a Σ-Δ ADC with a band-pass, rather than a low-pass characteristic. Such devices would appear to be useful in direct IF-to-digital conversion, digital radios, ultrasound, and other undersampling applications. However, the modulator and the digital BPF must be designed for the specific set of frequencies required by the system application, thereby somewhat limiting the flexibility of this approach.

In an undersampling application of a band pass Σ-Δ ADC, the minimum sampling frequency must be at least twice the signal bandwidth, BW. The signal is centered around a carrier frequency, f_c. A typical digital radio application using a 455 kHz center frequency and a signal bandwidth of 10 kHz is described in Reference 1. An oversampling frequency Kf_s = 2 MSPS and an output rate f_s = 20 kSPS yielded a dynamic range of 70 dB within the signal bandwidth.

ADCs for DSP Applications

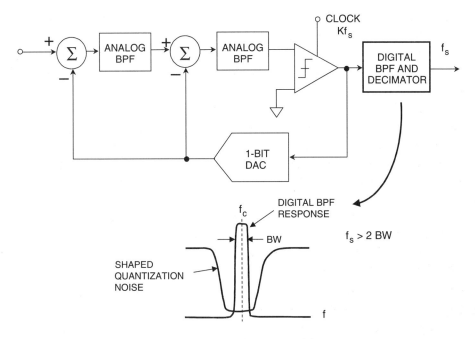

Figure 3-20: Replacing Integrators with Resonators Gives a Band-Pass Sigma-Delta ADC

Most sigma-delta ADCs generally have a fixed internal digital filter. The filter's cutoff frequency and the ADC output data rate scales with the master clock frequency. The AD7725 is a 16-bit sigma-delta ADC with a programmable internal digital filter. The modulator operates at a maximum oversampling rate of 19.2 MSPS. The modulator is followed by a preset FIR filter that decimates the modulator output by a factor of eight, yielding an output data rate of 2.4 MSPS. The output of the preset filter drives a programmable FIR filter. By loading the ROM with suitable coefficient values, this filter can be programmed for the desired frequency response.

The programmable filter is flexible with respect to number of taps and decimation rate. The filter can have up to 108 taps, up to five decimation stages, and a decimation factor between 2 and 256. Coefficient precision is 24 bits, and arithmetic precision is 30 bits.

The AD7725 contains Systolix's PulseDSP™ postprocessor, which permits the filter characteristics to be programmed through the parallel or serial microprocessor interface. Or, it may boot at power-on-reset from its internal ROM or from an external EPROM.

The postprocessor is a fully programmable core that provides processing power of up to 130 million multiply-accumulates (MACs) per second. To program the post processor, the user must produce a configuration file that contains the programming data for the filter function. This file is generated by a compiler available from Analog Devices. The AD7725 compiler accepts filter coefficient data as an input and automatically generates the required device programming data.

The coefficient file for the FIR filter response can be generated using a digital filter design package such as QEDesign from Momentum Data Systems. The response of the filter can be plotted so the user knows the response before generating the filter coefficients. The data is available to the processor at a 2.4 MSPS rate. When decimation is employed in a multistage filter, the first filter will be operated at 2.4 MSPS, and the user can then decimate between stages. The number of taps that can be contained in the processor is 108. Therefore, a single filter with 108 taps can be generated, or a multistage filter can be designed whereby the total number of taps adds up to 108. The filter characteristic can be low-pass, high-pass, band-stop, or band-pass.

The AD7725 operates on a single 5 V supply, has an on-chip 2.5 V reference, and is packaged in a 44-pin PQFP. Power dissipation is approximately 350 mW when operating at full power. A half-power mode is available with a master clock frequency of 10 MSPS maximum. Power consumption in the standby mode is 200 mW maximum. More details of the AD7725 operation can be found in Section 9.

Summary

A Σ-Δ ADC works by oversampling, wherein simple analog filters in the Σ-Δ modulator shape the quantization noise so that the SNR *in the bandwidth of interest* is much greater than would otherwise be the case, and by using high performance digital filters and decimation to eliminate noise outside the required pass band.

Oversampling has the added benefit of relaxing the requirements on the antialiasing filter. Because the analog circuitry is relatively undemanding, it may be built with the same digital VLSI process that is used to fabricate the DSP circuitry of the digital filter. Because the basic ADC is 1-bit (a comparator), the technique is inherently linear.

Although the detailed analysis of Σ-Δ ADCs involves quite complex mathematics, their basic design can be understood without the necessity of any mathematics at all. For further discussion on Σ-Δ ADCs, refer to References 1 through 18 at the end of this section.

- Inherently Excellent Linearity
- Oversampling Relaxes Analog Antialiasing Filter Requirements
- Ideal for Mixed-Signal IC Processes, No Trimming
- No SHA Required
- Added Functionality: On-Chip PGAs, Analog Filters, Autocalibration
- On-Chip Programmable Digital Filters (AD7725: Low-Pass, High-Pass, Band-Pass, Band-Stop)
- Upper Sampling Rate Currently Limits Applications to Measurement, Voice Band, and Audio, but Band-Pass Sigma-Delta Techniques May Change This
- Analog Multiplexer Switching Speed Limited by Internal Filter Settling Time. Consider One Sigma-Delta ADC per Channel.

Figure 3-21: Sigma-Delta Summary

Flash Converters

Flash ADCs (sometimes called *parallel* ADCs) are the fastest type of ADC and use large numbers of comparators. An N-bit flash ADC consists of 2^N resistors and 2^N-1 comparators arranged as in Figure 3-22. Each comparator has a reference voltage that is 1 LSB higher than that of the one below it in the chain. For a given input voltage, all the comparators below a certain point will have their input voltage larger than their reference voltage and a "1" logic output, and all the comparators above that point will have a reference voltage larger than the input voltage and a "0" logic output. The 2^N-1 comparator outputs therefore behave in a way analogous to a mercury thermometer, and the output code at this point is sometimes called a *thermometer* code. Since 2^N-1 data outputs are not really practical, they are processed by a decoder to an N-bit binary output.

The input signal is applied to all the comparators at once, so the thermometer output is delayed by only one comparator delay from the input, and the encoder N-bit output by only a few gate delays on top of that, so the process is very fast. However, the architecture uses large numbers of resistors and comparators and is limited to low resolutions and, if it is to be fast, each comparator must run at relatively high power levels. Hence, the problems of flash ADCs include limited resolution, high power dissipation because of the large number of high speed comparators (especially at sampling rates greater than 50 MSPS), and relatively large (and therefore expensive) chip sizes. In addition, the resistance of the reference resistor chain must be kept low to supply adequate bias current to the fast comparators, so the voltage reference has to source quite large currents (>10 mA).

Section Three

In practice, flash converters are available up to 10 bits, but more commonly they have 8 bits of resolution. Their maximum sampling rate can be as high as 1 GHz, with input full-power bandwidths in excess of 300 MHz.

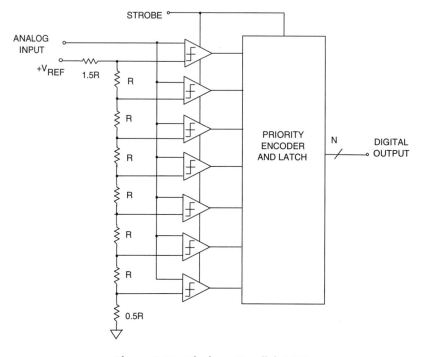

Figure 3-22: Flash or Parallel ADC

As mentioned earlier, full-power bandwidths are not necessarily full-resolution bandwidths. Ideally, the comparators in a flash converter are well matched both for dc and ac characteristics. Because the strobe is applied to all the comparators simultaneously, the flash converter is inherently a sampling converter. In practice, there are delay variations between the comparators and other ac mismatches that cause a degradation in ENOB at high input frequencies. This is because the inputs are slewing at a rate comparable to the comparator conversion time.

The input to a flash ADC is applied in parallel to a large number of comparators. Each has a voltage-variable junction capacitance, and this signal-dependent capacitance results in most flash ADCs having reduced ENOB and higher distortion at high input frequencies.

Adding 1 bit to the total resolution of a flash converter requires doubling the number of comparators. This limits the practical resolution of high speed flash converters to 8 bits because of excessive power dissipation.

ADCs for DSP Applications

However, in the AD9410 10-bit, 200 MSPS ADC, a technique called *interpolation* is used to minimize the number of preamplifiers in the flash converter comparators and also reduce the power (1.8 W). The method is shown in Figure 3-23.

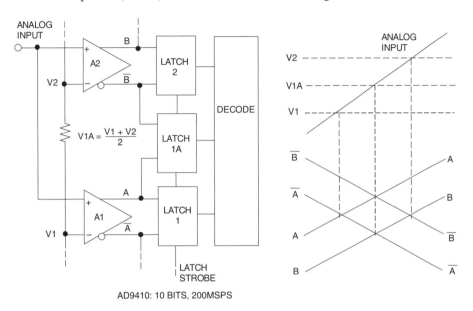

Figure 3-23: "Interpolating" Flash Reduces the Number of Preamplifiers by Factor of Two

The preamplifiers (labeled "A1," "A2,") are low gain g_m stages whose bandwidth is proportional to the tail currents of the differential pairs. Consider the case for a positive-going ramp input that is initially below the reference to AMP A1, V1. As the input signal approaches V1, the differential output of A1 approaches zero (i.e., A = $\overline{A}$), and the decision point is reached. The output of A1 drives the differential input of LATCH 1. As the input signal continues to go positive, A continues to go positive, and $\overline{B}$ begins to go negative. The interpolated decision point is determined when A = $\overline{B}$. As the input continues positive, the third decision point is reached when B = $\overline{B}$. This novel architecture reduces the ADC input capacitance and thereby minimizes its change with signal level and the associated distortion. The AD9410 also uses an input sample-and-hold circuit for improved ac linearity.

Subranging (Pipelined) ADCs

Although it is not practical to make flash ADCs with high resolution (greater than 10 bits), flash ADCs are often used as subsystems in "subranging" ADCs (sometimes known as "half-flash ADCs"), which are capable of much higher resolutions (up to 16 bits).

Section Three

A block diagram of an 8-bit subranging ADC based upon two 4-bit flash converters is shown in Figure 3-24. Although 8-bit flash converters are readily available at high sampling rates, this example will be used to illustrate the theory. The conversion process is done in two steps. The first four significant bits (MSBs) are digitized by the first flash (to better than 8-bit accuracy), and the 4-bit binary output is applied to a 4-bit DAC (again, better than 8-bits accurate). The DAC output is subtracted from the held analog input, and the resulting residue signal is amplified and applied to the second 4-bit flash. The outputs of the two 4-bit flash converters are then combined into a single 8-bit binary output word. If the residue signal range does not exactly fill the range of the second flash converter, nonlinearities and perhaps missing codes will result.

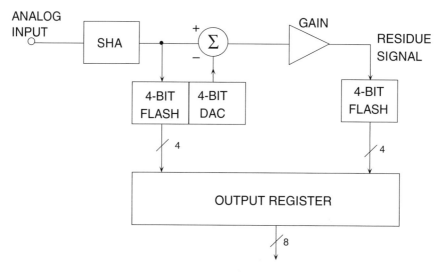

Figure 3-24: 8-Bit Subranging ADC

Modern subranging ADCs use a technique called *digital correction* to eliminate problems associated with the architecture of Figure 3-24. A simplified block diagram of a 12-bit digitally corrected subranging (DCS) ADC is shown in Figure 3-25. The architecture is similar to that used in the AD6640 12-bit, 65 MSPS ADC. Note that a 6-bit and a 7-bit ADC have been used to achieve an overall 12-bit output. These are not flash ADCs, but utilize a *magnitude-amplifier* (MagAmp™) architecture that will be described shortly.

If there were no errors in the first-stage conversion, the 6-bit "residue" signal applied to the 7-bit ADC by the summing amplifier would never exceed one-half of the range of the 7-bit ADC. The extra range in the second ADC is used in conjunction with the error correction logic (usually just a full adder) to correct the output data for most of the errors inherent in the traditional uncorrected subranging converter architecture. It is important to note that the 6-bit DAC must be better than 12 bits accurate, because the digital error correction does not correct for DAC errors. In practice, "thermometer" or "fully decoded" DACs using one current switch per level (63 switches in the case of a 6-bit DAC) are often used instead of a "binary" DAC to ensure excellent differential and integral linearity and minimum switching transients.

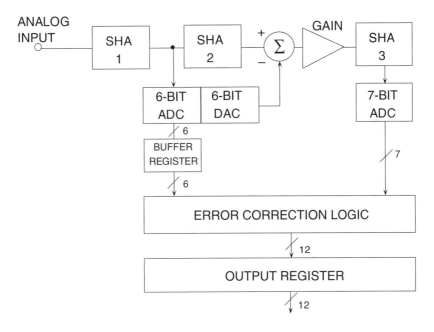

Figure 3-25: AD6640 12-Bit, 65 MSPS Pipelined Subranging ADC with Digital Error Correction

The second SHA delays the held output of the first SHA while the first-stage conversion occurs, thereby maximizing throughput. The third SHA serves to *deglitch* the residue output signal, thereby allowing a full conversion cycle for the 7-bit ADC to make its decision (the 6- and 7-bit ADCs in the AD6640 are bit-serial MagAmp ADCs that require more settling time than a flash converter).

This multistage conversion technique is sometimes referred to as "pipelining." Additional shift registers in series with the digital outputs of the first-stage ADC ensure that its output is ultimately time-aligned with the last 7 bits from the second ADC when their outputs are combined in the error correction logic. A pipelined ADC therefore has a specified number of clock cycles of *latency*, or *pipeline delay*, associated with the output data. The leading edge of the sampling clock (for sample N) is used to clock the output register, but the data that appears as a result of that clock edge corresponds to sample N – L, where L is the number of clock cycles of latency. In the case of the AD6640, there are two clock cycles of latency.

Section Three

The error correction scheme described above is designed to correct for errors made in the first conversion. Internal ADC gain, offset, and linearity errors are corrected as long as the residue signal falls within the range of the second-stage ADC. These errors will not affect the linearity of the overall ADC transfer characteristic. Errors made in the final conversion, however, do translate directly as errors in the overall transfer function. Also, linearity errors or gain errors either in the DAC or the residue amplifier will not be corrected and will show up as nonlinearities or nonmonotonic behavior in the overall ADC transfer function.

So far, we have considered only two-stage subranging ADCs, as these are easiest to analyze. There is no reason to stop at two stages, however. Three-pass and four-pass subranging pipelined ADCs are quite common, and can be made in many different ways, usually with digital error correction.

A simplified block diagram of the AD9220 12-bit, 10 MSPS single-supply, 250 mW CMOS ADC is shown in Figure 3-26. The AD9221 (1.25 MSPS, 60 mW) and the AD9223 (3 MSPS, 100 mW) ADCs use the identical architecture but operate at lower power and lower sampling rates. This is a four-stage pipelined architecture with an additional bit in the second, third, and fourth stage for error correction. Because of the pipelined architecture, these ADCs have a three-clock-cycle latency (see Figure 3-27).

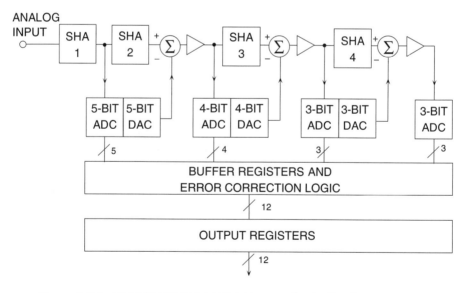

Figure 3-26: AD9220/AD9221/AD9223 12-Bit Pipelined CMOS ADC

ADCs for DSP Applications

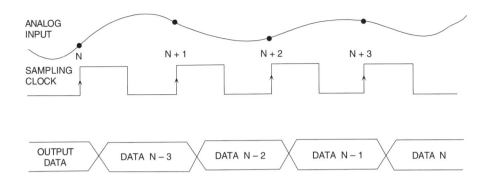

Figure 3-27: Latency (Pipeline Delay) of AD9220/AD9221/AD9223 ADC

Bit-Per-Stage (Serial or Ripple) ADCs

Various architectures exist for performing A/D conversion using one stage per bit. In fact, a multistage subranging ADC with one bit per stage and no error correction is one form. Figure 3-28 shows the overall concept. The SHA holds the input signal constant during the conversion cycle. There are N stages, each of which have a bit output and a residue output. The residue output of one stage is the input to the next. The last bit is detected with a single comparator as shown.

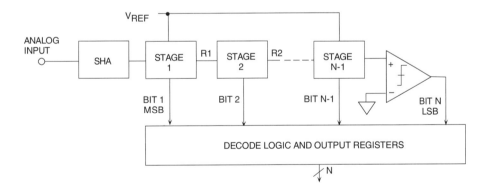

Figure 3-28: Bit-Per-Stage, Serial, or Ripple ADC

The basic stage for performing a single binary bit conversion is shown in Figure 3-29. It consists of a gain-of-two amplifier, a comparator, and a 1-bit DAC. Assume that this is the first stage of the ADC. The MSB is simply the polarity of the input, and that is detected with the comparator, which also controls the 1-bit DAC. The 1-bit DAC output is summed with the output of the gain-of-two amplifier. The resulting residue output is then applied to the next stage. In order to better understand how the circuit works, the diagram shows the residue output for the case of a linear ramp input voltage that traverses the entire ADC range, $-V_R$ to $+V_R$. Notice that the polarity of the residue output determines the binary bit output of the next stage.

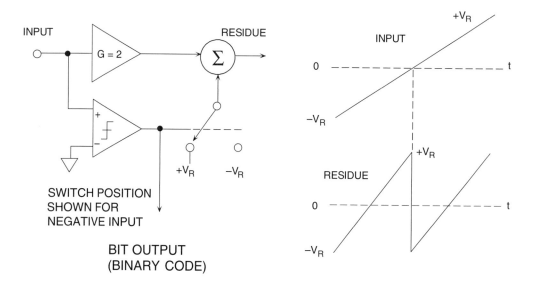

Figure 3-29: Single Stage of Binary ADC

A simplified 3-bit serial-binary ADC is shown in Figure 3-30, and the residue outputs are shown in Figure 3-31. Again, the case is shown for a linear ramp input voltage whose range is between $-V_R$ and $+V_R$. Each residue output signal has discontinuities that correspond to the point where the comparator changes state and causes the DAC to switch. The fundamental problem with this architecture is the discontinuity in the residue output waveforms. Adequate settling time must be allowed for these transients to propagate through all the stages and settle at the final comparator input. The prospects of making this architecture operate at high speed are therefore dismal.

ADCs for DSP Applications

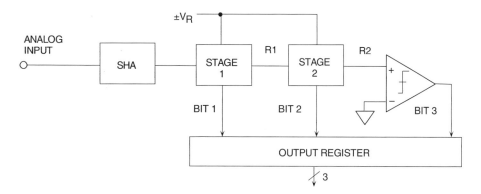

Figure 3-30: 3-Bit Serial ADC with Binary Output

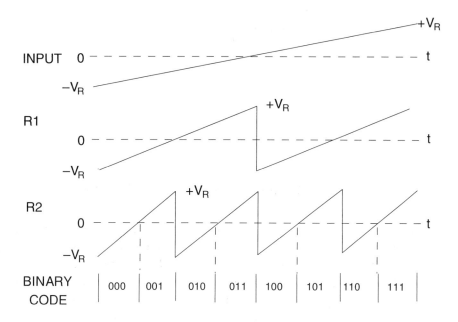

Figure 3-31: Input and Residue Waveforms of 3-Bit Binary Ripple ADC

A much better bit-per-stage architecture was developed by F.D. Waldhauer (Reference 21) based on absolute value amplifiers (magnitude amplifiers, or simply *MagAmps*). This scheme has often been referred to as *serial-Gray* (since the output coding is in Gray code), or *folding* converter (References 22, 23, 24). The basic stage is shown functionally in Figure 3-32 along with its transfer function. The input to the stage is assumed to be a linear ramp voltage whose range is between $-V_R$ and $+V_R$. The comparator detects the polarity of the input signal and provides the Gray bit output for the stage. It also determines whether the overall stage gain is +2 or −2. The reference voltage V_R is summed with the switch output to generate the residue signal that is applied to the next stage. The polarity of the residue signal determines the Gray bit for the next stage. The transfer function for the folding stage is also shown in Figure 3-32.

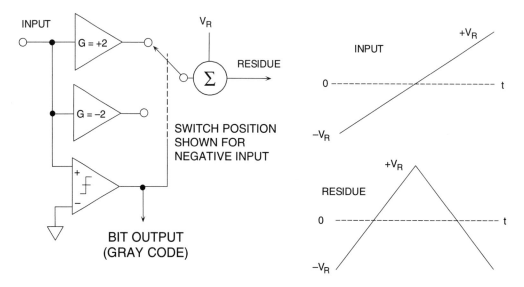

Figure 3-32: MagAmp Stage Functional Equivalent Circuit

A 3-bit MagAmp folding ADC is shown in Figure 3-33, and the corresponding residue waveforms in Figure 3-34. As in the case of the binary ripple ADC, the polarity of the residue output signal of a stage determines the value of the Gray bit for the next stage. The polarity of the input to the first stage determines the Gray MSB; the polarity of R1 output determines the Gray bit-2; and the polarity of R2 output determines the Gray bit-3. Notice that unlike the binary ripple ADC, there is no abrupt transition in any of the folding stage residue output waveforms. This makes operation at high speeds quite feasible.

The key to operating this architecture at high speeds is the folding stage. Early designs (see References 22, 23, 24) used discrete op amps with diodes inside the feedback loop to generate the folding transfer function. Modern IC circuit designs implement the transfer function using current-steering open-loop gain techniques which can be made to operate much faster. Fully differential stages (including the SHA) also provide speed, lower distortion, and yield 8-bit accurate folding stages with no requirement for thin film resistor laser trimming (see References 25, 26, 27).

ADCs for DSP Applications

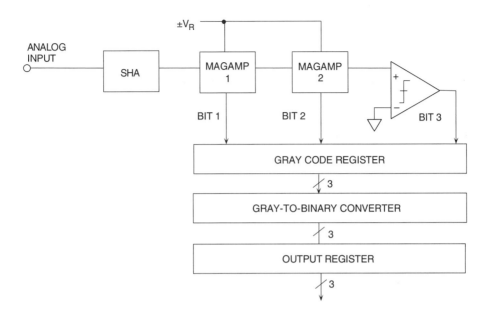

Figure 3-33: 3-Bit MagAmp (Folding) ADC Block Diagram

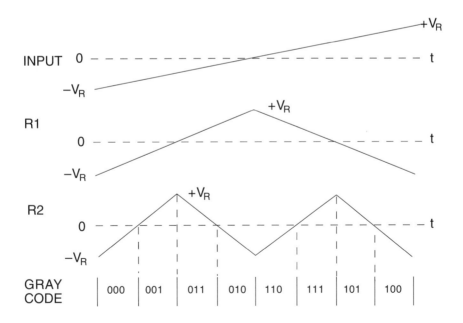

Figure 3-34: Input and Residue Waveforms for 3-Bit MagAmp ADC

Section Three

The MagAmp architecture can be extended to sampling rates previously dominated by flash converters. The AD9288-100 8-bit, 100 MSPS dual ADC is shown in Figure 3-35. The first five bits (Gray code) are derived from five differential MagAmp stages. The differential residue output of the fifth MagAmp stage drives a 3-bit flash converter, rather than a single comparator. The Gray-code output of the five MagAmps and the binary code output of the 3-bit flash are latched, all converted into binary, and latched again in the output data register.

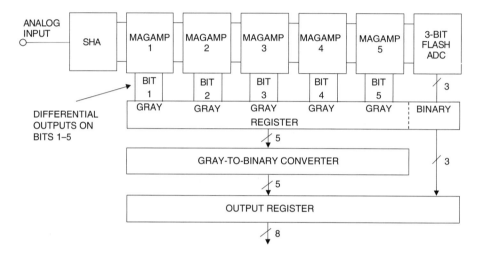

Figure 3-35: AD9288-100 Dual 8-Bit, 100 MSPS ADC Functional Diagram

References

1. S. A. Jantzi, M. Snelgrove and P. F. Ferguson Jr., *A 4th-Order Bandpass Sigma-Delta Modulator*, **IEEE Journal of Solid State Circuits,** Vol. 38, No. 3, March 1993, pp. 282–291.

2. **System Applications Guide**, Analog Devices, Inc., 1993, Section 14.

3. **Mixed Signal Design Seminar**, Analog Devices, Inc., 1991, Section 6.

4. **AD77XX-Series Data Sheets**, Analog Devices, http://www.analog.com.

5. **Linear Design Seminar**, Analog Devices, Inc., 1995, Section 8.

6. J. Dattorro, A. Charpentier, D. Andreas, *The Implementation of a One-Stage Multirate 64:1 FIR Decimator for Use in One-Bit Sigma-Delta A/D Applications*, **AES 7th International Conference**, May 1989.

7. W.L. Lee and C.G. Sodini, *A Topology for Higher-Order Interpolative Coders*, **ISCAS PROC. 1987**.

8. P.F. Ferguson, Jr., A. Ganesan and R. W. Adams, *One Bit Higher Order Sigma-Delta A/D Converters*, **ISCAS PROC. 1990**, Vol. 2, pp. 890–893.

9. R. Koch, B. Heise, F. Eckbauer, E. Engelhardt, J. Fisher, and F. Parzefall, *A 12-Bit Sigma-Delta Analog-to-Digital Converter with a 15 MHz Clock Rate*, **IEEE Journal of Solid-State Circuits**, Vol. SC-21, No. 6, December 1986.

10. Wai Laing Lee, *A Novel Higher Order Interpolative Modulator Topology for High Resolution Oversampling A/D Converters*, **MIT Masters Thesis,** June 1987.

11. D. R. Welland, B. P. Del Signore and E. J. Swanson, *A Stereo 16-Bit Delta-Sigma A/D Converter for Digital Audio*, **J. Audio Engineering Society**, Vol. 37, No. 6, June 1989, pp. 476–485.

12. R. W. Adams, *Design and Implementation of an Audio 18-Bit Analog-to-Digital Converter Using Oversampling Techniques*, **J. Audio Engineering Society**, Vol. 34, March 1986, pp. 153-166.

13. B. Boser and Bruce Wooley, *The Design of Sigma-Delta Modulation Analog-to-Digital Converters*, **IEEE Journal of Solid-State Circuits**, Vol. 23, No. 6, December 1988, pp. 1298–1308.

14. Y. Matsuya, et. al., *A 16-Bit Oversampling A/D Conversion Technology Using Triple-Integration Noise Shaping*, **IEEE Journal of Solid-State Circuits**, Vol. SC-22, No. 6, December 1987, pp. 921–929.

15. Y. Matsuya, et. al., *A 17-Bit Oversampling D/A Conversion Technology Using Multistage Noise Shaping*, **IEEE Journal of Solid-State Circuits**, Vol. 24, No. 4, August 1989, pp. 969–975.

16. P. Ferguson, Jr., A. Ganesan, R. Adams, et. al., *An 18-Bit 20 kHz Dual Sigma-Delta A/D Converter*, **ISSCC Digest of Technical Papers**, February 1991.

17. Steven Harris, *The Effects of Sampling Clock Jitter on Nyquist Sampling Analog-to-Digital Converters and on Oversampling Delta Sigma ADCs*, **Audio Engineering Society Reprint 2844 (F-4)**, October, 1989.

18. Max W. Hauser, *Principles of Oversampling A/D Conversion,* **Journal Audio Engineering Society**, Vol. 39, No. 1/2, January/February 1991, pp. 3-26.

19. Daniel H. Sheingold, **Analog-Digital Conversion Handbook**, Third Edition, Prentice-Hall, 1986.

20. Chuck Lane, *A 10-bit 60MSPS Flash ADC*, **Proceedings of the 1989 Bipolar Circuits and Technology Meeting**, IEEE Catalog No. 89CH2771-4, September 1989, pp. 44–47.

21. F.D. Waldhauer, *Analog to Digital Converter*, **U.S. Patent 3-187-325**, 1965.

22. J.O. Edson and H.H. Henning, *Broadband Codecs for an Experimental 224Mb/s PCM Terminal*, **Bell System Technical Journal**, 44, November 1965, pp. 1887–1940.

23. **J.S. Mayo,** *Experimental 224Mb/s PCM Terminals*, **Bell System Technical Journal**, 44, November 1965, pp. 1813–1941.

24. Hermann Schmid, **Electronic Analog/Digital Conversions**, Van Nostrand Reinhold Company, New York, 1970.

25. Carl Moreland, *An 8-Bit 150 MSPS Serial ADC*, 1**995 ISSCC Digest of Technical Papers**, Vol. 38, p. 272.

26. Roy Gosser and Frank Murden, *A 12-Bit 50 MSPS Two-Stage A/D Converter*, **1995 ISSCC Digest of Technical Papers**, p. 278.

27. Carl Moreland, **An Analog-to-Digital Converter Using Serial-Ripple Architecture**, Masters' Thesis, Florida State University College of Engineering, Department of Electrical Engineering, 1995.

28. **Practical Analog Design Techniques**, Analog Devices, 1995, Chapters 4, 5, and 8.

29. **Linear Design Seminar**, Analog Devices, 1995, Chapter 4, 5.

30. **System Applications Guide**, Analog Devices, 1993, Chapters 12, 13, 15, 16.

31. **Amplifier Applications Guide**, Analog Devices, 1992, Chapter 7.

32. Walt Kester, *Drive Circuitry is Critical to High-Speed Sampling ADCs*, **Electronic Design Special Analog Issue**, Nov. 7, 1994, pp. 43–50.

33. Walt Kester, *Basic Characteristics Distinguish Sampling A/D Converters*, **EDN**, Sept. 3, 1992, pp. 135–144.

34. Walt Kester, *Peripheral Circuits Can Make or Break Sampling ADC Systems*, **EDN**, Oct. 1, 1992, pp. 97–105.

35. Walt Kester, *Layout, Grounding, and Filtering Complete Sampling ADC System*, **EDN**, Oct. 15, 1992, pp. 127–134.

36. **High Speed Design Techniques**, Analog Devices, 1996, Chapters 4, 5.

Section 4
DACs for DSP Applications

- DAC Structures
- Low Distortion DAC Architectures
- DAC Logic
- Sigma-Delta DACs
- Direct Digital Synthesis (DDS)

Section 4
DACs for DSP Applications
Walt Kester and James Bryant

DAC Structures

The most commonly used DAC structures (other than a simple 1-bit DAC based on a single switch used with a reference voltage) are binary weighted DACs or ladder networks, but these, though relatively simple in structure, require quite complex analysis. We will start by examining one of the simplest structures of all, the Kelvin divider shown in Figure 4-1. An N-bit version of this DAC simply consists of 2^N equal resistors in series. The output is taken from the appropriate tap by closing one of the 2^N switches by decoding 1 of 2^N switches from the N-bit data. Recent DACs using this architecture are referred to as "string DACs."

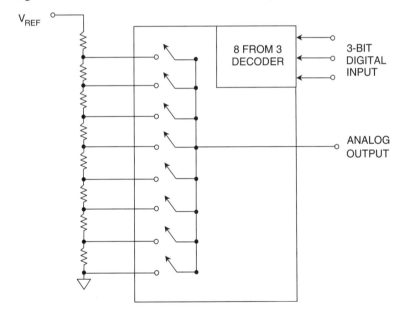

Figure 4-1: Simplest Voltage Output DAC: The Kelvin Divider ("String DAC")

This architecture is simple, has a voltage output (but a code-varying Z_{OUT}), and is inherently monotonic (even if a resistor is zero, $OUTPUT_N$ cannot exceed $OUTPUT_{N+1}$). It is linear if all the resistors are equal, but may be made deliberately nonlinear if a nonlinear DAC is required. Since only two switches operate during a transition, it is a low glitch architecture. Its major drawback is the large number of resistors required

Section Four

for high resolution, and as a result it is not commonly used—but, as we shall see later, it is used as a component in more complex DAC structures.

There is an analogous current output DAC that consists, again, of 2^N resistors (or current sources) but, in this case, they are all connected in parallel between the reference voltage input and the virtual ground output (see Figure 4-2).

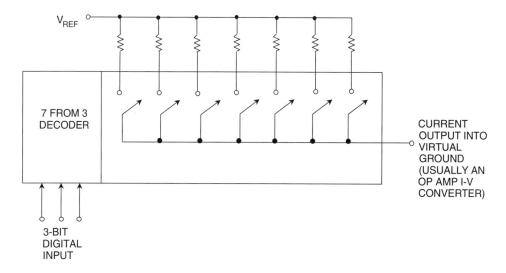

Figure 4-2: The Simplest Current Output DAC

In this DAC, once a resistor is switched into circuit by increasing digital code, any further increases do not switch it out again. The structure is thus inherently monotonic, regardless of inaccuracies in the resistors and, as in the previous case, may be made intentionally nonlinear where a specific nonlinearity is required. Again, as in the previous case, the architecture is rarely, if ever, used to fabricate a complete DAC because of the large numbers of resistors and switches required. However, it is often used as a component in a more complex DAC structure.

Unlike the Kelvin divider, this type of DAC does not have a unique name, although both types are referred to as *fully decoded DACs* or *thermometer DACs* or *string DACs*.

Fully decoded DACs are often used as components of more complex DACs. The most common are "segmented DACs" where part of the output of a fully decoded DAC is further subdivided. The structure is used because the fully decoded DAC is inherently monotonic, so if the subdivision is also monotonic, the whole resulting DAC is also monotonic.

DACs for DSP Applications

A voltage segmented DAC (see Figure 4-3) works by further subdividing the voltage across one resistor of a Kelvin divider. The subdivision may be done with a further Kelvin divider (in which case the whole structure is known as a "Kelvin-Varley divider") or with some other DAC structure.

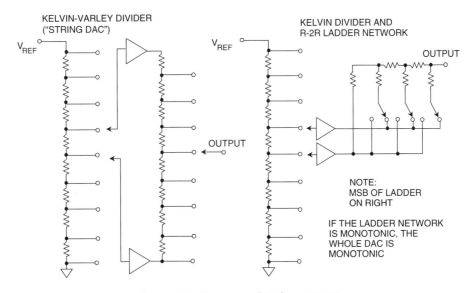

Figure 4-3: Segmented Voltage DACs

In all DACs, the output is the product of the reference voltage and the digital code, so in that sense, all DACs are multiplying DACs, but many DACs operate well only over a limited range of V_{REF}. True MDACs, however, are designed to operate over a wide range of V_{REF}. A strict definition of a multiplying DAC demands that its reference voltage range includes 0 V, and many, especially current mode ladder networks with CMOS switches, permit positive, negative, and ac V_{REF}. DACs that do not work down to 0 V. V_{REF} are still useful, however, and types where V_{REF} can vary by 10:1 or so are often called MDACs, although a more accurate description might be "semimultiplying" DACs.

Low Distortion DAC Architectures

Because of the emphasis in communications systems on DDS DACs with high SFDR, much effort has been placed on determining optimum DAC architectures. Practically all low distortion high speed DACs make use of some form of nonsaturating current-mode switching. A straight binary DAC with one current switch per bit produces code-dependent glitches as discussed above and is certainly not the most optimum architecture (Figure 4-4). A DAC with one current source per code level can be shown not to have code-dependent glitches, but it is not practical to implement for high resolutions. However, this performance can be approached by decoding the first few

Section Four

MSBs into a "thermometer" code and have one current switch per level. For example, a 5-bit thermometer DAC would have an architecture similar to that shown in Figure 4-5.

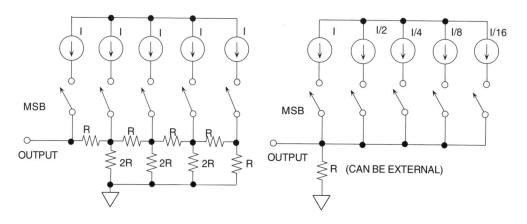

Figure 4-4: 5-Bit Binary DAC Architectures

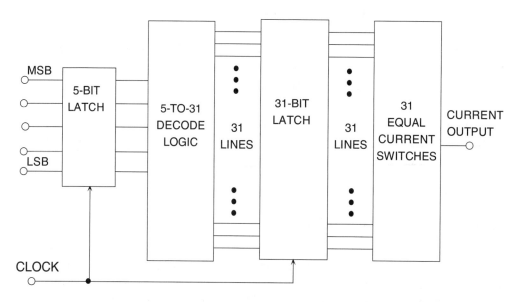

Figure 4-5: 5-Bit "Thermometer" or "Fully-Decoded" DAC Minimizes Code-Dependent Glitches

The input binary word is latched and then decoded into 31 outputs that drive a second latch. The output of the second latch drives 31 equally weighted current switches whose outputs are summed together. This scheme effectively removes nearly all the code dependence of the output glitch. The residual glitch that does occur at the output is equal, regardless of the output code change (it is code-independent) and can be filtered because it occurs at the DAC update frequency and its harmonics. The distortion mechanisms associated with the fully decoded architecture are primarily asymmetrical output slewing, finite switch turn-on and turn-off times, and integral nonlinearity.

The obvious disadvantage of this type of thermometer DAC is the large number of latches and switches required to make a 14-, 12-, 10-, or even 8-bit DAC. However, if this technique is used on the 5 MSBs of an 8-, 10-, 12-, or 14-bit DAC, a significant reduction in the code-dependent glitch is possible. This process is called *segmentation* and is quite common in low distortion DACs.

Figure 4-6 shows a scheme whereby the first five bits of a 10-bit DAC are decoded as described above and drive 31 equally weighted switches. The last five bits are derived from binarily weighted current sources. Equally weighted current sources driving an R/2R resistor ladder could be used to drive the LSBs; however, this approach requires thin film resistors, which are not generally available on a low cost CMOS process. Also, the use of R/2R networks lowers the DAC output impedance, thereby requiring more drive current to develop the same voltage across a fixed load resistance.

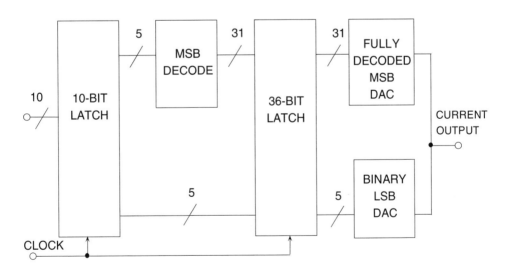

Figure 4-6: 10-Bit Segmented DAC

The AD9772 14-bit, 150 MSPS TxDAC uses three sections of segmentation as shown in Figure 4-7. Other members of the AD977x-family and the AD985x-family also use this same core.

The first five bits (MSBs) are fully decoded and drive 31 equally weighted current switches, each supplying 512 LSBs of current. The next four bits are decoded into 15 lines that drive 15 current switches, each supplying 32 LSBs of current. The five LSBs are latched and drive a traditional binary weighted DAC that supplies 1 LSB per output level. A total of 51 current switches and latches are required to implement this architecture.

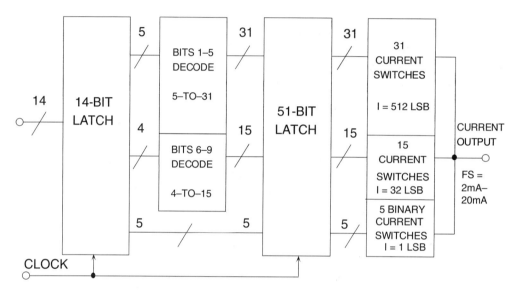

Figure 4-7: AD9772 TxDAC 14-Bit CMOS DAC Core

The basic current switching cell is made up of a differential PMOS transistor pair as shown in Figure 4-8. The differential pairs are driven with low level logic to minimize switching transients and time skew. The DAC outputs are symmetrical differential currents that help to minimize even-order distortion products (especially when driving a differential output such as a transformer or an op amp differential I/V converter).

The overall architecture of the AD977x TxDAC family and the AD985x DDS family is an excellent trade-off between power/performance and allows the entire DAC function to be implemented on a standard CMOS process with no thin film resistors. Single-supply operation on 3.3 V or 5 V make the devices extremely attractive for portable and low power applications.

DACs for DSP Applications

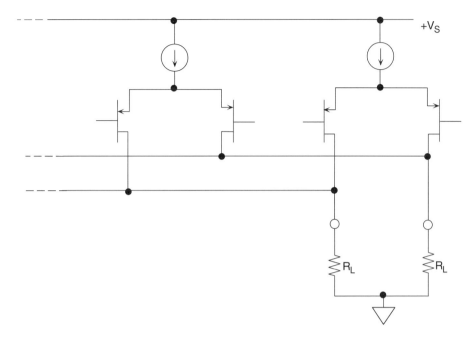

Figure 4-8: PMOS Transistor Current Switches

DAC Logic

The earliest monolithic DACs contained little, if any, logic circuitry, and parallel data had to be maintained on the digital input to maintain the digital output. Today almost all DACs have input latches, and data need only be written once, not maintained.

There are innumerable variations of DAC input structures, but the majority today are "double-buffered." A double-buffered DAC has two sets of latches. Data is initially latched in the first rank and subsequently transferred to the second as shown in Figure 4-9. There are three reasons why this arrangement is useful.

The first is that it allows data to enter the DAC in many different ways. A DAC without a latch, or with a single latch, must be loaded with all bits at once, in parallel, since otherwise its output during loading may be totally different from what it was or what it is to become. A double-buffered DAC, on the other hand, may be loaded with parallel data, serial data, or with 4-bit or 8-bit words, and the output will be unaffected until the new data is completely loaded and the DAC receives its update instruction.

Section Four

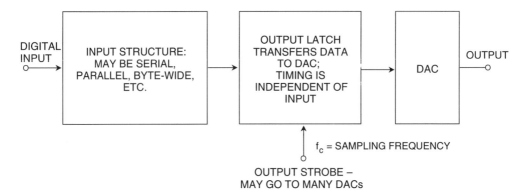

Figure 4-9: Double-Buffered DAC Permits Complex Input Structures and Simultaneous Update

The second feature of this type of input structure is that the output clock can operate at a fixed frequency (the DAC update rate), while the input latch can be loaded asynchronously. This is useful in real-time signal reconstruction applications.

The third convenience of the double-buffered structure is that many DACs may be updated simultaneously; data is loaded into the first rank of each DAC in turn, and when all is ready, the output buffers of all DACs are updated at once. There are many DAC applications where the output of several DACs must change simultaneously, and the double-buffered structure allows this to be done very easily.

Most early monolithic high resolution DACs had parallel or byte-wide data ports and tended to be connected to parallel data buses and address decoders and addressed by microprocessors as if they were very small write-only memories (some DACs are not write-only, but can have their contents read as well—this is convenient for some applications but is not very common). A DAC connected to a data bus is vulnerable to capacitive coupling of logic noise from the bus to the analog output, and many DACs today have serial data structures. These are less vulnerable to such noise (since fewer noisy pins are involved), use fewer pins and therefore take less space, and are frequently more convenient for use with modern microprocessors, many of which have serial data ports. Some, but not all, of such serial DACs have data outputs as well as data inputs so that several DACs may be connected in series and data clocked to them all from a single serial port. The arrangement is referred to as "daisy-chaining."

Another development in DACs is the ability to make several on a single chip, which is useful to reduce PCB sizes and assembly costs. Today it is possible to buy sixteen 8-bit, eight 12-bit, four 14-bit, or two 16-/18-/20-/22-/24-bit DACs in a single package. In the future, even higher densities are probable.

DACs for DSP Applications

Interpolating DACs

In ADC-based systems, oversampling can ease the requirements on the antialiasing filter, and a sigma-delta ADC has this inherent advantage. In a DAC-based system (such as DDS), the concept of interpolation can be used in a similar manner. This concept is common in digital audio CD players, where the basic update rate of the data from the CD is about 44 kSPS. "Zeros" are inserted into the parallel data, thereby increasing the effective update rate to four times, eight times, or 16 times the fundamental throughput rate. The 4×, 8×, or 16× data stream is passed through a digital interpolation filter, which generates the extra data points. The high oversampling rate moves the image frequencies higher, thereby allowing a less complex filter with a wider transition band. The sigma-delta 1-bit DAC architecture represents the ultimate extension of this concept and has become popular in modern CD players.

The same concept can be applied to a high speed DAC. Assume a traditional DAC is driven at an input word rate of 30 MSPS (see Figure 4-10A). Assume the DAC output frequency is 10 MHz. The image frequency component at 30 − 10 = 20 MHz must be attenuated by the analog antialiasing filter, and the transition band of the filter is 10 MHz to 20 MHz. Assume that the image frequency must be attenuated by 60 dB. The filter must therefore go from a pass band of 10 MHz to 60 dB stop-band attenuation over the transition band lying between 10 MHz and 20 MHz (one octave). A Butterworth filter design gives 6 dB attenuation per octave for each pole. Therefore, a minimum of 10 poles are required to provide the desired attenuation. Filters become even more complex as the transition band becomes narrower.

Assume that we increase the DAC update rate to 60 MSPS and insert a "zero" between each original data sample. The parallel data stream is now 60 MSPS, but we must now determine the value of the zero-value data points. This is done by passing the 60 MSPS data stream with the added zeros through a digital interpolation filter which computes the additional data points. The response of the digital filter relative to the 2× oversampling frequency is shown in Figure 4-10B. The analog antialiasing filter transition zone is now 10 MHz to 50 MHz (the first image occurs at $2f_c - f_o = 60 - 10 = 50$ MHz). This transition zone is a little greater than two octaves, implying that a five- or six-pole Butterworth filter is sufficient.

The AD9772 is a 2× oversampling interpolating 14-bit DAC, and a simplified block diagram is shown in Figure 4-11. The device is designed to handle 14-bit input word rates up to 150 MSPS. The output word rate is 300 MSPS maximum. For an output frequency of 60 MHz, an update rate of 150 MHz, and an oversampling ratio of 2, the image frequency occurs at 300 MHz − 60 MHz = 240 MHz. The transition band for the analog filter is therefore 60 MHz to 240 MHz. Without oversampling, the image frequency occurs at 150 MHz − 60 MHz = 90 MHz, and the filter transition band is 60 MHz to 90 MHz.

Section Four

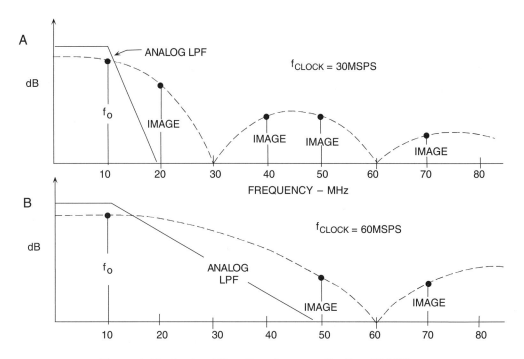

Figure 4-10: Analog Filter Requirements for f_o = 10 MHz:
f_c = 30 MSPS, and f_c = 60 MSPS

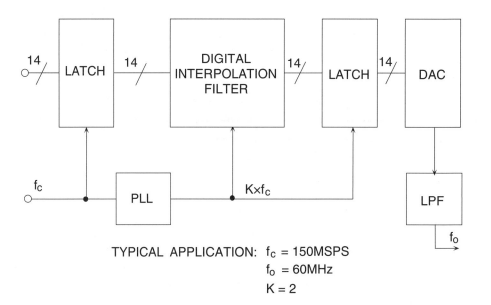

Figure 4-11: AD9772 14-Bit, 150 MSPS Interpolating TxDAC

Sigma-Delta DACs

Another way of obtaining high resolution is to use oversampling techniques and a 1-bit DAC. The technique, known as sigma-delta (S-D), is computation intensive, so has only recently become practical for the manufacture of high resolution DACs; since it uses a 1-bit DAC, it is intrinsically linear and monotonic.

A Σ-Δ DAC, unlike the Σ-Δ ADC, is mostly digital (see Figure 4-12A). It consists of an "interpolation filter" (a digital circuit that accepts data at a low rate, inserts zeros at a high rate, and then applies a digital filter algorithm and outputs data at a high rate), a Σ-Δ modulator (which effectively acts as a low-pass filter to the signal but as a high-pass filter to the quantization noise, and converts the resulting data to a high speed bit stream), and a 1-bit DAC whose output switches between equal positive and negative reference voltages. The output is filtered in an external analog LPF. Because of the high oversampling frequency, the complexity of the LPF is much less than the case of traditional Nyquist operation.

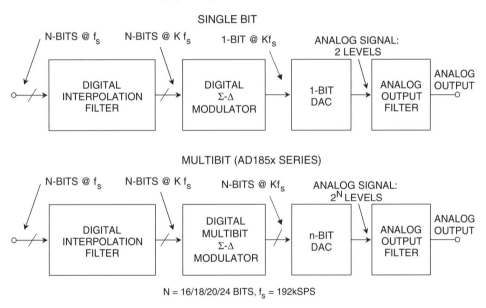

Figure 4-12: Sigma-Delta (Σ-Δ) DACs: 1-Bit and Multibit

It is possible to use more than one bit in the DAC, and this leads to the *multibit* architecture shown in Figure 4-12B. The concept is similar to that of the interpolating DAC previously discussed, with the addition of the digital sigma-delta modulator. In the past, multibit DACs have been difficult to design because of the accuracy requirement on the n-bit internal DAC (this DAC, although only n-bits, must have the linearity of the final number of bits, N). The AD185x series of audio DACs, however, use a proprietary *data scrambling* technique (*called direct data scrambling, or D^2S*)

Section Four

which overcomes this problem and produces excellent performance with respect to THD + N. For instance, the AD1853 dual 24-bit, 192 kSPS DAC has greater than 115 dB THD + N at a 48 kSPS sampling rate.

Direct Digital Synthesis (DDS)

A frequency synthesizer generates multiple frequencies from one or more frequency references. These devices have been used for decades, especially in communications systems. Many are based upon switching and mixing frequency outputs from a bank of crystal oscillators. Others have been based upon well-understood techniques utilizing phase-locked loops (PLLs). This mature technology is illustrated in Figure 4-13. A fixed-frequency reference drives one input of the phase comparator. The other phase comparator input is driven from a divide-by-N counter which, in turn, is driven by a voltage-controlled oscillator (VCO). Negative feedback forces the output of the internal loop filter to a value that makes the VCO output frequency N-times the reference frequency. The time constant of the loop is controlled by the loop filter. There are many trade-offs in designing a PLL, such as phase noise, tuning speed, and frequency resolution; and there are many good references on the subject (see References 1, 2, and 3).

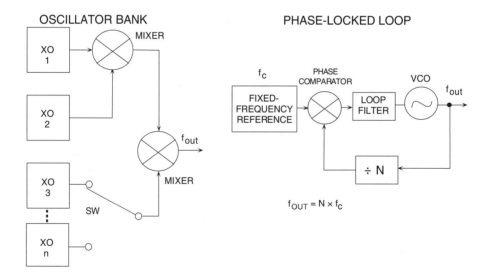

Figure 4-13: Frequency Synthesis Using Oscillators and Phase-Locked Loops

With the widespread use of digital techniques in instrumentation and communications systems, a digitally controlled method of generating multiple frequencies from a reference frequency source has evolved, called direct digital synthesis (DDS). The basic architecture is shown in Figure 4-14. In this simplified model, a stable clock drives a programmable read-only memory (PROM) that stores one or more integral number of cycles of a sine wave (or other arbitrary waveform, for that matter). As the address counter steps through each memory location, the corresponding digital amplitude of the signal at each location drives a DAC, which in turn generates the analog output signal. The spectral purity of the final analog output signal is determined primarily by the DAC. The phase noise is basically that of the reference clock.

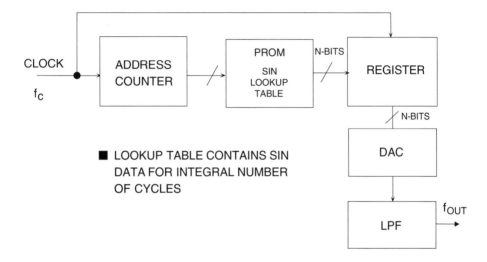

Figure 4-14: Fundamental Direct Digital Synthesis System

The DDS system differs from the PLL in several ways. Because a DDS system is a sampled data system, all the issues involved in sampling must be considered, including quantization noise, aliasing, and filtering. For instance, the higher order harmonics of the DAC output frequencies fold back into the Nyquist bandwidth, making them unfilterable, whereas the higher order harmonics of the output of PLL-based synthesizers can be filtered. Other considerations will be discussed shortly.

A fundamental problem with this simple DDS system is that the final output frequency can be changed only by changing the reference clock frequency or by reprogramming the PROM, making it rather inflexible. A practical DDS system implements this basic function in a much more flexible and efficient manner using digital hardware called a numerically controlled oscillator (NCO). A block diagram of such a system is shown in Figure 4-15.

Section Four

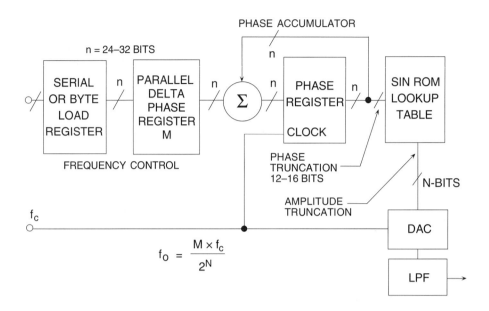

Figure 4-15: A Flexible DDS System

The heart of the system is the *phase accumulator* whose contents are updated once each clock cycle. Each time the phase accumulator is updated, the digital number, M, stored in the *delta phase register* is added to the number in the phase accumulator register. Assume that the number in the delta phase register is 00...01 and that the initial contents of the phase accumulator are 00...00. The phase accumulator is updated by 00...01 on each clock cycle. If the accumulator is 32 bits wide, 2^{32} clock cycles (over 4 billion) are required before the phase accumulator returns to 00...00, and the cycle repeats.

The truncated output of the phase accumulator serves as the address to a sine (or cosine) look-up table. Each address in the look-up table corresponds to a phase point on the sine wave from 0° to 360°. The look-up table contains the corresponding digital amplitude information for one complete cycle of a sine wave. (Actually, only data for 90° is required because the quadrature data is contained in the two MSBs.) The look-up table therefore maps the phase information from the phase accumulator into a digital amplitude word, which in turn drives the DAC.

Consider the case for n = 32, and M = 1. The phase accumulator steps through each of 2^{32} possible outputs before it overflows. The corresponding output sine wave frequency is equal to the clock frequency divided by 2^{32}. If M = 2, the phase accumulator register "rolls over" twice as fast, and the output frequency is doubled. This can be generalized as follows.

For an n-bit phase accumulator (n generally ranges from 24 to 32 in most DDS systems), there are 2^N possible phase points. The digital word in the delta phase register, M, represents the amount the phase accumulator is incremented each clock cycle. If f_c is the clock frequency, the frequency of the output sine wave is equal to:

$$f_o = \frac{M \times f_c}{2^N}$$

This equation is known as the DDS "tuning equation." Note that the frequency resolution of the system is equal to $f_c/2^n$. For n = 32, the resolution is greater than one part in four billion. In a practical DDS system, all the bits out of the phase accumulator are not passed on to the look-up table, but are truncated, leaving only the first 12 to 16 MSBs. This reduces the size of the lookup table and does not affect the frequency resolution. The phase truncation adds only a small but acceptable amount of phase noise to the final output; the bulk of the output distortion comes from the DAC itself.

The basic DDS system described above is extremely flexible and has high resolution. The frequency can be changed instantaneously with no phase discontinuity by simply changing the contents of the M register. However, practical DDS systems first require the execution of a serial, or byte-loading sequence to get the new frequency word into an internal buffer register that precedes the parallel-output M register. This is done to minimize package lead count. After the new word is loaded into the buffer register, the parallel-output delta-phase register is clocked, thereby changing all the bits simultaneously. The number of clock cycles required to load the delta-phase buffer register determines the maximum rate at which the output frequency can be changed.

The AD9850 125 MSPS DDS system (Figure 4-16) uses a 32-bit phase accumulator, which is truncated to 14 bits (MSBs) before being passed to the look-up table. The final digital output is 10 bits to the internal DAC. The AD9850 allows the output phase to be modulated using an additional register and an adder placed between the output of the phase accumulator register and the input to the look-up table. The AD9850 uses a 5-bit word to control the phase, which allows shifting the phase in increments of 180°, 90°, 45°, 22.5°, 11.25°, and any combination thereof. The device also contains an internal high speed comparator that can be configured to accept the (externally) filtered output of the DAC to generate a low jitter output pulse suitable for driving the sampling clock input of an ADC. The full-scale output current can be adjusted from 10 mA to 20 mA using a single external resistor, and the output voltage compliance is 1 V.

The frequency tuning (delta-phase register input word) and phase modulation words are loaded into the AD9850 via a parallel or serial loading format. The parallel load format consists of five consecutive loads of an 8-bit control word (byte). The first 8-bit byte controls phase modulation (five bits), power-down enable (one bit), and loading format (two bits). Bytes 2–5 comprise the 32-bit frequency tuning word. The maximum control register update frequency is 23 MHz. Serial loading of the AD9850 is accomplished via a 40-bit serial data stream on a single pin. Maximum update rate of the control register in the serial-load mode is 3 MHz.

Section Four

The AD9850 consumes only 380 mW of power on a single 5 V supply at a maximum 125 MSPS clock rate. The device is available in a 28-lead surface-mount SSOP (Shrink Small Outline Package).

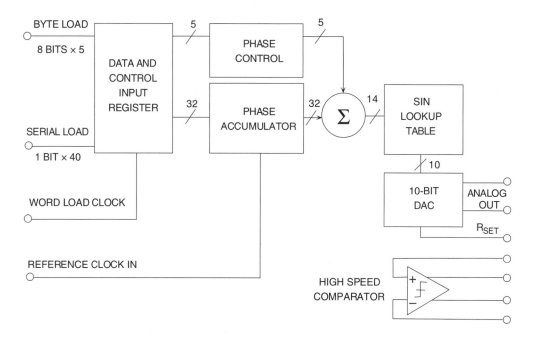

Figure 4-16: AD9850 CMOS 125 MSPS DDS/DAC Synthesizer

Analog Devices offers a number of DDS systems for a variety of applications. The AD983x family consists of low cost 10-bit systems with clock frequencies up to 50 MSPS. The AD985x family offers 10-bit and 12-bit systems with clock frequencies up to 300 MSPS and additional functions such as quadrature and phase modulation, chirp-mode capability, and programmable on-chip reference clock multipliers.

References

1. R.E. Best, **Phase-Locked Loops**, McGraw-Hill, New York, 1984.
2. F.M. Gardner, **Phaselock Techniques**, 2nd Edition, John Wiley, New York, 1979.
3. *Phase-Locked Loop Design Fundamentals*, Application Note AN-535, Motorola, Inc.
4. **The ARRL Handbook for Radio Amateurs**, American Radio Relay League, Newington, CT, 1992.
5. Richard J. Kerr and Lindsay A. Weaver, *Pseudorandom Dither for Frequency Synthesis Noise*, United States Patent Number 4,901,265, February 13, 1990.
6. Henry T. Nicholas, III and Henry Samueli, *An Analysis of the Output Spectrum of Direct Digital Frequency Synthesizers in the Presence of Phase-Accumulator Truncation*, **IEEE 41st Annual Frequency Control Symposium Digest of Papers**, 1987, pp. 495–502, IEEE Publication No. CH2427-3/87/0000-495.
7. Henry T. Nicholas, III and Henry Samueli, *The Optimization of Direct Digital Frequency Synthesizer Performance in the Presence of Finite Word Length Effects*, **IEEE 42nd Annual Frequency Control Symposium Digest of Papers**, 1988, pp. 357–363, IEEE Publication No. CH2588-2/88/0000-357.

Section 5
Fast Fourier Transforms

- The Discrete Fourier Transform
- The Fast Fourier Transform
- FFT Hardware Implementation and Benchmarks
- DSP Requirements for Real-Time FFT Applications
- Spectral Leakage and Windowing

Section 5
Fast Fourier Transforms
Walt Kester

The Discrete Fourier Transform

In 1807 the French mathematician and physicist Jean Baptiste Joseph Fourier presented a paper to the *Institut de France* on the use of sinusoids to represent temperature distributions. The paper made the controversial claim that *any continuous periodic signal could be represented by the sum of properly chosen sinusoidal waves.* Among the publication review committee were two famous mathematicians: Joseph Louis Lagrange, and Pierre Simon de Laplace. Lagrange objected strongly to publication on the basis that Fourier's approach would not work with signals having discontinuous slopes, such as square waves. Fourier's work was rejected, primarily because of Lagrange's objection, and was not published until the death of Lagrange, some 15 years later. In the meantime, Fourier's time was occupied with political activities, expeditions to Egypt with Napoleon, and trying to avoid the guillotine after the French Revolution. (This bit of history extracted from Reference 1, p. 141.)

It turns out that both Fourier and Lagrange were at least partially correct. Lagrange was correct that a summation of sinusoids cannot *exactly* form a signal with a corner. However, you can get *very* close if enough sinusoids are used. (This is described by the *Gibbs effect,* and is well understood by scientists, engineers, and mathematicians today.)

Fourier analysis forms the basis for much of digital signal processing. Simply stated, the Fourier transform (there are actually several members of this family) allows a time domain signal to be converted into its equivalent representation in the frequency domain. Conversely, if the frequency response of a signal is known, the inverse Fourier transform allows the corresponding time domain signal to be determined.

In addition to frequency analysis, these transforms are useful in filter design, since the frequency response of a filter can be obtained by taking the Fourier transform of its impulse response. Conversely, if the frequency response is specified, the required impulse response can be obtained by taking the inverse Fourier transform of the frequency response. Digital filters can be constructed based on their impulse response, because the coefficients of an FIR filter and its impulse response are identical.

The Fourier transform family (*Fourier Transform, Fourier Series, Discrete Time Fourier Series*, and *Discrete Fourier Transform*) is shown in Figure 5-2. These accepted definitions have evolved (not necessarily logically) over the years and depend upon whether the signal is *continuous–aperiodic, continuous–periodic, sampled–aperiodic, or sampled–periodic.* In this context, the term *sampled* is the same as *discrete* (i.e., a *discrete* number of time samples).

Section Five

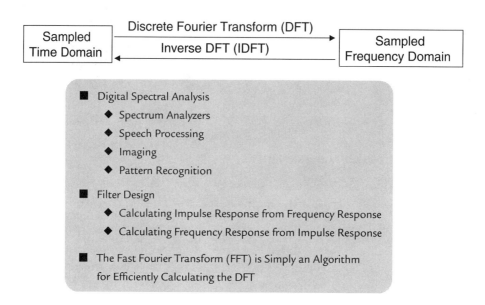

Figure 5-1: Applications of the Discrete Fourier Transform (DFT)

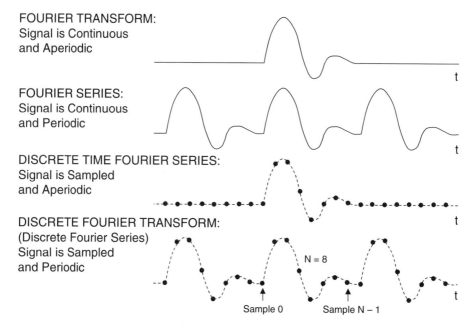

Figure 5-2: Fourier Transform Family
As a Function of Time Domain Signal Type

Fast Fourier Transforms

The only member of this family that is relevant to digital signal processing is the *Discrete Fourier Transform (DFT)*, which operates on a *sampled* time domain signal that is *periodic*. The signal must be periodic in order to be decomposed into the summation of sinusoids. However, only a finite number of samples (N) are available for inputting into the DFT. This dilemma is overcome by placing an infinite number of groups of the same N samples "end-to-end," thereby forcing mathematical (but not real-world) periodicity as shown in Figure 5-2.

The fundamental analysis equation for obtaining the N-point DFT is as follows:

$$X(k) = \frac{1}{N}\sum_{n=0}^{N-1} x(n) e^{-j2\pi nk/N} = \frac{1}{n}\sum_{n=0}^{N-1} x(n)\left[\cos(2\pi nk/N) - j\sin(2\pi nk/N)\right]$$

At this point, some terminology clarifications are in order regarding the above equation (also see Figure 5-3). X(k) (capital letter X) represents the DFT frequency output at the kth spectral point, where k ranges from 0 to N − 1. The quantity N represents the number of sample points in the DFT data frame.

Note that "N" should not be confused with ADC or DAC resolution, which is also given by the quantity N in other places in this book.

The quantity x(n) (lower case letter x) represents the nth time sample, where n also ranges from 0 to N − 1. In the general equation, x(n) can be real or complex.

Notice that the cosine and sine terms in the equation can be expressed in either polar or rectangular coordinates using Euler's equation:

$$e^{j\theta} = \cos + j\sin\theta$$

The DFT output spectrum, X(k), is the correlation between the input time samples and N cosine and N sine waves. The concept is best illustrated in Figure 5-4. In this figure, the real part of the first four output frequency points is calculated; therefore, only the cosine waves are shown. A similar procedure is used with sine waves in order to calculate the imaginary part of the output spectrum.

The first point, X(0), is simply the sum of the input time samples, because cos(0) = 1. The scaling factor, 1/N, is not shown, but must be present in the final result. Note that X(0) is the average value of the time samples, or simply the dc offset. The second point, ReX(1), is obtained by multiplying each time sample by each corresponding point on a cosine wave that makes one complete cycle in the interval N and summing the results. The third point, ReX(2), is obtained by multiplying each time sample by each corresponding point of a cosine wave that has two complete cycles in the interval N and then summing the results. Similarly, the fourth point, ReX(3), is obtained by multiplying each time sample by the corresponding point of a cosine wave that has three complete cycles in the interval N and summing the results. This process continues until all N outputs have been computed. A similar procedure is followed using sine waves in order to calculate the imaginary part of the frequency spectrum. The cosine and sine waves are referred to as *basis functions*.

Section Five

- A periodic signal can be decomposed into the sum of properly chosen cosine and sine waves (Jean Baptiste Joseph Fourier, 1807)

- The DFT operates on a finite number (N) of digitized time samples, x(n). When these samples are repeated and placed "end-to-end," they appear periodic to the transform.

- The complex DFT output spectrum X(k) is the result of correlating the input samples with sine and cosine basis functions:

$$X(k) = \frac{1}{N}\sum_{n=0}^{N-1} x(n)\, e^{\frac{-j2\pi nk}{N}} = \frac{1}{N}\sum_{n=0}^{N-1} x(n)\left[\cos\frac{2\pi nk}{N} - j\sin\frac{2\pi nk}{N}\right]$$

$$0 \leq k \leq N-1$$

Figure 5-3: The Discrete Fourier Transform (DFT)

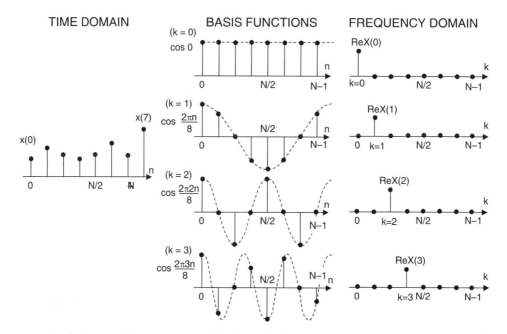

Figure 5-4: Correlation of Time Samples with Basis Functions Using the DFT for N = 8

Assume that the input signal is a cosine wave having a period of N, i.e., it makes one complete cycle during the data window. Also assume its amplitude and phase is identical to the first cosine wave of the basis functions, cos(2pn/8). The output ReX(1) contains a single point, and all the other ReX(k) outputs are zero. Assume that the input cosine wave is now shifted to the right by 90°. The correlation between it and the corresponding basis function is zero. However, there is an additional correlation required with the basis function sin(2pn/8) to yield ImX(1). This shows why both real and imaginary parts of the frequency spectrum need to be calculated in order to determine both the amplitude and phase of the frequency spectrum.

Notice that the correlation of a sine/cosine wave of any frequency other than that of the basis function produces a zero value for both ReX(1) and ImX(1).

A similar procedure is followed when using the *inverse* DFT (IDFT) to reconstruct the time domain samples, x(n), from the frequency domain samples X(k). The synthesis equation is given by:

$$x(n) = \sum_{k=0}^{N-1} X(k) e^{j2\pi nk/N} = \sum_{k=0}^{N-1} X(k)\left[\cos(2\pi nk/N) + j\sin(2\pi nk/N)\right]$$

There are two basic types of DFTs: *real* and *complex*. The equations shown in Figure 5-5 are for the complex DFT, where the input and output are both complex numbers. Since time domain input samples are real and have no imaginary part, the imaginary part of the input is always set to zero. The output of the DFT, X(k), contains a real and imaginary component that can be converted into amplitude and phase.

The *real* DFT, although somewhat simpler, is basically a simplification of the *complex* DFT. Most FFT routines are written using the complex DFT format, therefore understanding the complex DFT and how it relates to the real DFT is important. For instance, if you know the real DFT frequency outputs and want to use a complex inverse DFT to calculate the time samples, you need to know how to place the real DFT output points into the complex DFT format before taking the complex inverse DFT.

Figure 5-6 shows the input and output of a real and a complex FFT. Notice that the output of the real DFT yields real and imaginary X(k) values, where k ranges from only 0 to N/2. Note that the imaginary points ImX(0) and ImX(N/2) are always zero because sin(0) and sin(np) are both always zero.

The frequency domain output X(N/2) corresponds to the frequency output at one-half the sampling frequency, f_s. The width of each frequency bin is equal to f_s/N.

Section Five

Frequency Domain ← ← DFT ← ← Time Domain

$$X(k) = \frac{1}{N}\sum_{n=0}^{N-1} x(n)\, e^{\frac{-j2\pi nk}{N}} = \frac{1}{N}\sum_{n=0}^{N-1} x(n)\left[\cos\frac{2\pi nk}{N} - j\sin\frac{2\pi nk}{N}\right]$$

$$\boxed{W_N = e^{\frac{-j2\pi}{N}}} \qquad = \frac{1}{N}\sum_{n=0}^{N-1} x(n)\, W_N^{nk}, \quad 0 \le k \le N-1$$

Time Domain ← ← INVERSE DFT ← ← Frequency Domain

$$x(n) = \sum_{k=0}^{N-1} X(k)\, e^{\frac{j2\pi nk}{N}} = \sum_{k=0}^{N-1} X(k)\left[\cos\frac{2\pi nk}{N} + j\sin\frac{2\pi nk}{N}\right]$$

$$= \sum_{k=0}^{N-1} X(k)\, W_N^{-nk}, \quad 0 \le n \le N-1$$

Figure 5-5: The Complex DFT

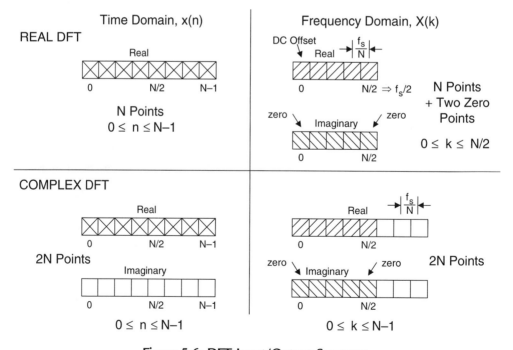

Figure 5-6: DFT Input/Output Spectrum

124

Fast Fourier Transforms

The complex DFT has real and imaginary values both at its input and output. In practice, the imaginary parts of the time domain samples are set to zero. If you are given the output spectrum for a complex DFT, it is useful to know how to relate them to the real DFT output and vice versa. The crosshatched areas in the diagram correspond to points that are common to both the real and complex DFT.

Figure 5-7 shows the relationship between the real and complex DFT in more detail. The real DFT output points are from 0 to N/2, with ImX(0) and ImX(N/2) always zero. The points between N/2 and N − 1 contain the negative frequencies in the complex DFT. Note that ReX(N/2 + 1) has the same value as Re(N/2 − 1), ReX(N/2 + 2) has the same value as ReX(N/2 − 2), and so on. Also, note that ImX(N/2 + 1) is the negative of ImX(N/2 − 1), ImX(N/2 + 2) is the negative of ImX(N/2 − 2), and so on. In other words, ReX(k) has *even symmetry* about N/2 and ImX(k) *odd symmetry* about N/2. In this way, the negative frequency components for the complex FFT can be generated if you are only given the real DFT components.

The equations for the complex and the real DFT are summarized in Figure 5-8. Note that the equations for the complex DFT work nearly the same whether taking the DFT, X(k) or the IDFT, x(n). The real DFT does not use complex numbers, and the equations for X(k) and x(n) are significantly different. Also, before using the x(n) equation, ReX(0) and ReX(N/2) must be divided by two. These details are explained in Chapter 31 of Reference 1, and the reader should study this chapter before attempting to use these equations.

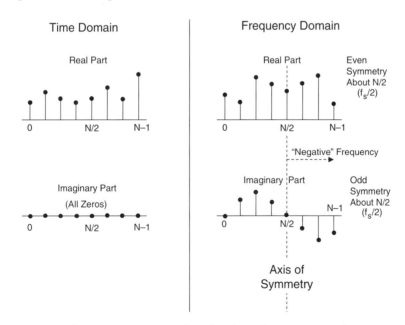

Figure 5-7: Constructing the Complex DFT Negative Frequency Components from the Real DFT

Section Five

The DFT output spectrum can be represented in either polar form (magnitude and phase) or rectangular form (real and imaginary) as shown in Figure 5-9. The conversion between the two forms is straightforward.

COMPLEX TRANSFORM

$$X(k) = \frac{1}{N} \sum_{n=0}^{N-1} x(n) e^{\frac{-j2\pi nk}{N}}$$

$$x(n) = \sum_{k=0}^{N-1} X(k) e^{\frac{j2\pi nk}{N}}$$

Time Domain: x(n) is complex, discrete, and periodic. n runs from 0 to N−1

Frequency Domain: X(k) is complex, discrete, and periodic. k runs from 0 to N−1
k = 0 to N/2 are positive frequencies.
k = N/2 to N−1 are negative frequencies

REAL TRANSFORM

$$ReX(k) = \frac{2}{N} \sum_{n=0}^{N-1} x(n) \cos(2\pi nk/N)$$

$$ImX(k) = \frac{-2}{N} \sum_{n=0}^{N-1} x(n) \sin(2\pi nk/N)$$

$$x(n) = \sum_{k=0}^{N/2} \left[ReX(k) \cos(2\pi nk/N) - ImX(k) \cos(2\pi nk/N) \right]$$

Time Domain: x(n) is real, discrete, and periodic.
n runs from 0 to N − 1

Frequency domain:
ReX(k) is real, discrete, and periodic.
ImX(k) is real, discrete, and periodic.
k runs from 0 to N/2

Before using x(n) equation, ReX(0) and ReX(N/2) must be divided by two.

Figure 5-8: Complex and Real DFT Equations

- $X(k) = ReX(k) + j\,ImX(k)$
- $MAG[X(k)] = \sqrt{ReX(k)^2 + ImX(k)^2}$
- $\varphi[X(k)] = \tan^{-1} \dfrac{ImX(k)}{ReX(k)}$

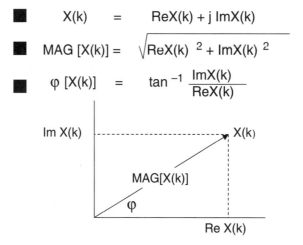

Figure 5-9: Converting Real and Imaginary DFT Outputs into Magnitude and Phase

Fast Fourier Transforms

The Fast Fourier Transform

In order to understand the development of the FFT, consider first the 8-point DFT expansion shown in Figure 5-10. In order to simplify the diagram, note that the quantity W_N is defined as:

$$W_N = e^{-j2\pi/N}$$

This leads to the definition of the *twiddle factors* as:

$$W_N^{nk} = e^{-j2\pi nk/N}$$

The twiddle factors are simply the sine and cosine basis functions written in polar form. Note that the 8-point DFT shown in the diagram requires 64 complex multiplications. In general, an N-point DFT requires N^2 complex multiplications. The number of multiplications required is significant because the multiplication function requires a relatively large amount of DSP processing time. In fact, the total time required to compute the DFT is directly proportional to the number of multiplications plus the required amount of overhead.

$$X(k) = \frac{1}{N}\sum_{n=0}^{N-1} x(n)\, e^{\frac{-j2\pi nk}{N}} = \frac{1}{N}\sum_{n=0}^{N-1} x(n)\, W_N^{nk} \qquad \boxed{W_N = e^{\frac{-j2\pi}{N}}}$$

$X(0) =$	$x(0)W_8^0 + x(1)W_8^0 + x(2)W_8^0 + x(3)W_8^0 + x(4)W_8^0 + x(5)W_8^0 + x(6)W_8^0 + x(7)W_8^0$
$X(1) =$	$x(0)W_8^0 + x(1)W_8^1 + x(2)W_8^2 + x(3)W_8^3 + x(4)W_8^4 + x(5)W_8^5 + x(6)W_8^6 + x(7)W_8^7$
$X(2) =$	$x(0)W_8^0 + x(1)W_8^2 + x(2)W_8^4 + x(3)W_8^6 + x(4)W_8^8 + x(5)W_8^{10} + x(6)W_8^{12} + x(7)W_8^{14}$
$X(3) =$	$x(0)W_8^0 + x(1)W_8^3 + x(2)W_8^6 + x(3)W_8^9 + x(4)W_8^{12} + x(5)W_8^{15} + x(6)W_8^{18} + x(7)W_8^{21}$
$X(4) =$	$x(0)W_8^0 + x(1)W_8^4 + x(2)W_8^8 + x(3)W_8^{12} + x(4)W_8^{16} + x(5)W_8^{20} + x(6)W_8^{24} + x(7)W_8^{28}$
$X(5) =$	$x(0)W_8^0 + x(1)W_8^5 + x(2)W_8^{10} + x(3)W_8^{15} + x(4)W_8^{20} + x(5)W_8^{25} + x(6)W_8^{30} + x(7)W_8^{35}$
$X(6) =$	$x(0)W_8^0 + x(1)W_8^6 + x(2)W_8^{12} + x(3)W_8^{18} + x(4)W_8^{24} + x(5)W_8^{30} + x(6)W_8^{36} + x(7)W_8^{42}$
$X(7) =$	$x(0)W_8^0 + x(1)W_8^7 + x(2)W_8^{14} + x(3)W_8^{21} + x(4)W_8^{28} + x(5)W_8^{35} + x(6)W_8^{42} + x(7)W_8^{49}$

NOTES: 1. N^2 Complex Multiplications

2. $\frac{1}{N}$ Scaling Factor Omitted

Figure 5-10: The 8-Point DFT (N = 8)

Section Five

The FFT is simply an algorithm to speed up the DFT calculation by reducing the number of multiplications and additions required. It was popularized by J. W. Cooley and J. W. Tukey in the 1960s and was actually a rediscovery of an idea of Runge (1903) and Danielson and Lanczos (1942), first occurring prior to the availability of computers and calculators, when numerical calculation could take many man-hours. In addition, the German mathematician Karl Friedrich Gauss (1777–1855) had used the method more than a century earlier.

In order to understand the basic concepts of the FFT and its derivation, note that the DFT expansion shown in Figure 5-10 can be greatly simplified by taking advantage of the symmetry and periodicity of the twiddle factors as shown in Figure 5-11. If the equations are rearranged and factored, the result is the Fast Fourier Transform (FFT) which requires only $(N/2) \log_2(N)$ complex multiplications. The computational efficiency of the FFT versus the DFT becomes highly significant when the FFT point size increases to several thousand as shown in Figure 5-12. However, notice that the FFT computes *all* the output frequency components (either all or none). If only a few spectral points need to be calculated, the DFT may actually be more efficient. Calculation of a single spectral output using the DFT requires only N complex multiplications.

Symmetry: $W_N^{r+N/2} = -W_N^r$, Periodicity: $W_N^{r+N} = W_N^r$

N = 8

$$W_8^4 = W_8^{0+4} = -W_8^0 = -1$$
$$W_8^5 = W_8^{1+4} = -W_8^1$$
$$W_8^6 = W_8^{2+4} = -W_8^2$$
$$W_8^7 = W_8^{3+4} = -W_8^3$$
$$W_8^8 = W_8^{0+8} = +W_8^0 = +1$$
$$W_8^9 = W_8^{1+8} = +W_8^1$$
$$W_8^{10} = W_8^{2+8} = +W_8^2$$
$$W_8^{11} = W_8^{3+8} = +W_8^3$$

Figure 5-11: Applying the Properties of Symmetry and Periodicity to W_N^r for N = 8

Fast Fourier Transforms

- The FFT is Simply an Algorithm for Efficiently Calculating the DFT
- Computational Efficiency of an N-Point FFT:
  - DFT: N^2 Complex Multiplications
  - FFT: $(N/2) \log_2(N)$ Complex Multiplications

N	DFT Multiplications	FFT Multiplications	FFT Efficiency
256	65,536	1,024	64 : 1
512	262,144	2,304	114 : 1
1,024	1,048,576	5,120	205 : 1
2,048	4,194,304	11,264	372 : 1
4,096	16,777,216	24,576	683 : 1

Figure 5-12: The Fast Fourier Transform (FFT) vs. the Discrete Fourier Transform (DFT)

The radix-2 FFT algorithm breaks the entire DFT calculation down into a number of 2-point DFTs. Each 2-point DFT consists of a multiply-and-accumulate operation called a *butterfly*, as shown in Figure 5-13. Two representations of the butterfly are shown in the diagram: the top diagram is the actual functional representation of the butterfly showing the digital multipliers and adders. In the simplified bottom diagram, the multiplications are indicated by placing the multiplier over an arrow, and addition is indicated whenever two arrows converge at a dot.

The 8-point decimation-in-time (DIT) FFT algorithm computes the final output in three stages as shown in Figure 5-14. The eight input time samples are first divided (or *decimated*) into four groups of 2-point DFTs. The four 2-point DFTs are then combined into two 4-point DFTs. The two 4-point DFTs are then combined to produce the final output X(k). The detailed process is shown in Figure 5-15, where all the multiplications and additions are shown. Note that the basic two-point DFT butterfly operation forms the basis for all computation. The computation is done in three stages. After the first stage computation is complete, there is no need to store any previous results. The first stage outputs can be stored in the same registers that originally held the time samples x(n). Similarly, when the second-stage computation is completed, the results of the first-stage computation can be deleted. In this way, *in-place* computation proceeds to the final stage. Note that in order for the algorithm to work properly, the order of the input time samples, x(n), must be properly reordered using a *bit reversal* algorithm.

Section Five

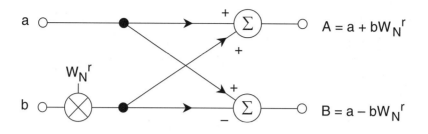

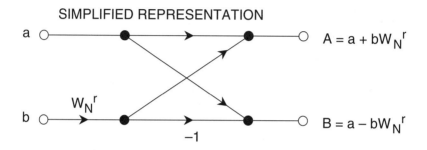

Figure 5-13: The Basic Butterfly Computation in the Decimation-in-Time FFT Algorithm

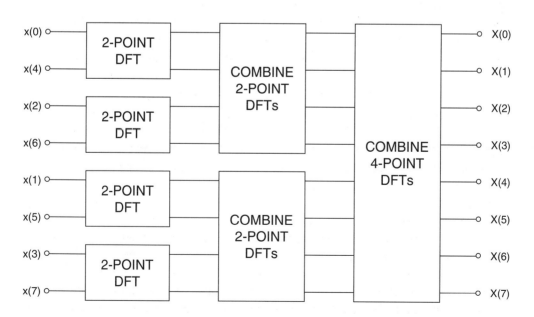

Figure 5-14: Computation of an 8-Point DFT in Three Stages Using Decimation-in-Time

Fast Fourier Transforms

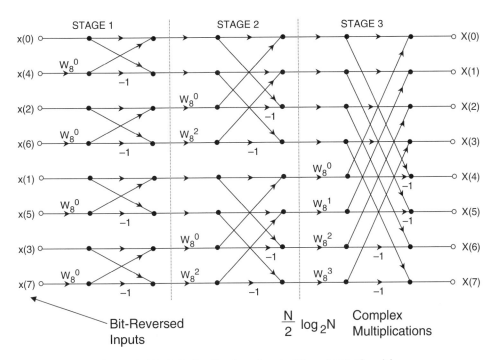

Figure 5-15: 8-Point Decimation-in-Time FFT Algorithm

The *bit reversal* algorithm used to perform this reordering is shown in Figure 5-16. The decimal index, n, is converted to its binary equivalent. The binary bits are then placed in reverse order, and converted back to a decimal number. Bit reversing is often performed in DSP hardware in the data address generator (DAG), thereby simplifying the software, reducing overhead, and speeding up the computations.

The computation of the FFT using *decimation-in-frequency* (DIF) is shown in Figures 5-17 and 5-18. This method requires that the bit reversal algorithm be applied to the output X(k). Note that the butterfly for the DIF algorithm differs slightly from the decimation-in-time butterfly as shown in Figure 5-19.

The use of decimation-in-time versus decimation-in-frequency algorithms is largely a matter of preference, as either yields the same result. System constraints may make one of the two a more optimal solution.

It should be noted that the algorithms required to compute the inverse FFT are nearly identical to those required to compute the FFT, assuming complex FFTs are used. In fact, a useful method for verifying a complex FFT algorithm consists of first taking the FFT of the x(n) time samples and then taking the inverse FFT of the X(k). At the end of this process, the original time samples, ReX(n), should be obtained and the imaginary part, ImX(n), should be zero (within the limits of the mathematical round-off errors).

Section Five

Decimal Number :	0	1	2	3	4	5	6	7
Binary Equivalent :	000	001	010	011	100	101	110	111
Bit-Reversed Binary :	000	100	010	110	001	101	011	111
Decimal Equivalent :	0	4	2	6	1	5	3	7

Figure 5-16: Bit Reversal Example for N = 8

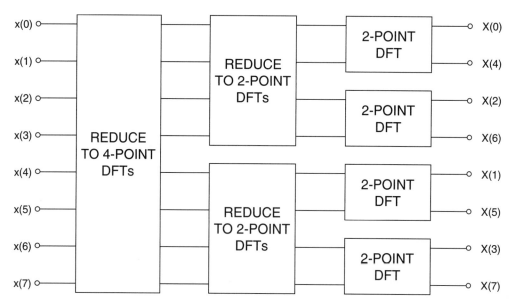

Figure 5-17: Computation of an 8-Point DFT in Three Stages Using Decimation-in-Frequency

Fast Fourier Transforms

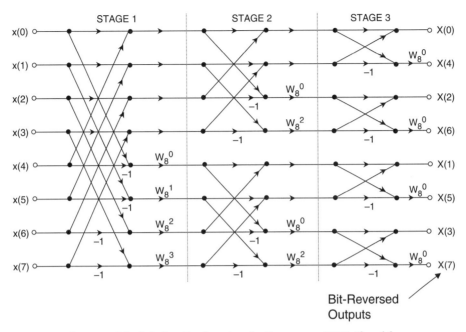

Figure 5-18: 8-Point Decimation-in-Frequency FFT Algorithm

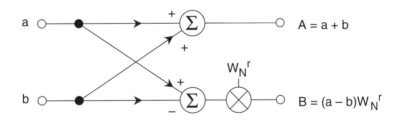

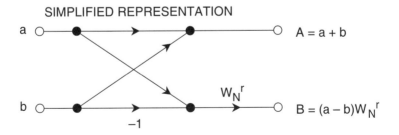

Figure 5-19: The Basic Butterfly Computation in the Decimation-in-Frequency FFT Algorithm

Section Five

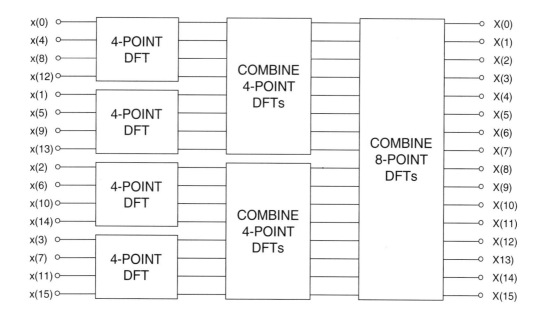

Figure 5-20: Computation of a 16-Point DFT in Three Stages Using Radix-4 Decimation-in-Time Algorithm

The FFTs discussed to this point are radix-2 FFTs, i.e., the computations are based on 2-point butterflies. This implies that the number of points in the FFT must be a power of two. If the number of points in an FFT is a power of four, however, the FFT can be broken down into a number of 4-point DFTs as shown in Figure 5-20. This is called a radix-4 FFT. The fundamental decimation-in-time butterfly for the radix-4 FFT is shown in Figure 5-21.

The radix-4 FFT requires fewer complex multiplications but more additions than the radix-2 FFT for the same number of points. Compared to the radix-2 FFT, the radix-4 FFT trades more complex data addressing and twiddle factors with less computation. The resulting savings in computation time varies between different DSPs but a radix-4 FFT can be as much as twice as fast as a radix-2 FFT for DSPs with optimal architectures.

Fast Fourier Transforms

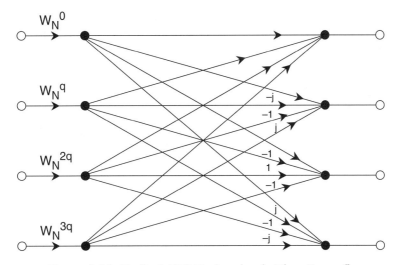

Figure 5-21: Radix-4 FFT Decimation-in-Time Butterfly

FFT Hardware Implementation and Benchmarks

In general terms, the memory requirements for an N-point FFT are N locations for the real data, N locations for the imaginary data, and N locations for the sinusoid data (sometimes referred to as twiddle factors). Additional memory locations will be required if windowing is used. Assuming the memory requirements are met, the DSP must perform the necessary calculations in the required time. Many DSP vendors will either give a performance benchmark for a specified FFT size or calculation time for a butterfly. When comparing FFT specifications, it is important to make sure that the same type of FFT is used in all cases. For example, the 1024-point FFT benchmark on one DSP derived from a radix-2 FFT should not be compared with the radix-4 benchmark from another DSP.

Another consideration regarding FFTs is whether to use a fixed-point or a floating-point processor. The results of a butterfly calculation can be larger than the inputs to the butterfly. This data growth can pose a potential problem in a DSP with a fixed number of bits. To prevent data overflow, the data needs to be scaled beforehand, leaving enough extra bits for growth. Alternatively, the data can be scaled after each stage of the FFT computation. The technique of scaling data after each pass of the FFT is known as *block floating point*. It is called this because a full array of data is scaled as a block, regardless of whether or not each element in the block needs to be scaled. The complete block is scaled so that the relative relationship of each data word remains the same. For example, if each data word is shifted right by one bit (divided by two), the absolute values have been changed but, relative to each other, the data stays the same.

Section Five

In a 16-bit fixed-point DSP, a 32-bit word is obtained after multiplication. The Analog Devices ADSP-21xx series DSPs have extended dynamic range by providing a 40-bit internal register in the multiply-accumulator (MAC).

The use of a floating-point DSP eliminates the need for data scaling and therefore results in a simpler FFT routine; however, the trade-off is the increased processing time required for the complex floating-point arithmetic. In addition, a 32-bit floating-point DSP will obviously have less round-off noise than a 16-bit fixed-point DSP. Figure 5-22 summarizes the FFT benchmarks for popular Analog Devices DSPs. Notice in particular that the ADSP-TS001 TigerSHARC® DSP offers both fixed-point and floating-point modes, thereby providing an exceptional degree of programming flexibility.

- ADSP-2189M, 16-Bit, Fixed-Point
  - 453 µs (1024-Point)
- ADSP-21160 SHARC®, 32-Bit, Floating-Point
  - 90 µs (1024-Point)
- ADSP-TS001 TigerSHARC @ 150 MHz,
  - 16-Bit, Fixed-Point Mode
    - 7.3 µs (256-Point FFT)
  - 32-Bit, Floating-Point Mode
    - 69 µs (1024-Point)

Figure 5-22: Radix-2 Complex FFT Hardware Benchmark Comparisons

DSP Requirements for Real-Time FFT Applications

There are two basic ways to acquire data from a real-world signal, either one sample at a time (continuous processing), or one frame at a time (batch processing). Sample-based systems, like a digital filter, acquire data one sample at a time. For each sample clock, a sample comes into the system and a processed sample is sent to the output. Frame-based systems, like an FFT-based digital spectrum analyzer, acquire a frame (or block of samples). Processing occurs on the entire frame of data and results in a frame of transformed output data.

In order to maintain real-time operation, the entire FFT must therefore be calculated during the frame period. This assumes that the DSP is collecting the data for the next frame while it is calculating the FFT for the current frame of data. Acquiring the data is one area where special architectural features of DSPs come into play. Seamless

Fast Fourier Transforms

data acquisition is facilitated by the DSP's flexible data addressing capabilities in conjunction with its direct memory accessing (DMA) channels.

Assume the DSP is the ADSP-TS001 TigerSHARC, which can calculate a 1024-point 32-bit complex floating-point FFT in 69 μs. The maximum sampling frequency is therefore 1024/69 μs = 14.8 MSPS. This implies a signal bandwidth of less than 7.4 MHz. It is also assumed that there is no additional FFT overhead or data transfer limitation.

- Assume 69 μs Execution Time for Radix-2, 1024-Point FFT (TigerSHARC, 32-Bit Mode)

- f_s (maximum) < $\dfrac{1024 \text{ Samples}}{69 \text{ μs}}$ = 14.8 MSPS

- Input Signal Bandwidth Therefore < 7.4 MHz

- This Assumes No Additional FFT Overhead and No Input/Output Data Transfer Limitations

Figure 5-23: Real-Time FFT Processing Example

The above example will give an estimate of the maximum bandwidth signal that can be handled by a given DSP using its FFT benchmarks. Another way to approach the issue is to start with the signal bandwidth and develop the DSP requirements. If the signal bandwidth is known, the required sampling frequency can be estimated by multiplying by a factor between 2 and 2.5 (the increased sampling rate may be required to ease the requirements on the antialiasing filter that precedes the ADC). The next step is to determine the required number of points in the FFT to achieve the desired frequency resolution. The frequency resolution is obtained by dividing the sampling rate f_s by N, the number of points in the FFT. These and other FFT considerations are shown in Figure 5-24.

The number of FFT points also determines the noise floor of the FFT with respect to the broadband noise level, and this may also be a consideration. Figure 5-25 shows the relationships between the system full-scale signal level, the broadband noise level (measured over the bandwidth dc to $f_s/2$), and the FFT noise floor. Notice that the FFT processing gain is determined by the number of points in the FFT. The FFT acts like an analog spectrum analyzer with a sweep bandwidth of f_s/N. Increasing the number of points increases the FFT resolution and narrows its bandwidth, thereby reducing the noise floor. This analysis neglects noise caused by the FFT round-off error. In practice, the ADC that is used to digitize the signal produces quantization noise, which is the dominant noise source.

At this point it is time to examine actual DSPs and their FFT processing times to make sure real-time operation can be achieved. This means that the FFT must be cal-

Section Five

culated during the acquisition time for one frame of data, which is N/f_s. Fixed-point versus floating-point, radix-2 versus radix-4, and processor power dissipation and cost may be other considerations.

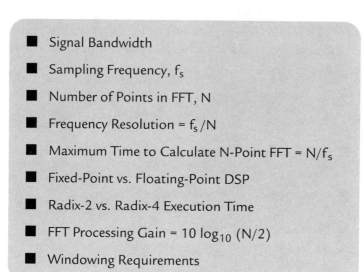

- Signal Bandwidth
- Sampling Frequency, f_s
- Number of Points in FFT, N
- Frequency Resolution = f_s/N
- Maximum Time to Calculate N-Point FFT = N/f_s
- Fixed-Point vs. Floating-Point DSP
- Radix-2 vs. Radix-4 Execution Time
- FFT Processing Gain = $10 \log_{10}(N/2)$
- Windowing Requirements

Figure 5-24: Real-Time FFT Considerations

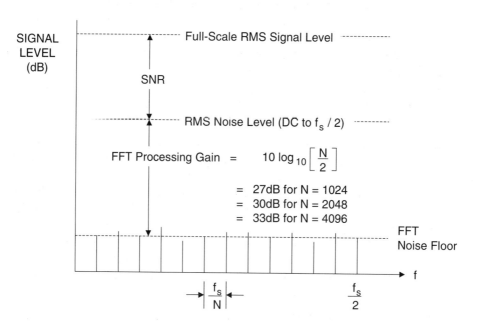

Figure 5-25: FFT Processing Gain Neglecting Round-Off Error

138

Fast Fourier Transforms

Spectral Leakage and Windowing

Spectral leakage in FFT processing can best be understood by considering the case of performing an N-point FFT on a pure sinusoidal input. Two conditions will be considered. In Figure 5-26, the ratio between the sampling frequency and the input sine wave frequency is such that precisely an integral number of cycles is contained within the data window (frame or record). Recall that the DFT assumes that an infinite number of these windows are placed end-to-end to form a periodic waveform as shown in the diagram as the periodic extensions. Under these conditions, the waveform appears continuous (no discontinuities), and the DFT or FFT output will be a single tone located at the input signal frequency.

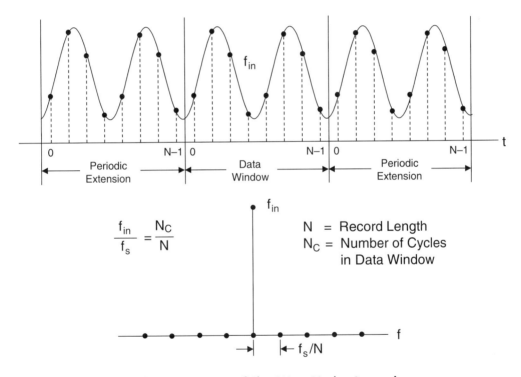

Figure 5-26: FFT of Sine Wave Having Integral Number of Cycles in Data Window

Figure 5-27 shows the condition where there is not an integral number of sine wave cycles within the data window. The discontinuities that occur at the endpoints of the data window result in leakage in the frequency domain because of the harmonics that are generated. In addition to the sidelobes, the main lobe of the sine wave is smeared over several frequency bins. This process is equivalent to multiplying the input sine wave by a rectangular window pulse that has the familiar $\sin(x)/x$ frequency response and associated smearing and sidelobes.

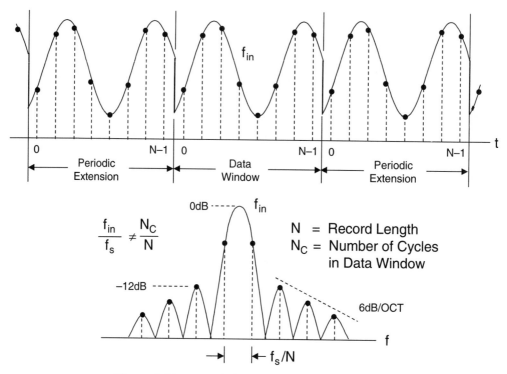

Figure 5-27: FFT of Sine Wave Having Nonintegral Number of Cycles in Data Window

Notice that the first sidelobe is only 12 dB below the fundamental, and that the sidelobes roll off at only 6 dB/octave. This situation would be unsuitable for most spectral analysis applications. Since in practical FFT spectral analysis applications the exact input frequencies are unknown, something must be done to minimize these sidelobes. This is done by choosing a window function other than the rectangular window. The input time samples are multiplied by an appropriate window function, which brings the signal to zero at the edges of the window as shown in Figure 5-28. The selection of a window function is primarily a trade-off between main lobe spreading and sidelobe roll-off. Reference 7 is highly recommended for an in-depth treatment of windows.

The mathematical functions that describe four popular window functions (Hamming, Blackman, Hanning, and Minimum 4-Term Blackman-Harris) are shown in Figure 5-29. The computations are straightforward, and the window function data points are usually precalculated and stored in the DSP memory to minimize their impact on FFT processing time. The frequency response of the rectangular, Hamming, and Blackman windows are shown in Figure 5-30. Figure 5-31 shows the trade-off between main lobe spreading and sidelobe amplitude and roll-off for the popular window functions.

Fast Fourier Transforms

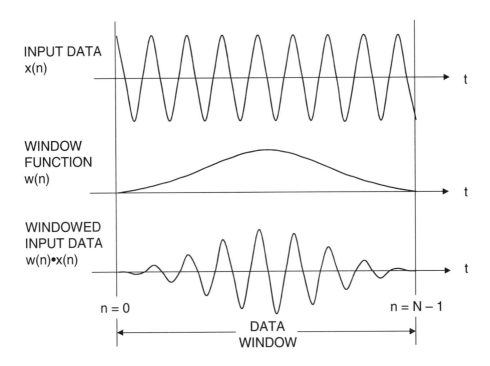

Figure 5-28: Windowing to Reduce Spectral Leakage

- Hamming: $w(n) = 0.54 - 0.46 \cos\left[\dfrac{2\pi n}{N}\right]$

- Blackman: $w(n) = 0.42 - 0.5 \cos\left[\dfrac{2\pi n}{N}\right] + 0.08 \cos\left[\dfrac{4\pi n}{N}\right]$

- Hanning: $w(n) = 0.5 - 0.5 \cos\left[\dfrac{2\pi n}{N}\right]$

- Minimum 4-Term Blackman Harris: $w(n) = 0.35875 - 0.48829 \cos\left[\dfrac{2\pi n}{N}\right]$
$\qquad + 0.14128 \cos\left[\dfrac{4\pi n}{N}\right]$
$\qquad - 0.01168 \cos\left[\dfrac{6\pi n}{N}\right]$

$\boxed{0 \leq n \leq N-1}$

Figure 5-29: Some Popular Window Functions

Section Five

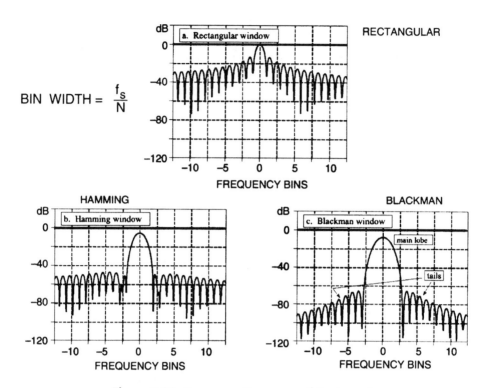

BIN WIDTH = $\frac{f_s}{N}$

Figure 5-30: Frequency Response of Rectangular, Hamming, and Blackman Windows For N = 256

WINDOW FUNCTION	3 dB BW (Bins)	6 dB BW (Bins)	HIGHEST SIDELOBE (dB)	SIDELOBE ROLL-OFF (dB/Octave)
Rectangle	0.89	1.21	−12	6
Hamming	1.3	1.81	−43	6
Blackman	1.68	2.35	−58	18
Hanning	1.44	2.00	−32	18
Minimum 4-Term Blackman-Harris	1.90	2.72	−92	6

Figure 5-31: Popular Windows and Figures of Merit

References

1. Steven W. Smith, **Digital Signal Processing: A Guide for Engineers and Scientists**. Newnes, 2002.

2. C. Britton Rorabaugh, **DSP Primer**, McGraw-Hill, 1999.

3. Richard J. Higgins, **Digital Signal Processing in VLSI,** Prentice-Hall, 1990.

4. A. V. Oppenheim and R. W. Schafer, **Digital Signal Processing**, Prentice-Hall, 1975.

5. L. R. Rabiner and B. Gold, **Theory and Application of Digital Signal Processing**, Prentice-Hall, 1975.

6. John G. Proakis and Dimitris G. Manolakis, **Introduction to Digital Signal Processing**, MacMillian, 1988.

7. Fredrick J. Harris, *On the Use of Windows for Harmonic Analysis with the Discrete Fourier Transform*, **Proc. IEEE**, Vol. 66, No. 1, 1978 pp. 51–83.

8. R. W. Ramirez, **The FFT: Fundamentals and Concepts**, Prentice-Hall, 1985.

9. J. W. Cooley and J. W. Tukey, *An Algorithm for the Machine Computation of Complex Fourier Series*, **Mathematics Computation, Vol. 19**, pp. 297–301, April 1965.

10. **Digital Signal Processing Applications Using the ADSP-2100 Family**, Vol. 1 and Vol. 2, Analog Devices, Free Download at: http://www.analog.com.

11. **ADSP-21000 Family Application Handbook**, Vol. 1, Analog Devices, Free Download at: http://www.analog.com.

Section 6
Digital Filters

- Finite Impulse Response (FIR) Filters
- Infinite Impulse Response (IIR) Filters
- Multirate Filters
- Adaptive Filters

Section 6
Digital Filters
Walt Kester

Digital filtering is one of the most powerful tools of DSP. Apart from the obvious advantages of virtually eliminating errors in the filter associated with passive component fluctuations over time and temperature, op amp drift (active filters), and other effects, digital filters are capable of performance specifications that would, at best, be extremely difficult, if not impossible, to achieve with an analog implementation. In addition, the characteristics of a digital filter can easily be changed under software control. Therefore, they are widely used in adaptive filtering applications in communications such as echo cancellation in modems, noise cancellation, and speech recognition.

The actual procedure for designing digital filters has the same fundamental elements as that for analog filters. First, the desired filter responses are characterized, and the filter parameters are then calculated. Characteristics such as amplitude and phase response are derived in the same way. The key difference between analog and digital filters is that instead of calculating resistor, capacitor, and inductor values for an analog filter, coefficient values are calculated for a digital filter. So for the digital filter, numbers replace the physical resistor and capacitor components of the analog filter. These numbers reside in a memory as filter coefficients and are used with the sampled data values from the ADC to perform the filter calculations.

The real-time digital filter, because it is a discrete time function, works with digitized data as opposed to a continuous waveform, and a new data point is acquired each sampling period. Because of this discrete nature, data samples are referenced as numbers such as sample 1, sample 2, and sample 3. Figure 6-1 shows a low frequency signal containing higher frequency noise which must be filtered out. This waveform must be digitized with an ADC to produce samples $x(n)$. These data values are fed to the digital filter, which in this case is a low-pass filter. The output data samples, $y(n)$, are used to reconstruct an analog waveform using a low glitch DAC.

Digital filters, however, are not the answer to all signal processing filtering requirements. In order to maintain real-time operation, the DSP processor must be able to execute all the steps in the filter routine within one sampling clock period, $1/f_s$. A fast general-purpose fixed-point DSP (such as the ADSP-2189M at 75 MIPS) can execute a complete filter tap multiply-accumulate instruction in 13.3 ns. The ADSP-2189M requires $N + 5$ instructions for an N-tap filter. For a 100-tap filter, the total execution time is approximately 1.4 μs. This corresponds to a maximum possible sampling frequency of 714 kHz, thereby limiting the upper signal bandwidth to a few hundred kHz.

Section Six

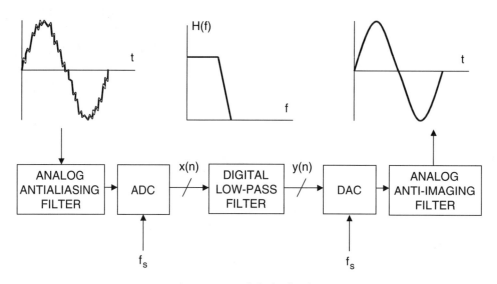

Figure 6-1: Digital Filtering

However, it is possible to replace a general-purpose DSP chip and design special hardware digital filters that will operate at video-speed sampling rates. In other cases, the speed limitations can be overcome by first storing the high speed ADC data in a buffer memory. The buffer memory is then read at a rate that is compatible with the speed of the DSP-based digital filter. In this manner, pseudo-real-time operation can be maintained as in a radar system, where signal processing is typically done on bursts of data collected after each transmitted pulse.

Another option is to use a third-party dedicated DSP filter engine like the Systolix PulseDSP filter core. The AD7725 16-bit sigma-delta ADC has an on-chip PulseDSP filter that can do 125 million multiply-accumulates per second.

Even in highly oversampled sampled data systems, an analog antialiasing filter is still required ahead of the ADC and a reconstruction (anti-imaging) filter after the DAC. Finally, as signal frequencies increase sufficiently, they surpass the capabilities of available ADCs, and digital filtering then becomes impossible. Active analog filtering is not possible at extremely high frequencies because of op amp bandwidth and distortion limitations, and filtering requirements must then be met using purely passive components. The primary focus of the following discussions will be on filters that can run in real-time under DSP program control.

As an example, consider the comparison between an analog and a digital filter shown in Figure 6-3. The cutoff frequency of both filters is 1 kHz. The analog filter is realized as a 6-pole Chebyshev Type 1 filter (ripple in pass-band, no ripple in stop-band). In practice, this filter would probably be realized using three 2-pole stages, each of which requires an op amp, and several resistors and capacitors. The 6-pole design is certainly not trivial, and maintaining the 0.5 dB ripple specification requires accurate component selection and matching.

On the other hand, the digital FIR filter shown has only 0.002 dB pass-band ripple, linear phase, and a much sharper roll-off. In fact, it could not be realized using analog techniques. In a practical application, there are many other factors to consider when evaluating analog versus digital filters. Most modern signal processing systems use a combination of analog and digital techniques in order to accomplish the desired function and take advantage of the best of both the analog and the digital world.

DIGITAL FILTERS	ANALOG FILTERS
High Accuracy	Less Accuracy – Component Tolerances
Linear Phase (FIR Filters)	Nonlinear Phase
No Drift Due to Component Variations	Drift Due to Component Variations
Flexible, Adaptive Filtering Possible	Adaptive Filters Difficult
Easy to Simulate and Design	Difficult to Simulate and Design
Computation Must be Completed in Sampling Period – Limits Real-Time Operation	Analog Filters Required at High Frequencies and for Antialiasing Filters
Requires High Performance ADC, DAC, and DSP	No ADC, DAC, or DSP Required

Figure 6-2: Digital vs. Analog Filtering

There are many applications where digital filters must operate in real-time. This places specific requirements on the DSP, depending upon the sampling frequency and the filter complexity. The key point is *that the DSP must finish all computations during the sampling period so it will be ready to process the next data sample*. Assume that the analog signal bandwidth to be processed is f_a. This requires the ADC sampling frequency f_s to be at least $2f_a$. The sampling period is $1/f_s$. All DSP filter computations (including overhead) must be completed during this interval. The computation time depends on the number of taps in the filter and the speed and efficiency of the DSP. Each tap on the filter requires one multiplication and one addition (multiply-accumulate). DSPs are generally optimized to perform fast multiply-accumulates, and many DSPs have additional features such as circular buffering and zero-overhead looping to minimize the "overhead" instructions that otherwise would be needed.

Section Six

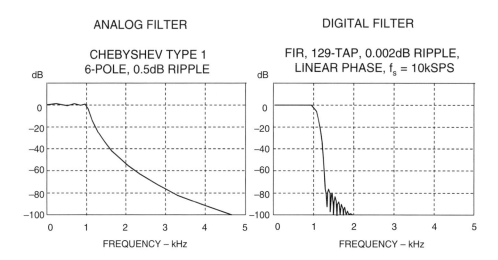

Figure 6-3: Analog vs. Digital Filter
Frequency Response Comparison

- Signal Bandwidth = f_a
- Sampling Frequency $f_s > 2f_a$
- Sampling Period = $1 / f_s$
- Filter Computational Time + Overhead < Sampling Period
  - Depends on Number of Taps
  - Speed of DSP Multiplication-Accumulates (MACs)
  - Efficiency of DSP
    - Circular Buffering
    - Zero-Overhead Looping

Figure 6-4: Processing Requirements
For Real-Time Digital Filtering

Finite Impulse Response (FIR) Filters

There are two fundamental types of digital filters: finite impulse response (FIR) and infinite impulse response (IIR). As the terminology suggests, these classifications refer to the filter's impulse response. By varying the weight of the coefficients and the number of filter taps, virtually any frequency response characteristic can be realized with a FIR filter. As has been shown, FIR filters can achieve performance levels that are not possible with analog filter techniques (such as perfect linear phase response). However, high performance FIR filters generally require a large number of multiply-accumulates and therefore require fast and efficient DSPs. On the other hand, IIR filters tend to mimic the performance of traditional analog filters and make use of feedback, so their impulse response extends over an infinite period of time. Because of feedback, IIR filters can be implemented with fewer coefficients than for a FIR filter. Lattice filters are simply another way to implement either FIR or IIR filters and are often used in speech processing applications. Finally, digital filters lend themselves to adaptive filtering applications simply because of the speed and ease with which the filter characteristics can be changed by varying the filter coefficients.

- Moving Average
- Finite Impulse Response (FIR)
  - Linear Phase
  - Easy to Design
  - Computationally Intensive
- Infinite Impulse Response (IIR)
  - Based on Classical Analog Filters
  - Computationally Efficient
- Lattice Filters (Can be FIR or IIR)
- Adaptive Filters

Figure 6-5: Types of Digital Filters

The most elementary form of a FIR filter is a *moving average* filter as shown in Figure 6-6. Moving average filters are popular for smoothing data, such as in the analysis of stock prices. The input samples, x(n) are passed through a series of buffer registers (labeled z^{-1}, corresponding to the z-transform representation of a delay element). In the example shown, there are four taps corresponding to a 4-point moving average. Each sample is multiplied by 0.25, and these results are added to yield the final moving average output y(n). The figure also shows the general equation of a moving average filter with N taps. Note again that N refers to the number of filter taps, and not the ADC or DAC resolution as in previous sections.

Section Six

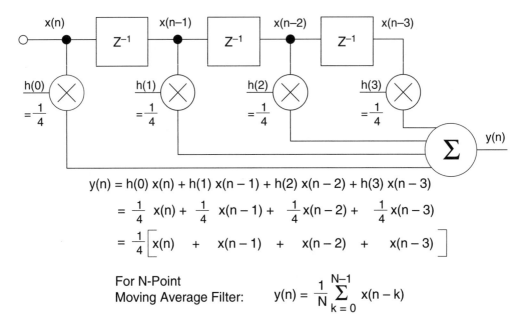

Figure 6-6: 4-Point Moving Average Filter

Since the coefficients are equal, an easier way to perform a moving average filter is shown in Figure 6-7. Note that the first step is to store the first four samples, $x(0)$, $x(1)$, $x(2)$, $x(3)$ in a register. These quantities are added and then multiplied by 0.25 to yield the first output, $y(3)$. Note that the initial outputs $y(0)$, $y(1)$, and $y(2)$ are not valid because all registers are not full until sample $x(3)$ is received.

When sample $x(4)$ is received, it is added to the result, and sample $x(0)$ is subtracted from the result. The new result must then be multiplied by 0.25. Therefore, the calculations required to produce a new output consist of one addition, one subtraction, and one multiplication, regardless of the length of the moving average filter.

The step function response of a 4-point moving average filter is shown in Figure 6-8. Notice that the moving average filter has no overshoot. This makes it useful in signal processing applications where random white noise must be filtered but pulse response preserved. Of all the possible linear filters that could be used, the moving average produces the lowest noise for a given edge sharpness. This is illustrated in Figure 6-9, where the noise level becomes lower as the number of taps are increased. Notice that the 0% to 100% rise time of the pulse response is equal to the total number of taps in the filter multiplied by the sampling period.

Digital Filters

$$y(3) = 0.25 \left[x(3) + x(2) + x(1) + x(0) \right]$$

$$y(4) = 0.25 \left[x(4) + x(3) + x(2) + x(1) \right]$$

$$y(5) = 0.25 \left[x(5) + x(4) + x(3) + x(2) \right]$$

$$y(6) = 0.25 \left[x(6) + x(5) + x(4) + x(3) \right]$$

$$y(7) = 0.25 \left[x(7) + x(6) + x(5) + x(4) \right]$$

-
- Each Output Requires:
- 1 Multiplication, 1 Addition, 1 Subtraction

Figure 6-7: Calculating Output of 4-Point Moving Average Filter

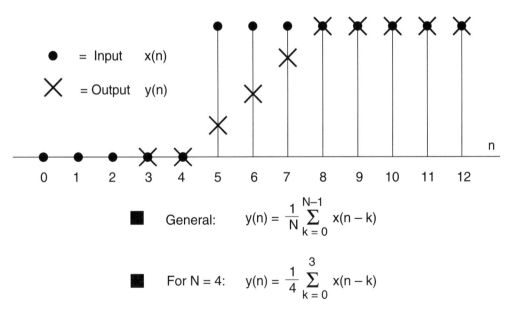

■ General: $\quad y(n) = \dfrac{1}{N} \sum_{k=0}^{N-1} x(n-k)$

■ For N = 4: $\quad y(n) = \dfrac{1}{4} \sum_{k=0}^{3} x(n-k)$

Figure 6-8: 4-Tap Moving Average Filter Step Response

Section Six

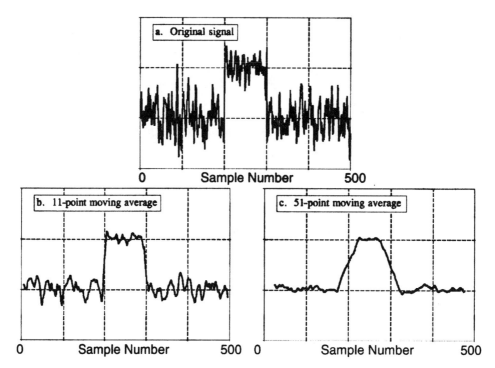

Figure 6-9: Moving Average Filter Response to Noise Superimposed on Step Input

The frequency response of the simple moving average filter is SIN (x)/x and is shown on a linear amplitude scale in Figure 6-10. Adding more taps to the filter sharpens the roll-off, but does not significantly reduce the amplitude of the sidelobes which are approximately 14 dB down for the 11- and 31-tap filter. These filters are definitely not suitable where high stop-band attenuation is required.

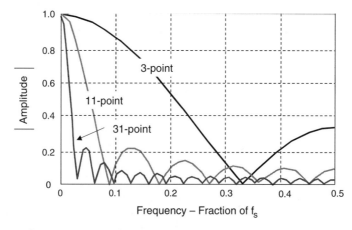

Figure 6-10: Moving Average Filter Frequency Response

Digital Filters

It is possible to dramatically improve the performance of the simple FIR moving average filter by properly selecting the individual weights or coefficients rather than giving them equal weight. The sharpness of the roll-off can be improved by adding more stages (taps), and the stop-band attenuation characteristics can be improved by properly selecting the filter coefficients. Note that unlike the moving average filter, one multiply-accumulate cycle is now required per tap for the generalized FIR filter. The essence of FIR filter design is the appropriate selection of the filter coefficients and the number of taps to realize the desired transfer function H(f). Various algorithms are available to translate the frequency response H(f) into a set of FIR coefficients. Most of this software is commercially available and can be run on PCs. *The key theorem of FIR filter design is that the coefficients h(n) of the FIR filter are simply the quantized values of the impulse response of the frequency transfer function H(f).* Conversely, the impulse response is the discrete Fourier transform of H(f).

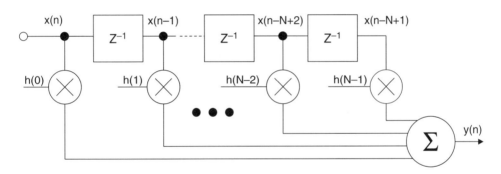

- $y(n) = h(n) * x(n) = \sum_{k=0}^{N-1} h(k) x(n-k)$
- $*$ = Symbol for Convolution
- Requires N multiply-accumulates for each output

Figure 6-11: N-Tap Finite Impulse Response (FIR) Filter

The generalized form of an N-tap FIR filter is shown in Figure 6-11. As has been discussed, an FIR filter must perform the following convolution equation:

$$y(n) = h(k)*x(n) = \sum_{k=0}^{N-1} h(k) x(n-k).$$

where h(k) is the filter coefficient array and x(n-k) is the input data array to the filter. The number N, in the equation, represents the number of taps of the filter and relates to the filter performance as has been discussed above. An N-tap FIR filter requires N multiply-accumulate cycles.

Section Six

FIR filter diagrams are often simplified as shown in Figure 6-12. The summations are represented by arrows pointing into the dots, and the multiplications are indicated by placing the h(k) coefficients next to the arrows on the lines. The z^{-1} delay element is often shown by placing the label above or next to the appropriate line.

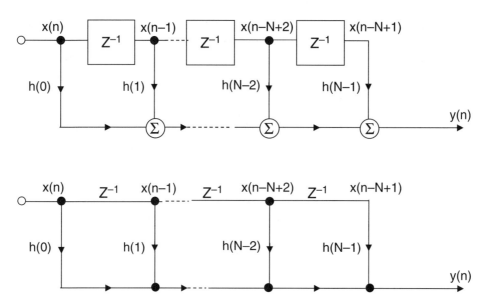

Figure 6-12: Simplified Filter Notations

FIR Filter Implementation in DSP Hardware Using Circular Buffering

In the series of FIR filter equations, the N coefficient locations are always accessed sequentially from h(0) to h(N–1). The associated data points circulate through the memory; new samples are added, replacing the oldest each time a filter output is computed. A fixed-boundary RAM can be used to achieve this circulating buffer effect as shown in Figure 6-13 for a four-tap FIR filter. The oldest data sample is replaced by the newest after each convolution. A "time history" of the four most recent data samples is always stored in RAM.

To facilitate memory addressing, old data values are read from memory starting with the value one location after the value that was just written. For example, x(4) is written into memory location 0, and data values are then read from locations 1, 2, 3, and 0. This example can be expanded to accommodate any number of taps. By addressing data memory locations in this manner, the address generator need only supply sequential addresses, regardless of whether the operation is a memory read or write. This data memory buffer is called *circular* because when the last location is reached, the memory pointer is reset to the beginning of the buffer.

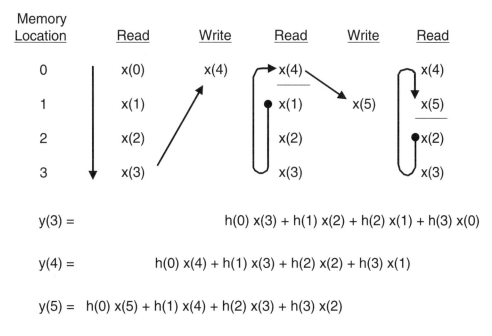

Figure 6-13: Calculating Outputs of 4-Tap FIR Filter Using a Circular Buffer

The coefficients are fetched simultaneously with the data. Due to the addressing scheme chosen, the oldest data sample is fetched first. Therefore, the last coefficient must be fetched first. The coefficients can be stored backwards in memory: h(N–1) is the first location, and h(0) is the last, with the address generator providing incremental addresses. Alternatively, coefficients can be stored in a normal manner with the accessing of coefficients starting at the end of the buffer, and the address generator being decremented. In the example shown in Figure 6-13, the coefficients are stored in a reverse manner.

A simple summary flowchart for these operations is shown in Figure 6-14. For Analog Devices DSPs, *all operations within the filter loop are completed in one instruction cycle*, thereby greatly increasing efficiency. This is referred to as *zero-overhead looping*. The actual FIR filter assembly code for the ADSP-21xx family of fixed-point DSPs is shown in Figure 6-15. The arrows in the diagram point to the actual executable instructions, and the rest of the code are simply comments added for clarification.

Section Six

The first instruction (labeled *fir:*) sets up the computation by clearing the MR register and loading the MX0 and MY0 registers with the first data and coefficient values from data and program memory. The multiply-accumulate with dual data fetch in the *convolution* loop is then executed N–1 times in N cycles to compute the sum of the first N–1 products. The final multiply-accumulate instruction is performed with the rounding mode enabled to round the result to the upper 24 bits of the MR register. The MR1 register is then conditionally saturated to its most positive or negative value, based on the status of the overflow flag contained in the MV register. In this manner, results are accumulated to the full 40-bit precision of the MR register, with saturation of the output only if the final result overflowed beyond the least significant 32 bits of the MR register.

The limit on the number of filter taps attainable for a real-time implementation of the FIR filter subroutine is primarily determined by the processor cycle time, the sampling rate, and the number of other computations required. The FIR subroutine presented here requires a total of N + 5 cycles for a filter of length N. For the ADSP-2189M 75 MIPS DSP, one instruction cycle is 13.3 ns, so a 100-tap filter would require 13.3 ns × 100 + 5 × 13.3 ns = 1330 ns + 66.5 ns = 1396.5 ns = 1.4 μs.

1. Obtain sample from ADC (typically interrupt-driven)
2. Move sample into input signal's circular buffer
3. Update the pointer for the input signal's circular buffer
4. Zero the accumulator
5. Implement filter (control the loop through each of the coefficients)
    6. Fetch the coefficient from the coefficient's circular buffer
    7. Update the pointer for the coefficient's circular buffer
    8. Fetch the sample from the input signal's circular buffer
    9. Update the pointer for the input signal's circular buffer
    10. Multiply the coefficient by the sample
    11. Add the product to the accumulator
12. Move the filtered sample to the DAC

```
ADSP-21xx Example code:

CNTR = N-1;
DO convolution UNTIL CE;
convolution:
    MR = MR + MX0 * MY0(SS), MX0 = DM(I0,M1), MY0 = PM(I4,M5);
```

Figure 6-14: Pseudocode for FIR Filter Program Using a DSP with Circular Buffering

Digital Filters

```
                .MODULE         fir_sub;
                {               FIR Filter Subroutine
                                Calling Parameters
                                        I0 --> Oldest input data value in delay line
                                        I4 --> Beginning of filter coefficient table
                                        L0 = Filter length (N)
                                        L4 = Filter length (N)
                                        M1,M5 = 1
                                        CNTR = Filter length - 1 (N-1)
                                Return Values
                                        MR1 = Sum of products (rounded and saturated)
                                        I0 --> Oldest input data value in delay line
                                        I4 --> Beginning of filter coefficient table
                                Altered Registers
                                        MX0,MY0,MR
                                Computation Time
                                        (N - 1) + 6 cycles = N + 5 cycles
                                All coefficients are assumed to be in 1.15 format. }
                .ENTRY          fir;
⟶       fir:            MR=0, MX0=DM(I0,M1), MY0=PM(I4,M5)
⟶                       CNTR = N-1;
⟶                       DO convolution UNTIL CE;
⟶       convolution:      MR=MR+MX0*MY0(SS), MX0=DM(I0,M1), MY0=PM(I4,M5);
⟶                       MR=MR+MX0*MY0(RND);
⟶                       IF MV SAT MR;
⟶                       RTS;
                .ENDMOD;
```

Figure 6-15: ADSP-21xx FIR Filter Assembly Code (Single Precision)

Designing FIR Filters

FIR filters are relatively easy to design using modern CAD tools. Figure 6-16 summarizes the characteristics of FIR filters as well as the most popular design techniques. *The fundamental concept of FIR filter design is that the filter frequency response is determined by the impulse response, and the quantized impulse response and the filter coefficients are identical.* This can be understood by examining Figure 6-17. The input to the FIR filter is an impulse, and as the impulse propagates through the delay elements, the filter output is identical to the filter coefficients. The FIR filter design process therefore consists of determining the impulse response from the desired frequency response, and then quantizing the impulse response to generate the filter coefficients.

It is useful to digress for a moment and examine the relationship between the time domain and the frequency domain to better understand the principles behind digital filters such as the FIR filter. In a sampled data system, a convolution operation can be carried out by performing a series of multiply-accumulates. The convolution operation in the time or frequency domain is equivalent to point-by-point multiplication in the opposite domain. For example, convolution in the time domain is equivalent to multiplication in the frequency domain. This is shown graphically in Figure 6-18. It can be seen that filtering in the frequency domain can be accomplished by multiplying all frequency components in the pass band by a 1 and all frequencies in the stop band by 0. Conversely, convolution in the frequency domain is equivalent to point-by-point multiplication in the time domain.

Section Six

- Impulse Response has a Finite Duration (N Cycles)
- Linear Phase, Constant Group Delay (N Must be Odd)
- No Analog Equivalent
- Unconditionally Stable
- Can be Adaptive
- Computational Advantages when Decimating Output
- Easy to Understand and Design
    - Windowed-Sinc Method
    - Fourier Series Expansion with Windowing
    - Frequency Sampling Using Inverse FFT – Arbitrary Frequency Response
    - Parks-McClellan Program with Remez Exchange Algorithm

Figure 6-16: Characteristics of FIR Filters

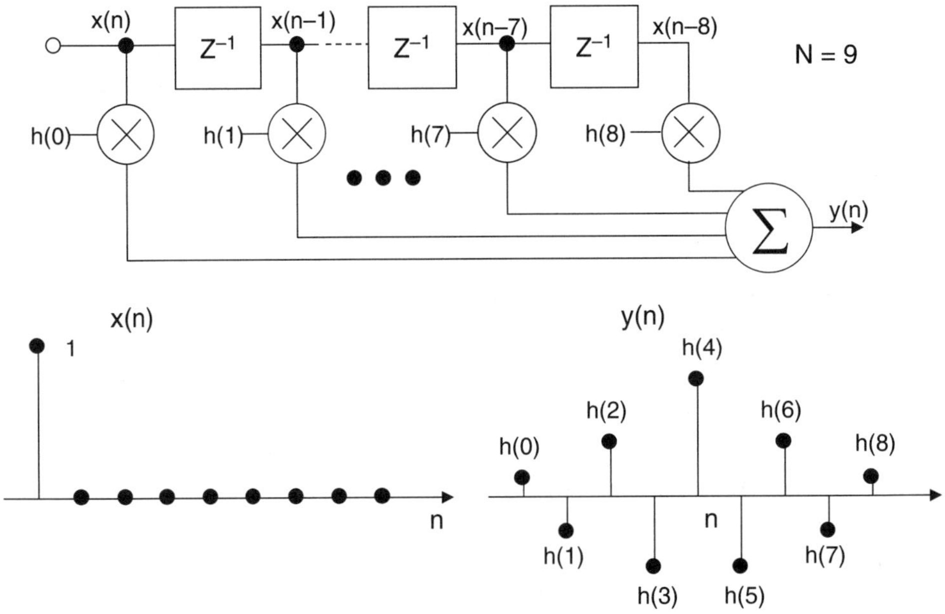

Figure 6-17: FIR Filter Impulse Response Determines the Filter Coefficients

Digital Filters

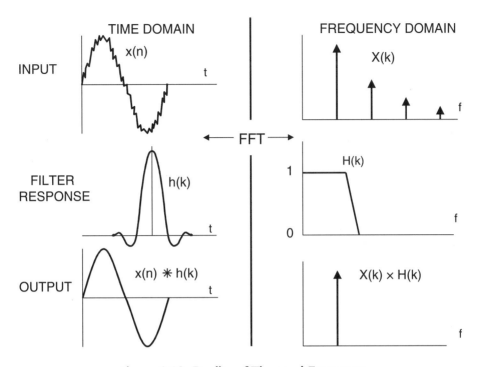

Figure 6-18: Duality of Time and Frequency

The transfer function in the frequency domain (either a 1 or a 0) can be translated to the time domain by the discrete Fourier transform (in practice, the fast Fourier transform is used). This transformation produces an impulse response in the time domain. Since the multiplication in the frequency domain (signal spectrum times the transfer function) is equivalent to convolution in the time domain (signal convolved with impulse response), the signal can be filtered by convolving it with the impulse response. The FIR filter is exactly this process. Since it is a sampled data system, the signal and the impulse response are quantized in time and amplitude, yielding discrete samples. The discrete samples comprising the desired impulse response are the FIR filter coefficients.

The mathematics involved in filter design (analog or digital) generally make use of transforms. In continuous-time systems, the Laplace transform can be considered to be a generalization of the Fourier transform. In a similar manner, it is possible to generalize the Fourier transform for discrete-time sampled data systems, resulting in what is commonly referred to as the z-transform. Details describing the use of the z-transform in digital filter design are given in References 1 through 6, but the theory is not necessary for the rest of this discussion.

FIR Filter Design Using the Windowed-Sinc Method

An ideal low-pass filter frequency response is shown in Figure 6-19A. The corresponding impulse response in the time domain is shown in Figure 6-19B, and follows the sin(x)/x (sinc) function. If a FIR filter is used to implement this frequency response, an infinite number of taps are required. The windowed-sinc method is used to implement the filter as follows. First, the impulse response is truncated to a reasonable number of N taps as in Figure 6-19C. As has been discussed in Section 5, the frequency response corresponding to Figure 6-19C has relatively poor sidelobe performance because of the end-point discontinuities in the truncated impulse response. The next step in the design process is to apply an appropriate window function as shown in Figure 6-19D to the truncated impulse. This forces the endpoints to zero. The particular window function chosen determines the roll-off and sidelobe performance of the filter. Window functions have been discussed in detail in Section 5, and there are several good choices depending upon the desired frequency response. The frequency response of the truncated and windowed-sinc impulse response of Figure 6-19E is shown in Figure 6-19F.

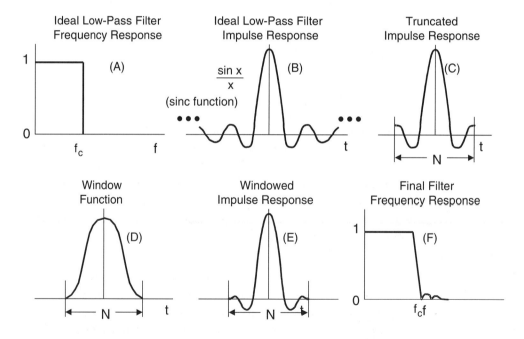

Figure 6-19: FIR Filter Design Using the Windowed-Sinc Method

FIR Filter Design Using the Fourier Series Method with Windowing

The Fourier series with windowing method (Figure 6-20) starts by defining the transfer function H(f) mathematically and expanding it in a Fourier series. The Fourier series coefficients define the impulse response and therefore the coefficients of the FIR filter. However, the impulse response must be truncated and windowed as in the previous method. After truncation and windowing, an FFT is used to generate the corresponding frequency response. The frequency response can be modified by choosing different window functions, although precise control of the stop-band characteristics is difficult in any method that uses windowing.

- Specify H(f)
- Expand H(f) in a Fourier Series: The Fourier Series Coefficients are the Coefficients of the FIR Filter, h(k), and its Impulse Response
- Truncate the Impulse Response to N Points (Taps)
- Apply a Suitable Window Function to h(k) to Smooth the Effects of Truncation
- Lacks Precise Control of Cutoff Frequency; Highly Dependent on Window Function

Figure 6-20: FIR Filter Design Using Fourier Series Method with Windowing

FIR Filter Design Using the Frequency Sampling Method

This method is extremely useful in generating a FIR filter with an arbitrary frequency response. H(f) is specified as a series of amplitude and phase points in the frequency domain. The points are then converted into real and imaginary components. Next, the impulse response is obtained by taking the complex inverse FFT of the frequency response. The impulse response is then truncated to N points, and a window function is applied to minimize the effects of truncation. The filter design should then be tested by taking its FFT and evaluating the frequency response. Several iterations may be required to achieve the desired response.

- Specify H(k) as a Finite Number of Spectral Points Spread Uniformly between 0 and $0.5f_s$ (512 Usually Sufficient)
- Specify Phase Points (Can Make Equal to Zero)
- Convert Rectangular Form (Real + Imaginary)
- Take the Complex Inverse FFT of H(f) Array to Obtain the Impulse Response
- Truncate the Impulse Response to N Points
- Apply a Suitable Window Function to h(k) to Smooth the Effects of Truncation
- Test Filter Design and Modify if Necessary
- CAD Design Techniques More Suitable for Low-Pass, High-Pass, Band-Pass, or Band-Stop Filters

Figure 6-21: Frequency Sampling Method for FIR Filters with Arbitrary Frequency Response

FIR Filter Design Using the Parks-McClellan Program

Historically, the design method based on the use of windows to truncate the impulse response and to obtain the desired frequency response was the first method used for designing FIR filters. The frequency-sampling method was developed in the 1970s and is still popular where the frequency response is an arbitrary function.

Modern CAD programs are available today that greatly simplify the design of low-pass, high-pass, band-pass, or band-stop FIR filters. A popular one was developed by Parks and McClellan and uses the Remez exchange algorithm. The filter design begins by specifying the parameters shown in Figure 6-22: pass-band ripple, stop-band ripple (same as attenuation), and the transition region. For this design example, the QED1000 program from Momentum Data Systems was used (a demo version is free and downloadable from http://www.mds.com).

For this example, we will design an audio low-pass filter that operates at a sampling rate of 44.1 kHz. The filter is specified as shown in Figure 6-22: 18 kHz pass-band frequency, 21 kHz stop-band frequency, 0.01 dB pass-band ripple, 96 dB stop-band ripple (attenuation). We must also specify the word length of the coefficients, which in this case is 16 bits, assuming a 16-bit fixed-point DSP is to be used.

Digital Filters

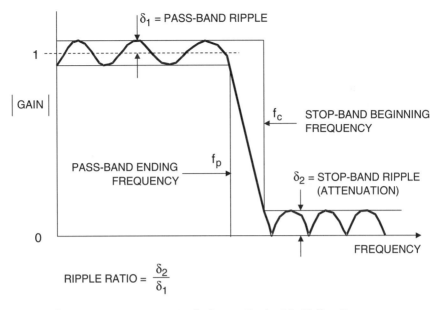

Figure 6-22: FIR CAD Techniques: Parks McClellan Program with Remez Exchange Algorithm

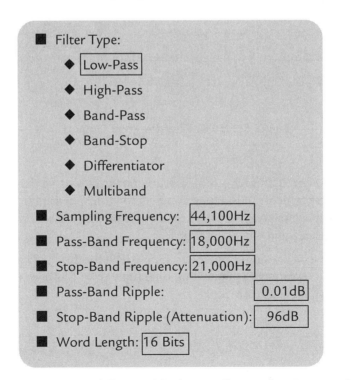

Figure 6-23: Parks McClellan Equiripple FIR Filter Design: Program Inputs

165

Section Six

The program allows us to choose between a window-based design or the equiripple Parks-McClellan program. We will choose the latter. The program now estimates the number of taps required to implement the filter based on the above specifications. In this case, it is 69 taps. At this point, we can accept this and proceed with the design or decrease the number of taps and see what degradation in specifications occur.

We will accept this number and let the program complete the calculations. The program outputs the frequency response (Figure 6-25), step function response (Figure 6-26), s- and z-plane analysis data, and the impulse response (Figure 6-27). The QED1000 program then outputs the quantized filter coefficients to a program that generates the actual DSP assembly code for a number of popular DSPs, including Analog Devices'. The program is quite flexible and allows the user to perform a number of scenarios to optimize the filter design.

- Estimated Number of Taps Required: 69
    ◆ Accept? Change? Accept
- Frequency Response (Linear and Log Scales)
- Step Response
- s- and z-Plane Analysis
- Impulse Response: Filter Coefficients (Quantized)
- DSP FIR Filter Assembly Code

Figure 6-24: FIR Filter Program Outputs

The 69-tap FIR filter requires 69 + 5 = 74 instruction cycles using the ADSP-2189M 75 MIPS processor, which yields a total computation time per sample of 74 × 13.3 ns = 984 ns. The sampling interval is 1/44.1 kHz, or 22.7 µs. This allows 22.7 µs – 0.984 µs = 21.7 µs for overhead and other operations.

Other options are to use a slower processor (3.3 MIPS) for this application, a more complex filter that takes more computation time (up to N = 1700), or increase the sampling frequency to about 1 MSPS.

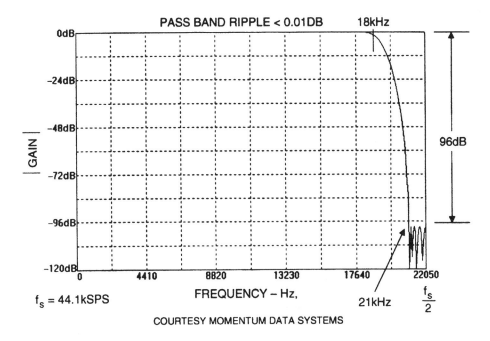

Figure 6-25: FIR Design Example: Frequency Response

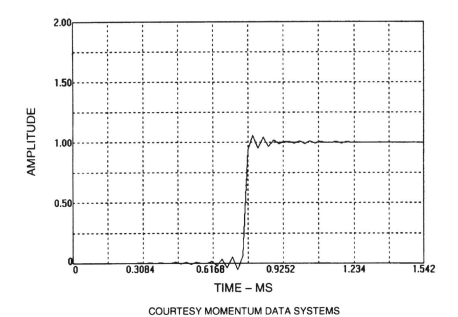

Figure 6-26: FIR Filter Design Example: Step Response

Section Six

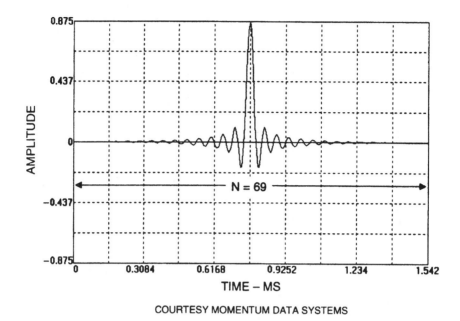

Figure 6-27: FIR Design Example: Impulse Response (Filter Coefficients)

- Sampling Frequency f_s = 44.1 kSPS
- Sampling Interval = $1/f_s$ = 22.7 μs
- Number of Filter Taps, N = 69
- Number of Required Instructions = N + 5 = 74
- Processing Time/Instruction = 13.3 ns (75 MIPS) (ADSP-2189M)
- Total Processing Time = 74 × 13.3 ns = 984 ns
- Total Processing Time < Sampling Interval with 22.7 μs − 0.984 μs = 21.7 μs for Other Operations
  - ◆ Increase Sampling Frequency to 1 MHz
  - ◆ Use Slower DSP (3.3 MIPS)
  - ◆ Add More Filter Taps (Up to N = 1700)

Figure 6-28: Design Example Using ADSP-2189M: Processor Time for 69-Tap FIR Filter

Designing High-Pass, Band-Pass, and Band-Stop Filters Based on Low-Pass Filter Design

Converting a low-pass filter design impulse response into a high-pass filter impulse response can be accomplished in one of two ways. In the *spectral inversion method*, the sign of each filter coefficient in the low-pass filter impulse response is changed. Next, 1 is added to the center coefficient. In the *spectral reversal method*, the sign of every other coefficient is changed. This reverses the frequency domain plot. In other words, if the cutoff of the low-pass filter is $0.2f_s$, the resulting high-pass filter will have a cutoff frequency of $0.5f_s - 0.2f_s = 0.3f_s$. This must be considered when doing the original low-pass filter design.

- **Spectral Inversion Technique:**
  - Design Low-Pass Filter (Linear Phase, N Odd)
  - Change the Sign of Each Coefficient in the Impulse Response, h(n)
  - Add 1 to the Coefficient at the Center of Symmetry

- **Spectral Reversal Technique:**
  - Design Low-Pass Filter
  - Change the Sign of Every Other Coefficient in the Impulse Response, h(n)
  - This Reverses the Frequency Domain Left-for-Right: 0 Becomes 0.5, and 0.5 Becomes 0; i.e., if the Cut-Off Frequency of the Low-Pass Filter is 0.2, the Cut-Off of the Resulting High-Pass Filter is 0.3

Figure 6-29: Designing High-Pass Filters Using Low-Pass Filter Impulse Response

Band-pass and band-stop filters can be designed by combining individual low-pass and high-pass filters in the proper manner. Band-pass filters are designed by placing the low-pass and high-pass filters in cascade. The equivalent impulse response of the cascaded filters is then obtained by *convolving* the two individual impulse responses.

A band-stop filter is designed by connecting the low-pass and high-pass filters in parallel and adding their outputs. The equivalent impulse response is then obtained by *adding* the two individual impulse responses.

Section Six

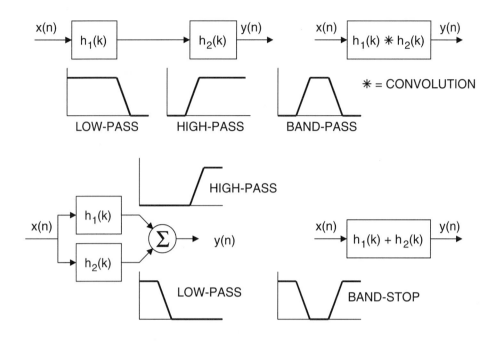

Figure 6-30: Band-Pass and Band-Stop Filters Designed from Low-Pass and High-Pass Filters

Infinite Impulse Response (IIR) Filters

As was mentioned previously, FIR filters have no real analog counterparts, the closest analogy being the weighted moving average. In addition, FIR filters have only zeros and no poles. On the other hand, IIR filters have traditional analog counterparts (Butterworth, Chebyshev, Elliptic, and Bessel) and can be analyzed and synthesized using more familiar traditional filter design techniques.

Infinite impulse response filters get their name because their impulse response extends for an infinite period of time. This is because they are recursive, i.e., they utilize feedback. Although they can be implemented with fewer computations than FIR filters, IIR filters do not match the performance achievable with FIR filters, and do not have linear phase. Also, there is no computational advantage achieved when the output of an IIR filter is decimated, because each output value must always be calculated.

Digital Filters

- Uses Feedback (Recursion)
- Impulse Response has an Infinite Duration
- Potentially Unstable
- Nonlinear Phase
- More Efficient than FIR Filters
- No Computational Advantage when Decimating Output
- Usually Designed to Duplicate Analog Filter Response
- Usually Implemented as Cascaded Second-Order Sections (Biquads)

Figure 6-31: Infinite Impulse Response (IIR) Filters

IIR filters are generally implemented in 2-pole sections called biquads because they are described with a biquadratic equation in the z-domain. Higher order filters are designed using cascaded biquad sections, e.g., a 6-pole filter requires three biquad sections.

The basic IIR biquad is shown in Figure 6-32. The zeros are formed by the feed-forward coefficients b_0, b_1, and b_2; the poles are formed by the feedback coefficients a_1, and a_2.

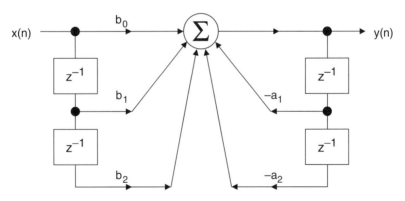

- $y(n) = b_0 x(n) + b_1 x(n-1) + b_2 x(n-2) - a_1 y(n-1) - a_2 y(n-2)$

- $y(n) = \sum_{k=0}^{M} b_k x(n-k) - \sum_{k=1}^{N} a_k x(n-k)$

- $H(z) = \dfrac{\sum_{k=0}^{M} b_k z^{-k}}{1 + \sum_{k=1}^{N} a_k z^{-k}}$ (Zeros) (Poles)

Figure 6-32: Hardware Implementation of Second-Order IIR Filter (Biquad) Direct Form 1

Section Six

The general digital filter equation is shown in Figure 6-32, which gives rise to the general transfer function H(z), which contains polynomials in both the numerator and the denominator. The roots of the denominator determine the pole locations of the filter, and the roots of the numerator determine the zero locations. Although it is possible to construct a high order IIR filter directly from this equation (called the *direct form* implementation), accumulation errors due to quantization errors (finite word-length arithmetic) may give rise to instability and large errors. For this reason, it is common to cascade several biquad sections with appropriate coefficients rather than use the direct form implementation. The biquads can be scaled separately and then cascaded in order to minimize the coefficient quantization and the recursive accumulation errors. Cascaded biquads execute more slowly than their direct form counterparts, but are more stable and minimize the effects of errors due to finite arithmetic errors.

The Direct Form 1 biquad section shown in Figure 6-32 requires four registers. This configuration can be changed into an equivalent circuit shown in Figure 6-33 that is called the Direct Form 2 and requires only two registers. It can be shown that the equations describing the Direct Form 2 IIR biquad filter are the same as those for Direct Form 1. As in the case of FIR filters, the notation for an IIR filter is often simplified as shown in Figure 6-34.

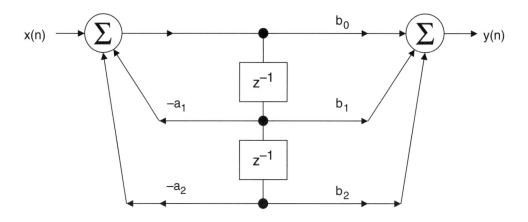

- REDUCES TO THE SAME EQUATION AS DIRECT FORM 1:

$$y(n) = b_0 x(n) + b_1 x(n-1) + b_2 x(n-2) - a_1 y(n-1) - a_2 y(n-2)$$

- REQUIRES ONLY TWO DELAY ELEMENTS (REGISTERS)

Figure 6-33: IIR Biquad Filter Direct Form 2

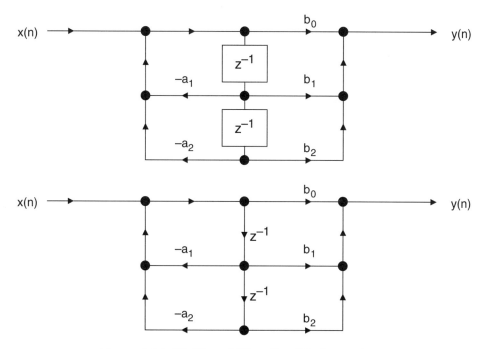

Figure 6-34: IIR Biquad Filter Simplified Notations

IIR Filter Design Techniques

A popular method for IIR filter design is to first design the analog-equivalent filter and then mathematically transform the transfer function H(s) into the z-domain, H(z). Multiple pole designs are implemented using cascaded biquad sections. The most popular analog filters are the Butterworth, Chebyshev, Elliptical, and Bessel (see Figure 6-35). There are many CAD programs available to generate the Laplace transform, H(s), for these filters.

The all-pole Butterworth (also called maximally flat) has no ripple in the pass band or stop band and has monotonic response in both regions. The all-pole Type 1 Chebyshev filter has a faster roll-off than the Butterworth (for the same number of poles) and has ripple in the pass band. The Type 2 Chebyschev filter is rarely used, but has ripple in the stop band rather than the pass band.

The Elliptical (Cauer) filter has poles and zeros and ripple in both the pass band and stop band. This filter has even faster roll-off than the Chebyshev for the same number of poles. The Elliptical filter is often used where degraded phase response can be tolerated.

Finally, the Bessel (Thompson) filter is an all-pole filter optimized for pulse response and linear phase but has the poorest roll-off of any of the types discussed for the same number of poles.

Section Six

- **Butterworth**
  - All Pole, No Ripples in Pass Band or Stop Band
  - Maximally Flat Response (Fastest Roll-Off with No Ripple)
- **Chebyshev (Type 1)**
  - All Pole, Ripple in Pass Band, No Ripple in Stop Band
  - Shorter Transition Region than Butterworth for Given Number of Poles
  - Type 2 has Ripple in Stop Band, No Ripple in Pass Band
- **Elliptical (Cauer)**
  - Has Poles and Zeros, Ripple in Both Pass Band and Stop Band
  - Shorter Transition Region than Chebyshev for Given Number of Poles
  - Degraded Phase Response
- **Bessel (Thompson)**
  - All Pole, No Ripples in Pass Band or Stop Band
  - Optimized for Linear Phase and Pulse Response
  - Longest Transition Region of All for Given Number of Poles

Figure 6-35: Review of Popular Analog Filters

All of the above types of analog filters are covered in the literature, and their Laplace transforms, H(s), are readily available, either from tables or CAD programs. There are three methods used to convert the Laplace transform into the z-transform: *impulse invariant* transformation, *bilinear* transformation, and the *matched z-transform*. The resulting z-transforms can be converted into the coefficients of the IIR biquad. These techniques are highly mathematically intensive and will not be discussed further.

A CAD approach for IIR filter design is similar to the Parks-McClellan program used for FIR filters. This technique uses the Fletcher-Powell algorithm.

In calculating the throughput time of a particular DSP IIR filter, one should examine the benchmark performance specification for a biquad filter section. For the ADSP-21xx family, seven instruction cycles are required to execute a biquad filter output sample. For the ADSP-2189M, 75 MIPS DSP, this corresponds to 7×13.3 ns = 93 ns, corresponding to a maximum possible sampling frequency of 10 MSPS (neglecting overhead).

Digital Filters

- **Impulse Invariant Transformation Method**
  - Start with H(s) for Analog Filter
  - Take Inverse Laplace Transform to Get Impulse Response
  - Obtain z-Transform H(z) from Sampled Impulse Response
  - z-Transform Yields Filter Coefficients
  - Aliasing Effects Must Be Considered
- **Bilinear Transformation Method**
  - Another Method for Transforming H(s) into H(z)
  - Performance Determined by the Analog System's Differential Equation
  - Aliasing Effects Do Not Occur
- **Matched z-Transform Method**
  - Maps H(s) into H(z) for Filters with Both Poles and Zeros
- **CAD Methods**
  - Fletcher-Powell Algorithm
  - Implements Cascaded Biquad Sections

Figure 6-36: IIR Filter Design Techniques

- Determine How Many Biquad Sections are Required to Realize the Desired Frequency Response
- Multiply This by the Execution Time per Biquad for the DSP (7 Instruction Cycles × 13.3 ns = 93 ns for the 75 MIPS ADSP-2189M, for example)
- The Result (Plus Overhead) is the Minimum Allowable Sampling Period ($1/f_s$) for Real-Time Operation

Figure 6-37: Throughput Considerations for IIR Filters

Section Six

Summary: FIR Versus IIR Filters

Choosing between FIR and IIR filter designs can be somewhat of a challenge, but a few basic guidelines can be given. Typically, IIR filters are more efficient than FIR filters because they require less memory and fewer multiply-accumulates are needed. IIR filters can be designed based upon previous experience with analog filter designs. IIR filters may exhibit instability problems, but this is much less likely to occur if higher order filters are designed by cascading second-order systems.

On the other hand, FIR filters require more taps and multiply-accumulates for a given cut off frequency response, but have linear phase characteristics. Since FIR filters operate on a finite history of data, if some data is corrupted (ADC sparkle codes, for example) the FIR filter will ring for only $N-1$ samples. Because of the feedback, however, an IIR filter will ring for a considerably longer period of time.

If sharp cut-off filters are needed, and processing time is at a premium, IIR elliptic filters are a good choice. If the number of multiply/accumulates is not prohibitive, and linear phase is a requirement, the FIR should be chosen.

IIR FILTERS	FIR FILTERS
More Efficient	Less Efficient
Analog Equivalent	No Analog Equivalent
May Be Unstable	Always Stable
Nonlinear Phase Response	Linear Phase Response
More Ringing on Glitches	Less Ringing on Glitches
CAD Design Packages Available	CAD Design Packages Available
No Efficiency Gained by Decimation	Decimation Increases Efficiency

Figure 6-38: Comparison Between FIR and IIR Filters

Digital Filters

Multirate Filters

There are many applications in which it is desirable to change the effective sampling rate in a sampled data system. In many cases, this can be accomplished simply by changing the sampling frequency to the ADC or DAC. However, it is often desirable to accomplish the sample rate conversion after the signal has been digitized. The most common techniques used are *decimation* (reducing the sampling rate by a factor of M), and *interpolation* (increasing the sampling rate by a factor of L). The decimation and interpolation factors (M and L) are normally integer numbers. In a generalized sample-rate converter, it may be desirable to change the sampling frequency by a noninteger number. In the case of converting the CD sampling frequency of 44.1 kHz to the digital audio tape (DAT) sampling rate of 48 kHz, interpolating by L = 160 followed by decimation by M = 147 accomplishes the desired result.

The concept of decimation is illustrated in Figure 6-39. The top diagram shows the original signal, f_a, which is sampled at a frequency f_s. The corresponding frequency spectrum shows that the sampling frequency is much higher than required to preserve information contained in f_a, i.e., f_a is oversampled. Notice that there is no information contained between the frequencies f_a and $f_s - f_a$. The bottom diagram shows the same signal where the sampling frequency has been reduced (decimated) by a factor of M. Notice that even though the sampling rate has been reduced, there is no aliasing and loss of information. Decimation by a larger factor than shown in Figure 6-39 will cause aliasing.

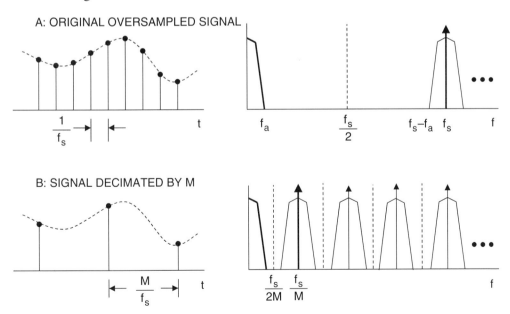

Figure 6-39: Decimation of a Sampled Signal by a Factor of M

Section Six

Figure 6-40A shows how to decimate the output of an FIR filter. The filtered data y(n) is stored in a data register that is clocked at the decimated frequency f_s/M. This does not change the number of computations required of the digital filter; i.e., it still must calculate each output sample y(n).

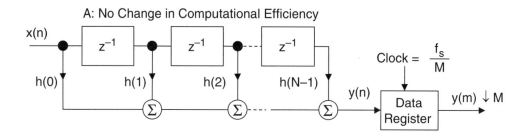

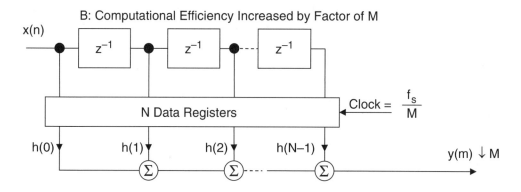

Figure 6-40: Decimation Combined with FIR Filtering

Figure 6-40B shows a method for increasing the computational efficiency of the FIR filter by a factor of M. The data from the delay registers are simply stored in N data registers that are clocked at the decimated frequency f_s/M. The FIR multiply-accumulates now only have to be done every Mth clock cycle. This increase in efficiency could be utilized by adding more taps to the FIR filter, doing other computations in the extra time, or using a slower DSP.

Figure 6-41 shows the concept of interpolation. The original signal in 6-41A is sampled at a frequency f_s. In 6-41B, the sampling frequency has been increased by a factor of L, and zeros have been added to fill in the extra samples. The signal with added zeros is passed through an interpolation filter, which provides the extra data values.

Digital Filters

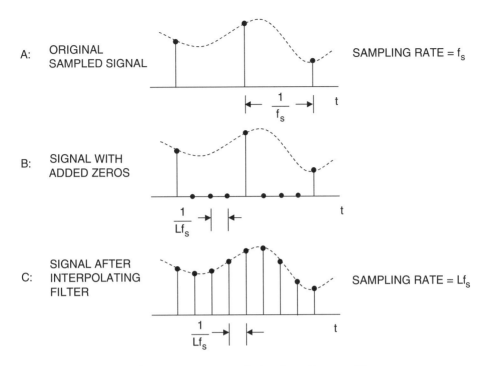

Figure 6-41: Interpolation by a Factor of L

The frequency domain effects of interpolation are shown in Figure 6-42. The original signal is sampled at a frequency f_s and is shown in 6-42A. The interpolated signal in 6-42B is sampled at a frequency Lf_s. An example of interpolation is a CD player DAC, where the CD data is generated at a frequency of 44.1 kHz. If this data is passed directly to a DAC, the frequency spectrum shown in Figure 6-42A results, and the requirements on the anti-imaging filter that precedes the DAC are extremely stringent to overcome this. An oversampling interpolating DAC is normally used, and the spectrum shown in Figure 6-42B results. Notice that the requirements on the analog anti-imaging filter are now easier to realize. This is important in maintaining relatively linear phase and also reducing the cost of the filter.

The digital implementation of interpolation is shown in Figure 6-43. The original signal x(n) is first passed through a rate expander that increases the sampling frequency by a factor of L and inserts the extra zeros. The data then passes through an interpolation filter that smoothes the data and interpolates between the original data points. The efficiency of this filter can be improved by using a filter algorithm that takes advantage of the fact that the zero-value input samples do not require multiply-accumulates. Using a DSP that allows circular buffering and zero-overhead looping also improves efficiency.

Section Six

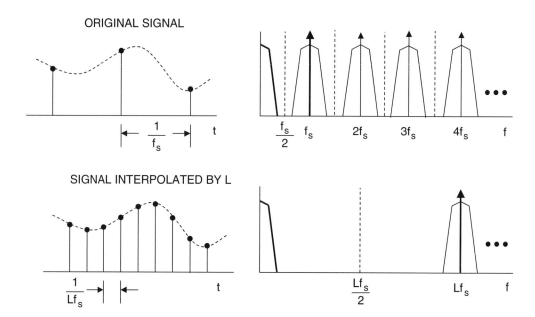

Figure 6-42: Effects of Interpolation on Frequency Spectrum

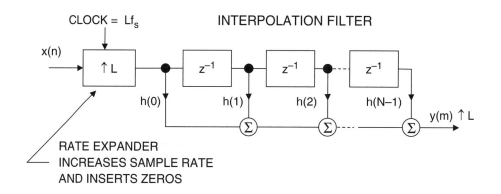

Efficient DSP algorithms take advantage of:
- Multiplications by Zero
- Circular Buffers
- Zero-Overhead Looping

Figure 6-43: Typical Interpolation Implementation

Digital Filters

Interpolators and decimators can be combined to perform fractional sample rate conversion as shown in Figure 6-44. The input signal x(n) is first interpolated by a factor of L and then decimated by a factor of M. The resulting output sample rate is Lf_s/M. To maintain the maximum possible bandwidth in the intermediate signal, the interpolation must come before the decimation; otherwise, some of the desired frequency content in the original signal would be filtered out by the decimator.

An example is converting from the CD sampling rate of 44.1 kHz to the digital audio tape (DAT) sampling rate of 48.0 kHz. The interpolation factor is 160, and the decimation factor, 147. In practice, the interpolating filter h´(k) and the decimating filter h´´(k) are combined into a single filter, h(k).

The entire sample rate conversion function is integrated into the AD1890, AD1891, AD1892, and AD1893 family which operates at frequencies between 8 kHz and 56 kHz (48 kHz for the AD1892). The AD1895 and AD1896 operate at up to 192 kHz.

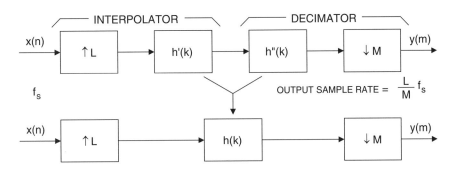

- Example: Convert CD Sampling Rate = 44.1kHz to DAT Sampling Rate = 48.0kHz
- Use L = 160, M = 147
- $f_{out} = \frac{L}{M} f_s = \frac{160}{147} \times 44.1 \text{kHz} = 48.0 \text{kHz}$
- AD189X - Family of Sample Rate Converters

Figure 6-44: Sample Rate Converters

Adaptive Filters

Unlike analog filters, the characteristics of digital filters can easily be changed by modifying the filter coefficients. This makes digital filters attractive in communications applications such as adaptive equalization, echo cancellation, noise reduction, speech analysis, and speech synthesis. The basic concept of an adaptive filter is shown in Figure 6-45. The objective is to filter the input signal, x(n), with an adaptive filter in such a manner that it matches the desired signal, d(n). The desired signal, d(n), is subtracted from the filtered signal, y(n), to generate an error signal. The error signal drives an adaptive algorithm that generates the filter coefficients in a manner that minimizes the error signal. The least-mean-square (LMS) or recursive-least-squares (RLS) algorithms are two of the most popular.

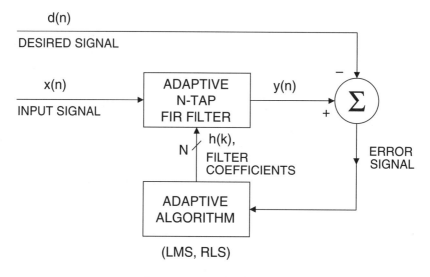

Figure 6-45: Adaptive Filter

Adaptive filters are widely used in communications to perform such functions as equalization, echo cancellation, noise cancellation, and speech compression. Figure 6-46 shows an application of an adaptive filter used to compensate for the effects of amplitude and phase distortion in the transmission channel. The filter coefficients are determined during a training sequence where a known data pattern is transmitted. The adaptive algorithm adjusts the filter coefficients to force the receive data to match the training sequence data. In a modem application, the training sequence occurs after the initial connection is made. After the training sequence is completed, the switches are put in the other position, and the actual data is transmitted. During this time, the error signal is generated by subtracting the input from the output of the adaptive filter.

Speech compression and synthesis also makes extensive use of adaptive filtering to reduce data rates. The linear predictive coding (LPC) model shown in Figure 6-47 models the vocal tract as a variable frequency impulse generator (for voiced portions of speech) and a random noise generator (for unvoiced portions of speech such as consonant sounds). These these two generators drive a digital filter that in turn generates the actual voice signal.

Digital Filters

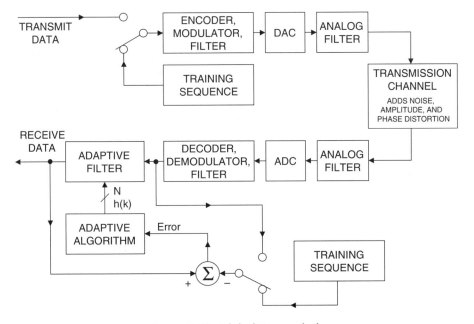

Figure 6-46: Digital Transmission
Using Adaptive Equalization

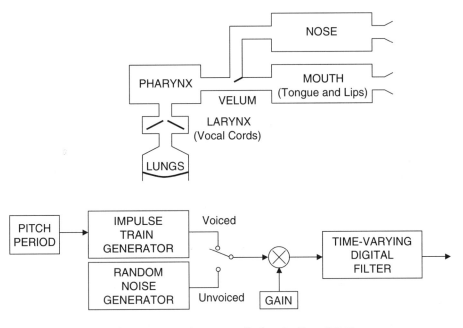

Figure 6-47: Linear Predictive Coding (LPC)
Model of Speech Production

Section Six

The application of LPC in a communication system such as GSM is shown in Figure 6-48. The speech input is first digitized by a 16-bit ADC at a sampling frequency of 8 kSPS. This produces output data at 128 Kbps, which is much too high to be transmitted directly. The transmitting DSP uses the LPC algorithm to break the speech signal into digital filter coefficients and pitch. This is done in 20 ms windows, which have been found to be optimum for most speech applications. The actual transmitted data is only 2.4 Kbps, which represents a 53.3 compression factor. The receiving DSP uses the LPC model to reconstruct the speech from the coefficients and the excitation data. The final output data rate of 128 Kbps then drives a 16-bit DAC for final reconstruction of the speech data.

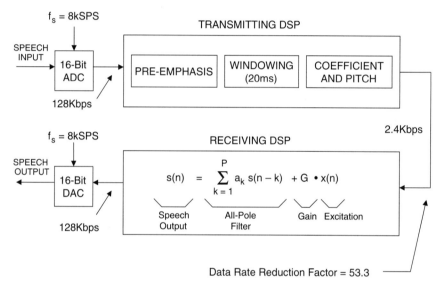

Figure 6-48: LPC Speech Companding System

The digital filters used in LPC speech applications can either be FIR or IIR, although all-pole IIR filters are the most widely used. Both FIR and IIR filters can be implemented in a lattice structure as shown in Figure 6-49 for a recursive all-pole filter. This structure can be derived from the IIR structure, but the advantage of the lattice filter is that the coefficients are more directly related to the outputs of algorithms that use the vocal tract model shown in Figure 6-47 than the coefficients of the equivalent IIR filter.

The parameters of the all-pole lattice filter model are determined from the speech samples by means of linear prediction as shown in Figure 6-50. Due to the nonstationary nature of speech signals, this model is applied to short segments (typically 20 ms) of the speech signal. A new set of parameters is usually determined for each time segment unless there are sharp discontinuities, in which case the data may be smoothed between segments.

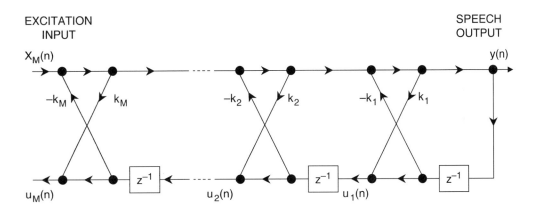

Figure 6-49: All Pole Lattice Filter

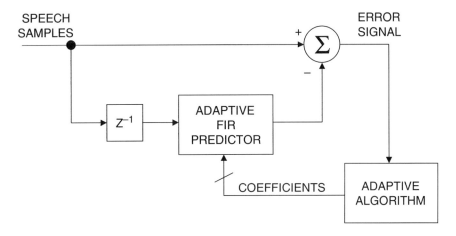

Figure 6-50: Estimation of Lattice Filter
Coefficients in Transmitting DSP

References

1. Steven W. Smith, **Digital Signal Processing: A Guide for Engineers and Scientists**, Newnes, 2002.

2. C. Britton Rorabaugh, **DSP Primer**, McGraw-Hill, 1999.

3. Richard J. Higgins, **Digital Signal Processing in VLSI,** Prentice-Hall, 1990.

4. A. V. Oppenheim and R. W. Schafer, **Digital Signal Processing**, Prentice-Hall, 1975.

5. L. R. Rabiner and B. Gold, **Theory and Application of Digital Signal Processing**, Prentice-Hall, 1975.

6. John G. Proakis and Dimitris G. Manolakis, **Introduction to Digital Signal Processing**, MacMillian, 1988.

7. J.H. McClellan, T.W. Parks, and L.R. Rabiner, *A Computer Program for Designing Optimum FIR Linear Phase Digital Filters*, **IEEE Trasactions on Audio and Electroacoustics**, Vol. AU-21, No. 6, December, 1973.

8. Fredrick J. Harris, *On the Use of Windows for Harmonic Analysis with the Discrete Fourier Transform*, **Proc. IEEE**, Vol. 66, No. 1, 1978 pp. 51–83.

9. Momentum Data Systems, Inc., 17330 Brookhurst St., Suite 140, Fountain Valley, CA 92708, http://www.mds.com.

10. **Digital Signal Processing Applications Using the ADSP-2100 Family**, Vol. 1 and Vol. 2, Analog Devices, Free Download at: http://www.analog.com.

11. **ADSP-21000 Family Application Handbook**, Vol. 1, Analog Devices, Free Download at: http://www.analog.com.

12. B. Widrow and S.D. Stearns, **Adaptive Signal Processing**, Prentice-Hall, 1985.

13. S. Haykin, **Adaptive Filter Theory**, 3rd Edition, Prentice-Hall, 1996.

14. Michael L. Honig and David G. Messerschmitt, **Adaptive Filters – Structures, Algorithms, and Applications**, Kluwer Academic Publishers, Hingham, MA 1984.

15. J.D. Markel and A.H. Gray, Jr., **Linear Prediction of Speech**, Springer-Verlag, New York, NY, 1976.

16. L.R. Rabiner and R.W. Schafer, **Digital Processing of Speech Signals**, Prentice-Hall, 1978.

Section 7
DSP Hardware

- Microcontrollers, Microprocessors, and Digital Signal Processors (DSPs)
- DSP Requirements
- ADSP-21xx 16-Bit Fixed-Point DSP Core
- Fixed-Point Versus Floating-Point
- ADI SHARC Floating-Point DSPs
- ADSP-2116x Single-Instruction, Multiple-Data (SIMD) Core Architecture
- TigerSHARC: The ADSP-TS001 Static Superscalar DSP
- DSP Benchmarks
- DSP Evaluation and Development Tools

Section 7
DSP Hardware
*Dan King, Greg Geerling, Ken Waurin,
Noam Levine, Jesse Morris, Walt Kester*

Microcontrollers, Microprocessors, and Digital Signal Processors (DSPs)

Computers are extremely capable in two broad areas: (1) *data manipulation*, such as word processing and database management, and (2) *mathematical calculation*, used in science, engineering, and digital signal processing. However, most computers are not optimized to perform *both* functions. In computing applications such as word processing, data must be stored, sorted, compared, and moved, and the time to execute a particular instruction is not critical, as long as the program's overall response time to various commands and operations is adequate enough to satisfy the end user. Occasionally, mathematical operations may also be performed, as in a spreadsheet or database program, but speed of execution is generally not the governing factor. In most general-purpose computing applications there is no concentrated attempt by software companies to make the code efficient. Application programs are loaded with "features" that require more memory and faster processors with every new release or upgrade.

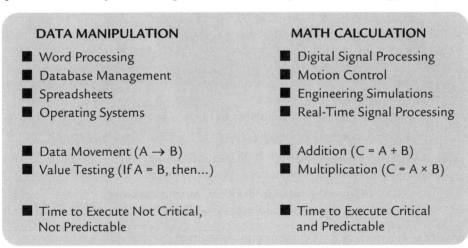

Figure 7-1: General Computing Applications

On the other hand, digital signal processing applications require that mathematical operations be performed quickly, and the time to execute a given instruction must be known precisely, and it must be predictable. Both code and hardware must be extremely efficient to accomplish this. As shown in the last two sections of this book, the most fundamental mathematical operation or *kernel* in all of DSP is the sum of products (or *dot product*). Fast execution of the dot product is critical to fast Fourier transforms (FFTs), real-time digital filters, matrix multiplications, graphics pixel manipulation, and other high speed algorithms.

Based on this introductory discussion of DSP requirements, it is important to understand the differences between *microcontrollers*, *microprocessors*, and *DSPs*. While microcontrollers used in industrial process control applications can perform functions such as multiplication, addition, and division, they are much more suited to applications in which I/O capability and control are more important than speed. Microcontrollers such as the 8051 family typically contain a CPU, RAM, ROM, serial/parallel interfaces, timers, and interrupt circuitry. The MicroConverter™ series from Analog Devices contains not only the 8051 core but also high performance ADC and DAC functions along with flash memory.

- **Microcontrollers**
  - CPU, RAM, ROM, Serial/Parallel Interface, Timer, Interrupt Circuitry
  - Well-Suited for Toasters as Well as Industrial Process Control
  - Speed is Not Generally a Requirement
  - Compact Instruction Sets
  - Examples: 8051, 68HC11, PIC
- **Microprocessors**
  - Single-Chip CPU – Requires Additional External Circuitry
  - RISC: Reduced Instruction Set Computer
  - CISC: Complex Instruction Set Computer
  - Examples: Pentium Series, PowerPC, MIPS
- **Digital Signal Processors (DSPs)**
  - RAM, ROM, Serial/Parallel Interface, Interrupt Circuitry
  - CPU Optimized for Fast Repetitive Math for Real-Time Processing
  - Examples: ADSP-21xx, ADSP-21K

Figure 7-2: Microcontrollers, Microprocessors, and Digital Signal Processors (DSPs)

Microprocessors, such as the Pentium series from Intel, are basically single-chip CPUs that require additional circuitry to make up the total computing function. Microprocessor instruction sets can be either complex instruction set computer (CISC) or reduced instruction set computer (RISC). The complex instruction set computer (CISC) includes instructions for basic processor operations, plus single instructions that are highly sophisticated; for example, to evaluate a high order polynomial. But CISC has a price: many of the instructions execute via microcode in the CPU and require numerous clock cycles plus silicon real estate for code storage memory.

DSP Hardware

In contrast, the reduced instruction set computer (RISC) recognizes that, in many applications, basic instructions such as LOAD and STORE—with simple addressing schemes—are used much more frequently than the advanced instructions, and should not incur an execution penalty. These simpler instructions are hardwired in the CPU logic to execute in a single clock cycle, reducing execution time and CPU complexity.

Although the RISC approach offers many advantages in general-purpose computing, it is not well-suited to DSP. For example, most RISCs do not support single-instruction multiplication, a very common and repetitive operation in DSP. The DSP is optimized to accomplish these tasks fast enough to maintain real-time operation in the context of the application. This requires single-cycle arithmetic operations and accumulations.

DSP Requirements

The most fundamental mathematical operation in DSP is shown in Figure 7-3: the sum of products (dot product). It is common to digital filters, FFTs, and many other DSP applications. A digital signal processor (DSP) is optimized to perform repetitive mathematical operations such as the dot product. There are five basic requirements for a DSP to optimize this performance: *fast arithmetic*, *extended precision*, *dual operand fetch*, *circular buffering*, and *zero-overhead looping*.

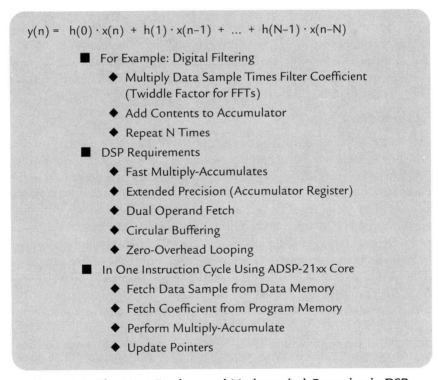

Figure 7-3: The Most Fundamental Mathematical Operation in DSP: The Sum of Products

Fast Arithmetic

Fast arithmetic is the simplest of these requirements to understand. Since real-time DSP applications are driven by performance, the multiply-accumulate, or MAC time, is a basic requirement; faster MACs mean potentially higher bandwidth. It is critical to remember that MAC time alone does not define DSP performance. This often forgotten fact leads to an inadequate measure of processor performance by simply examining its MIPS (million instructions per second) rating. Since most DSP and DSP-like architectures feature MACs that can execute an instruction every cycle, most processors are given a MIPS rating equal to its MAC throughput. This does not necessarily account for the other factors that can degrade a processor's overall performance in real-world applications. The other four criteria can wipe out MAC gains if they are not satisfied.

In addition to the requirement for fast arithmetic, a DSP should be able to support other general-purpose math functions and should therefore have an appropriate arithmetic logic unit (ALU) and a programmable shifter function for bit manipulation.

Extended Precision

Apart from the obvious need for fast multiplication and addition (MAC), there is also a requirement for extended precision in the accumulator register. For example, when two 16-bit words are multiplied, the result is a 32-bit word. The Analog Devices ADSP-21xx 16-bit fixed-point core architecture has an internal 40-bit accumulator that provides a high degree of overflow protection. While floating-point DSPs eliminate most of the problems associated with precision and overflow, fixed-point processors are still popular for many applications, and therefore overflow, underflow, and data scaling issues must be dealt with properly.

Dual Operand Fetch

Regardless of the nature of a processor, performance limitations are generally based on bus bandwidth. In the case of general-purpose microprocessors or microcontrollers, code is dominated by single memory fetch instructions, usually addressed as a base plus offset value. This leads architects to embed fixed data into the instruction set so that this class of memory access is fast and memory efficient. DSPs, on the other hand, are dominated by instructions requiring two independent memory fetches. This is driven by the basic form of the convolution (kernel or dot product) $Sh(i)x(i)$. The goal of fast dual operand fetches is to keep the MAC fully loaded. We saw in the discussion on MACs that the performance of a DSP is first limited by MAC time. Assuming an adequate MAC cycle time, two data values need to be supplied at the same rate; increases in operand fetch cycle time will result in corresponding increases in MAC cycle time. Ideally, the operand fetches occur simultaneously with the MAC instruction so that the combination of the MAC and memory addressing occurs in one cycle.

Dual operand fetch is implemented in DSPs by providing separate buses for program memory data and data memory data. In addition, separate program memory address and data memory address buses are also provided. The MAC can therefore receive inputs from each data bus simultaneously. This architecture is often referred to as the Harvard architecture.

Circular Buffering

If we examine the kernel equation more carefully, the advantages of circular buffering in DSP applications become apparent. A finite impulse response (FIR) filter is used to demonstrate the point. First, coefficients or tap values for FIR filters are periodic in nature. Second, the FIR filter uses the newest real-world signal value and discards the oldest value when calculating each output.

In the series of FIR filter equations, the N coefficient locations are always accessed sequentially from $h(0)$ to $h(N-1)$. The associated data points circulate through the memory as follows: new samples are stored, replacing the oldest data each time a filter output is computed. A fixed-boundary RAM can be used to achieve this circulating buffer effect. The oldest data sample is replaced by the newest with each convolution. A "time history" of the N most recent samples is kept in RAM.

This delay line can be implemented in fixed-boundary RAM in a DSP chip if new data values are written into memory, overwriting the oldest value. To facilitate memory addressing, old data values are read from memory, starting with the value one location after the value that was just written. In a 4-tap FIR filter, for example, $x(4)$ is written into memory location 0, and data values are then read from locations 1, 2, 3, and 0. This example can be expanded to accommodate any number of taps. By addressing data memory locations in this manner, the address generator need only supply sequential addresses regardless of whether the operation is a memory read or write. This data memory buffer is called *circular* because when the last location is reached, the memory pointer must be reset to the beginning of the buffer.

The coefficients are fetched simultaneously with the data. Due to the addressing scheme chosen, the oldest data sample is fetched first. Therefore, the last coefficient must be fetched first. The coefficients can be stored backwards in memory: $h(N-1)$ is the first location, and $h(0)$ is the last, with the address generator providing incremental addresses. Alternatively, coefficients can be stored in a normal manner with the accessing of coefficients starting at the end of the buffer, and the address generator being decremented.

This allows direct support of the FIR filter unit delay taps without software overhead. These data characteristics are DSP algorithm-specific and must be supported in hardware to achieve the best DSP performance. Implementing circular buffers in hardware allows buffer parameters (e.g., start and length) to be set up outside of the core instruction loop. This eliminates the need for extra instructions within the loop body. Lack of a hardware implementation for circular buffering can significantly affect MAC performance.

Zero-Overhead Looping

Zero-overhead looping is required by the repetitive nature of the kernel equation. The multiply-accumulate function and the data fetches required are repeated N times every time the kernel function is calculated. Traditional microprocessors implement loops that have one instruction execution time or more of overhead associated with repeating the loop. Analog Devices' DSP architectures provide hardware support that eliminates the need for looping instructions within the loop body. For true DSP architectures, the difference of zero-overhead body looping and programmed looping can easily exceed 20% cycle time.

Summary

Any processor can accomplish any software task, given enough time. However, DSPs are optimized for the unique computational requirements of real-time, real-world signal processing. Traditional computers are better suited for tasks that can be performed in non-real-time. In the following section, we will examine the architecture of a high performance 16-bit fixed-point DSP microcomputers, the Analog Devices ADSP-21xx family.

ADSP-21xx 16-Bit Fixed-Point DSP Core

Traditional microprocessors use the *Von Neumann architecture* (named after the American mathematician John Von Neumann) as shown in Figure 7-4A. The Von Neumann architecture consists of a single memory that contains data and instructions and a single bus for transferring data and instructions into and out of the CPU. Multiplying two numbers requires at least three cycles: two cycles are required to transfer the two numbers into the CPU, and one cycle to transfer the instruction. This architecture is satisfactory when all the required tasks can be executed serially. In fact, most general-purpose computers today use the Von Neumann architecture.

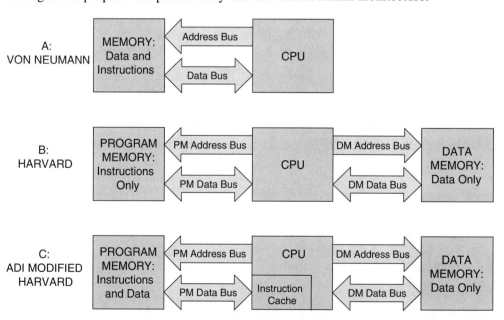

Figure 7-4: Microprocessor Architectures

For faster processing, however, the *Harvard architecture* shown in Figure 7-4B is more suitable. This is named for the work done at Harvard University under the leadership of Howard Aiken. Data and program instructions each have separate memories and buses as shown. Since the buses operate independently, program instructions and data can be fetched at the same time, thereby improving the speed over the single-bus Von Neumann design. In order to perform a single FIR filter multiply-accumulate, an instruction is fetched from the program memory, and, during the same cycle, a coefficient can be fetched from data memory. A second cycle is required to fetch the data word from the data memory.

Figure 7-4C illustrates Analog Devices' modified Harvard architecture where *instructions and data are allowed in the program memory*. For example, in the case of a digital filter, the coefficients are stored in the program memory, and the data samples in the data memory. A coefficient and a data sample can thus be fetched in a single cycle. In addition to fetching the coefficient from program memory and a data sample from data memory, an instruction must also be fetched from program memory. Analog Devices' DSPs handle this in one of two ways. In the first method, the program memory is accessed twice (double pumped) in an instruction cycle. The ADSP-218x series uses this method. In the second method, a program memory cache is provided. If an algorithm requires dual data fetches, the programmer places one buffer in program memory and the other in data memory. The first time the processor executes an instruction, there is a one-cycle stall because it must fetch the instruction and the coefficient over the program memory data bus. However, whenever this conflict occurs, the DSP "caches" the instruction in a cache memory. The next time this instruction is required, the program sequencer obtains it from the cache, while the coefficient is obtained over the program memory data bus. The cache method is used in the ADSP-219x family as well as in the SHARC family.

Digital Filter Example

Now that the fundamental architecture of the ADSP-21xx family has been presented, a simple FIR filter design will illustrate the ease of programming the family. Pseudocode for an FIR filter design is shown in Figure 7-5. For Analog Devices' DSPs, *all operations within the filter loop are completed in one instruction cycle*, thereby greatly increasing efficiency. Extra instructions are not required to repeat the loop. This is referred to as *zero-overhead looping*. The actual FIR filter assembly code for the ADSP-21xx family of fixed-point DSPs is shown in Figure 7-6. The arrows in the diagram point to the actual executable instructions (seven lines), the rest of the code consists simply of comments added for clarification. The first instruction (labeled *fir:*) sets up the computation by clearing the MR register and loading the MX0 and MY0 registers with the first data and coefficient values from data and program memory. The multiply-accumulate with dual data fetch in the *convolution* loop is then executed N–1 times in N–1 cycles to compute the sum of the first N–1 products. The final multiply-accumulate instruction is performed with the rounding mode enabled to round the result to the upper 24 bits of the MR register. The MR1 register is then conditionally saturated to its most positive or negative value based on the status of the overflow flag contained in the MV register. In this manner, results are accumulated to the full 40-bit precision of the MR register, with saturation of the output only if the final result overflowed beyond the least significant 32 bits of the MR register.

Section Seven

1. Obtain sample from ADC (typically interrupt driven)
2. Move sample into input signal's circular buffer
3. Update the pointer for the input signal's circular buffer
4. Zero the accumulator
5. Implement filter (control the loop through each of the coefficients)
   6. Fetch the coefficient from the coefficient's circular buffer
   7. Update the pointer for the coefficient's circular buffer
   8. Fetch the sample from the input signal's circular buffer
   9. Update the pointer for the input signal's circular buffer
   10. Multiply the coefficient by the sample
   11. Add the product to the accumulator
12. Move the filtered sample to the DAC

```
ADSP21xx Example code:

CNTR = N-1;
DO convolution UNTIL CE;
convolution:
   MR = MR + MX0 * MY0(SS), MX0 = DM(I0,M1), MY0 = PM(I4,M5);
```

Figure 7-5: Pseudocode for FIR Filter Program Using a DSP with Circular Buffering

```
         .MODULE         fir_sub;
         {               FIR Filter Subroutine
                         Calling Parameters
                                 I0 --> Oldest input data value in delay line
                                 I4 --> Beginning of filter coefficient table
                                 L0 = Filter length (N)
                                 L4 = Filter length (N)
                                 M1,M5 = 1
                                 CNTR = Filter length - 1 (N-1)
                         Return Values
                                 MR1 = Sum of products (rounded and saturated)
                                 I0 --> Oldest input data value in delay line
                                 I4 --> Beginning of filter coefficient table
                         Altered Registers
                                 MX0,MY0,MR
                         Computation Time
                                 (N - 1) + 6 cycles = N + 5 cycles
                         All coefficients are assumed to be in 1.15 format. }
         .ENTRY          fir;
         fir:            MR=0, MX0=DM(I0,M1), MY0=PM(I4,M5)
                         CNTR = N-1;
                         DO convolution UNTIL CE;
         convolution:        MR=MR+MX0*MY0(SS), MX0=DM(I0,M1), MY0=PM(I4,M5);
                         MR=MR+MX0*MY0(RND);
                         IF MV SAT MR;
                         RTS;
         .ENDMOD;
```

Figure 7-6: ADSP-21xx FIR Filter Assembly Code (Single Precision)

DSP Hardware

The ADSP-21xx family architecture (Figure 7-7) is optimized for digital signal processing and other high speed numeric processing applications. This family of DSPs combines the complete ADSP-2100 core architecture (three computational units, data address generators, and a program sequencer) with two serial ports, a programmable timer, extensive interrupt capabilities, and on-board program and data memory RAM. ROM-based versions are also available.

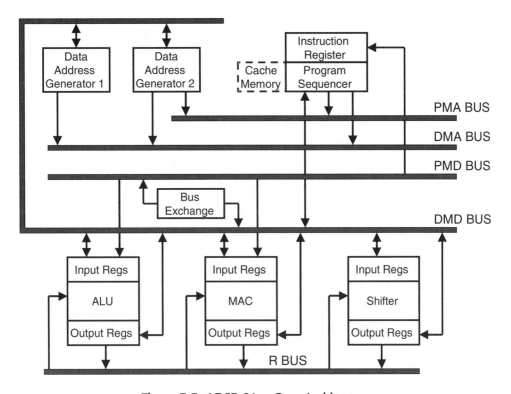

Figure 7-7: ADSP-21xx Core Architecture

The ADSP-21xx's flexible architecture and comprehensive instruction set support a high degree of operational parallelism. In one cycle the ADSP-21xx can generate the next program address, fetch the next instruction, perform one or two data moves, update one or two data address pointers, perform a computational operation, receive and transmit data via the two serial ports, and update a timer.

Section Seven

Figure 7-8:
ADSP-21xx Core Architecture

- **Buses**
  - Program Memory Address (PMA)
  - Data Memory Address (DMA)
  - Program Memory Data (PMD)
  - Data Memory Data (DMD)
  - Result (R)
- **Computational Units**
  - Arithmetic Logic Unit (ALU)
  - Multiply-Accumulator (MAC)
  - Shifter
- **Data Address Generators**
- **Program Sequencer**
- **On-Chip Peripheral Options**
  - Program Memory RAM or ROM
  - Data Memory RAM
  - Serial Ports
  - Timer
  - Host Interface Port
  - DMA Port

Buses

The ADSP-21xx processors have five internal buses to ensure efficient data transfer. The program memory address (PMA) and data memory address (DMA) buses are used internally for the addresses associated with program and data memory. The program memory data (PMD) and data memory data (DMD) buses are used for the data associated with the memory spaces. Off chip, the buses are multiplexed into a single external address bus and a single external data bus; the address spaces are selected by the appropriate control signals. The result (R) bus transfers intermediate results directly between the various computational units.

The PMA bus is 14 bits wide, allowing direct access of up to 16 Kwords of data. The data memory data (DMD) bus is 16 bits wide. The DMD bus provides a path for the contents of any register in the processor to be transferred to any other register or to any data memory location in a single cycle. The data memory address comes from two sources: an absolute value specified in the instruction code (direct addressing) or the output of a data address generator (indirect addressing). Only indirect addressing is supported for data fetches from program memory.

The program memory data (PMD) bus can also be used to transfer data to and from the computational units through direct paths or via the PMD-DMD bus exchange unit. The PMD-DMD bus exchange unit permits data to be passed from one bus to the other. It contains hardware to overcome the 8-bit width discrepancy between the two buses, when necessary.

Program memory can store both instructions and data, permitting the ADSP-21xx to fetch two data operands in a single cycle, one from program memory and one from data memory. The corresponding instruction is obtained directly from program memory by "double pumping" (ADSP-218x series) or from a cache memory (ADSP-219x and SHARC series).

Computational Units (ALU, MAC, Shifter)

The processor contains three independent computational units: the arithmetic logic unit (ALU), the multiplier-accumulator (MAC), and the barrel shifter. The computational units process 16-bit data directly and have provisions to support multiprecision computations. The ALU has a carry-in (CI) bit which allows it to support 32-bit arithmetic.

The ALU provides a standard set of arithmetic and logic functions: add, subtract, negate, increment, decrement, absolute value, AND, OR, EXCLUSIVE OR, and NOT. Two divide primitives are also provided.

- Add, Subtract, Negate, Increment, Decrement, Absolute Value, AND, OR, EXCLUSIVE OR, NOT
- Bitwise Operators, Constant Operators
- Multiprecision Math Capabilities
- Divide Primitives
- Saturation Mode for Overflow Support
- Background Registers for Single-Cycle Context Switch
- Example Instructions:
    - ◆ IF EQ AR = AX0 + AY0;
    - ◆ AF = MR1 XOR AY1;
    - ◆ AR = TGLBIT 7 OF AX1;

Figure 7-9: Arithmetic Logic Unit (ALU) Features

The MAC performs single-cycle multiply, multiply/add, and multiply/subtract operations. It also contains a 40-bit accumulator that provides 8 bits of overflow in successive additions to ensure that no loss of data occurs; 256 overflows would have to occur before any data is lost. Special instructions are provided for implementing block floating-point scaling of data. A set of background registers is also available in the MAC for interrupt service routines. If after a DSP routine is finished and the MV flag has been set, this means that the register contains a word greater than 32 bits. The register can be "saturated" using the saturation routine, which normalizes the 40-bit word to either a negative full-scale or positive full-scale 32-bit word in 1.32 format.

- Single-Cycle Multiply, Multiply-Add, Multiply-Subtract
- 40-Bit Accumulator for Overflow Protection (219x Adds Second 40-Bit Accumulator)
- Saturation Instruction Performs Single-Cycle Overflow Cleanup
- Background Registers for Single-Cycle Context Switch
- Example MAC Instructions:
    - `MR = MX0 * MY0(US);`
    - `IF MV SAT MR;`
    - `MR = MR - AR * MY1(SS);`
    - `MR = MR + MX1 * MY0(RND);`
    - `IF LT MR = MX0 * MX0(UU);`

Figure 7-10: Multiply-Accumulator (MAC) Features

The shifter performs logical and arithmetic shifts, normalization, denormalization, and derive-exponent operations. The shifter can be used to efficiently implement numeric format control including multiword floating-point representations.

- Normalize (Fixed-Point to Floating-Point Conversion)
- Denormalize (Floating-Point to Fixed-Point Conversion)
- Arithmetic and Logical Shifts
- Block Floating Point Support
- Derive Exponent
- Background Registers for Single-Cycle Context Switch
- Example Shifter Instructions:
    - `SR = ASHIFT SI BY -6(LO);`   {Arithmetic Shift}
    - `SR = SR OR LSHIFT SI BY 3(HI);`   {Logical Shift}
    - `SR = NORM MR1(LO);`   {Normalization}

Figure 7-11: Shifter Features

DSP Hardware

The computational units are arranged side-by-side instead of serially so that the output of any unit may be the input of any unit on the next cycle. The internal result (R) bus directly connects the computational units to make this possible.

Data Address Generators and Program Sequencer

Two dedicated data address generators and a powerful program sequencer ensure efficient use of the computational units. The data address generators (DAGs) provide memory addresses when memory data is transferred to or from the input or output registers. Each DAG keeps track of up to four address pointers. Whenever the pointer is used to access data (indirect addressing), it is postmodified by the value of a specified modify register. A length value may be associated with each pointer to implement automatic modulo addressing for circular buffers. With two independent DAGs, the processor can generate two addresses simultaneously for dual operand fetches.

DAG1 can supply addresses to data memory only; DAG2 can supply addresses to either data memory or program memory. When the appropriate mode bit is set in the mode status register (MSTAT), the output address of DAG1 is bit-reversed before being driven onto the address bus. This feature facilitates addressing in radix-2 FFT algorithms.

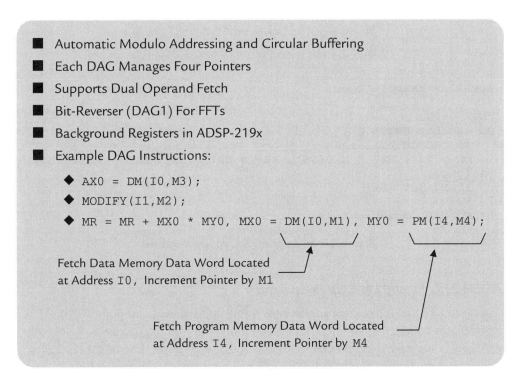

Figure 7-12: Data Address Generator Features

The program sequencer supplies instruction addresses to the program memory. The sequencer is driven by the instruction register that holds the currently executing instruction. The instruction register introduces a single level of pipelining into the program flow. Instructions are fetched and loaded into the instruction register during one processor cycle, and executed during the following cycle while the next instruction is pre-fetched. To minimize overhead cycles, the sequencer supports conditional jumps, subroutine calls, and returns in a single cycle. With an internal loop counter and loop stack, the processor executes looped code with zero overhead. No explicit jump instructions are required to loop. The sequencer also efficiently processes interrupts with its interrupt controller for fast interrupt response with minimum latency. When an interrupt occurs, it causes a jump to a known specified location in memory. Short interrupt service routines can be coded in place. For interrupt service routines with more than four instructions, program control is transferred to the service routine by means of a JUMP instruction placed at the interrupt vector location.

- Generates Next Instruction Address
- Low Latency Interrupt Handling
- Hardware Stacks
- Single-Cycle Conditional Branch (218x)
- Supports Zero-Overhead Looping

```
ADSP21xx Example code:

CNTR = 10;
DO endloop UNTIL CE;
   IO(DACCONTROL) = AX0;
   MR = MR + MX0 * MY0(SS), MX0 = DM(I0,M1), MY0 = PM(I4,M5);
endloop:
   IF MV SET FL1;

IF EQ CALL mysubroutine;
```

Figure 7-13: Program Sequencer Features

ADSP-21xx Family On-Chip Peripherals

The discussion so far has involved the core architecture of the fixed-point ADSP-21xx DSPs that is common to all members of the family. This section discusses the on-chip peripherals that have different configurations and options depending on the particular processor in the family. The ADSP-218x architecture is shown in Figure 7-14.

DSP Hardware

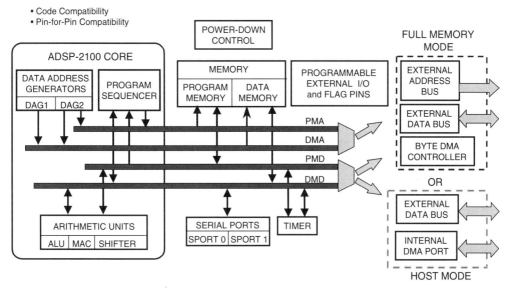

External bus features are multiplexed for 100-lead parts. All are available on 128-lead parts

Figure 7-14: ADSP-218x Family Architecture

- All Family Members Use an Enhanced Harvard Architecture
  - ◆ Separate Program Memory and Data Memory Spaces
  - ◆ Can Access Data Values in Program Memory
- Different Family Members Have Different Memory Configurations
- External Memory Interface Supports Both Fast and Slow Memories with Programmable Wait States
- Supports DSP Boot Loading from Byte Memory Space or from a Host Processor
- Supports Memory-Mapped Peripherals Through I/O Space
- Bus Request/Grant Mechanism for Shared External Bus

Figure 7-15: ADSP-21xx On-Chip Peripherals: Memory Interface

205

Section Seven

The 21xx family comes with a variety of on-chip memory options, and the newer ADSP-218x family has up to 48 Kwords of program memory and 56 Kwords of data memory. All family members use the modified Harvard architecture, which provides separate program and data memory and allows data to be stored in program memory. The external memory interface supports both fast and slow memories with programmable wait states. The ADSP-218x family also supports memory-mapped peripherals through I/O space.

All ADSP-21xx parts (except the ADSP-2105) have two double-buffered serial ports (SPORTs) for transmitting or receiving serial data. Each SPORT is bidirectional, full duplex, and has its own programmable serial clock. The SPORT word length can be configured from 3 bits to 16 bits. Data can be framed or unframed. Each SPORT generates an interrupt and supports A-law and μ-law companding.

- ADSP-21xx SPORTs Are Used For Synchronous Communication
- Full Duplex
- Fully Programmable
- Autobuffer/DMA Capability
- TDM Multichannel Capability
- A-Law and μ-Law Companding
- Data Rates of 25 Mbits/sec and above
- Glueless Interface to a Wide Range of Serial Peripherals or Processors
- 219x DSPs Add SPI and UART Serial Ports (with Boot Capability)

Figure 7-16: ADSP-21xx On-Chip Peripherals: Serial Ports (SPORTS)

The ADSP-218x family internal direct memory access (IDMA) port supports booting from and runtime access by a host processor. This feature allows data to be transferred to and from internal memory in the background while continuing foreground processing. The IDMA port allows a host processor to access all of the DSP's internal memory without using mailbox registers. The IDMA port supports 16- and 24-bit words, and 24-bit transfers take two cycles to complete.

DSP Hardware

- Allows an External System to Access DSP Internal Memory
- External Device or DSP Can Specify Internal Starting Address
- Address Automatically Increments to Speed Throughput
- 16-Bit Bus Supports Both Data and Instruction Transfers (ADSP-219x has 8-Bit Bus Support)
- Single DSP Processor-Cycle Transfers
- Supports Power-On Booting

Figure 7-17: ADSP-21xx On-Chip Peripherals: Internal DMA (IDMA)

The ADSP-218x family also has a byte memory interface that supports booting from, and runtime access to, 8-bit memories. It can access up to 4 MB. This memory space takes the place of the boot memory space found on other ADSP-21xx family processors. Byte memory consists of 256 pages of 16K × 8 locations. This memory can be written and read in 24-bit, 16-bit, or 8-bit left- or right-justified transfers. Transfers happen in the background to the DSP internal memory by stealing cycles.

- Provides Bulk Storage for Both Data and Program Code
- Can Access up to 4 Mbytes of External Code and Data
- Supports Multiple Data Formats
  - Automatic Data Packing/Unpacking to 16 Bits and 24 Bits
  - 8-Bit Transfers, Left- or Right-Justified
- Background Transfer to DSP Internal Memory
  - One Cycle per Word
  - DSP Specifies Destination/Source and Word Count
- Supports Power-On Booting
- Allows Multiple Code Segments
  - DSP Can Overlay Code Sections
  - Processor Can Run During Transfer, or Halt and Restart

Figure 7-18: ADSP-21xx On-Chip Peripherals: Byte DMA Port (BDMA)

Section Seven

The ADSP-218x, ADSP-219x devices provide a power-down feature that allows the processor to enter a very low power state (less than 1 mW) through hardware or software control. This feature is extremely useful for battery-powered applications. During some of the power-down modes, the internal clocks are disabled, but the processor registers and memory are maintained.

- Nonmaskable Interrupt
  - Hardware Pin (PWD), or Software Forced
- Holds Processor in CMOS Standby
- Rapid CLKIN-Cycle Recovery
- Acknowledge Handshake (PWDACK)
- Ideal for Battery-Powered Applications
- ADSP-219x is Fully Static

Figure 7-19: ADSP-21xx Internal Peripherals: Power-Down

From the above discussions it should be obvious that ADI DSPs are designed for maximum efficiency when performing typical DSP functions such as FFTs or digital filtering. ADI DSPs can perform several operations in an instruction cycle as has been shown in the above filter example. DSPs are often rated in terms of millions of instructions per second, or MIPS. However, the MIPS rating does not tell the entire story. For example if processor A has an instruction rate of 50 MIPS and can perform one operation per instruction, it can perform 50 million operations per second, or 50 MOPS. Now assume that processor B has an instruction rate of 20 MIPS, but can perform four operations per instruction. Processor B can therefore perform 80 million operations per second, or 80 MOPS, and is actually more efficient than processor A. A better way to evaluate DSP performance is to use well-defined benchmarks such as an FIR filter with a prescribed number of taps or an FFT of known size. Benchmark comparisons eliminate the confusion often associated with MIPS or MOPS ratings alone and are discussed later in this section. Even this form of benchmarking does not give a true comparison of performance between two processors. Analysis of the target system requirements, processor architecture, memory needs, and other factors must be considered.

The ADSP-219x family is ADI's latest generation of ADSP-21xx code compatible DSPs. The ADSP-219x family builds on the popular ADSP-218x platform and includes many enhancements to the DSP core and peripheral set over the previous generation. This provides customers with a smooth upgrade path to higher performance. Many of the ADSP-219x core enhancements are also designed to improve C-compiler efficiency. A block diagram of the ADSP-219x core is shown in Figure 7-20.

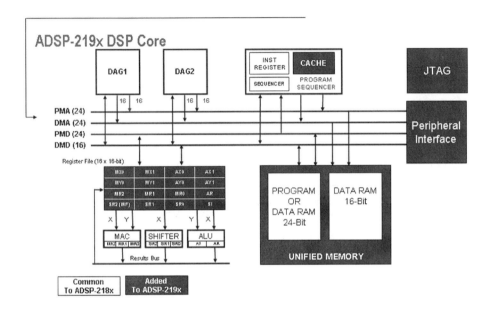

Figure 7-20: ADSP-219x Series Architecture

A global register allocator and support for register file-like operand access reduce spills and reduce reliance on the local stack. The data register file consists of a set of 16-bit data registers that transfers data between the data buses and the computational units. These registers also provide local storage for operands and results. The register file on ADSP-219x DSPs enables any register in the data register file to provide input to any of the computational units. This drastically increases compiler efficiency and differs from the data registers in the ADSP-218x core where only the dedicated registers for each computational unit can provide inputs.

ADSP-219x DSPs balance a high performance processor core with high performance buses (PM, DM, DMA). The core contains two 40-bit accumulators and a 40-bit shifter, which help minimize data overflow during complex operations. The address bus has been increased from 14 bits to 24 bits to support 64 Kword direct memory addressing or 16 Mword paged memory addressing. Two independent 16-bit address generators (DAGs) can generate addresses for both the PMA or DMA buses to support a unified PM and DM memory space. This allows programmers greater code flexibility and more efficient use of on-chip memory. The ADSP-219x core supports all of the ADSP-21xx addressing modes and adds five additional modes; register, indirect-postmodify, immediate-modify, and direct- and indirect-offset addressing. Each address generator supports as many as four circular buffers, each with three registers. The ADSP-219x core also supports as many as 16 circular buffers using a DAG shadow register set and a set of base registers for additional circular-buffering flexibility.

The ADSP-219x core supports up to eight-deep nesting via its loop hardware, and the program sequencer features a six-deep pipeline and supports delayed branching. The instruction cache provides for "third-bus" performance for multifunction instructions that require fetching an operand from the program memory block. Power consumption is reduced through local access of instruction from cache—not from memory.

In 2001, Analog Devices introduced a number of new DSPs based on the ADSP-219x core, that operate at 160 MHz and integrate a variety of peripherals and memory sizes to address the communications and industrial markets. Models are available with up to 2 Mbits of on-chip SRAM to increase overall system performance. All of the processors integrate a programmable DMA controller to support maximum I/O throughput and processor efficiency. The ADSP-21xx road map including the ADSP-219x processors is shown in Figure 7-22.

For more information on this family of products please visit: http://www.analog.com/technology/dsp/index.html.

Many of the enhancements in the ADSP-219x are designed to improve compiler efficiency. A global register allocator and support for register file-like operand access reduce spills and reduce reliance on the local stack. The compiler features DSP intrinsic support including fractional and complex. On-chip cache memory has also been added.

The ADSP-219x core will serve as a key DSP technology for ADI's 16-bit general-purpose DSP offerings and embedded DSP solutions, where application-specific circuitry and software are custom-designed to a customer's precise requirements. For performance-driven applications, multiple cores will be integrated together on a single die.

DSP Hardware

- Code Compatible
  - Compatible with ADSP-218x Series
  - Single-Cycle Instruction Execution, Zero-Overhead Looping, Single-Cycle Context Switch
- Performance
  - Architectural Performance Beyond 300 MIPs
  - Fully Transparent Instruction Cache
- Compiler-Friendly and Tool-Friendly
  - 64 Kword Direct and 16 Mword Paged Memory Support
  - 5 New DAG Addressing Modes
  - Register File-Like Operand Access
  - JTAG Interface Support

Figure 7-21: ADSP-219x Family Key Specifications

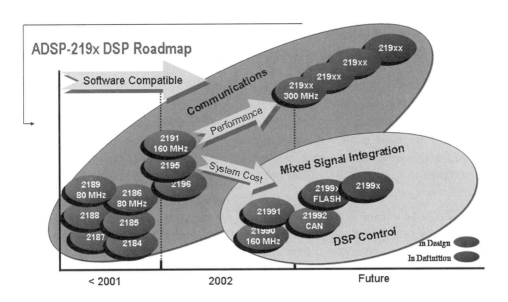

Figure 7-22: 16-Bit DSP Family Tree: History of Improvements in Packaging, Power, Performance

Section Seven

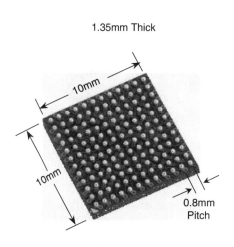

- Up to 160 MIPS in 1 cm^2
- Up to 2 Mbits on-chip SRAM
- 0.3 mA per MIP for low power applications

miniBGA package
144-ball grid array (BGA)

Figure 7-23: 'M' Series Offers the Largest MIPS/Memory Density with miniBGA Packaging

Fixed-Point Versus Floating-Point

DSP arithmetic can be divided into two catagories: *fixed-point* and *floating-point*. These refer to the format used to store and manipulate the numbers within the devices. The Analog Devices fixed-point DSPs, such as those discussed so far, represent each number with 16 bits. There are four common ways that $2^{16} = 65,536$ possible bit patterns can represent a number. In *unsigned integer format*, the stored number takes on any integer value from 0 to 65,536. In *signed integer format*, twos complement is used to make the range include negative numbers, from –32,768 to +32,767. Using *unsigned fractional format*, the 65,536 levels are spread uniformly between 0 and +1. Finally, *signed fractional format* allows negative numbers, with 65,536 levels equally spaced between –1 and +1.

The ADSP-21xx family arithmetic is optimized for the signed fractional format denoted by 1.15 ("one dot fifteen"). In the 1.15 format, there is one sign bit (the MSB) and fifteen fractional bits representing values from –1 up to 1 LSB less than +1 as shown in Figure 7-24.

DSP Hardware

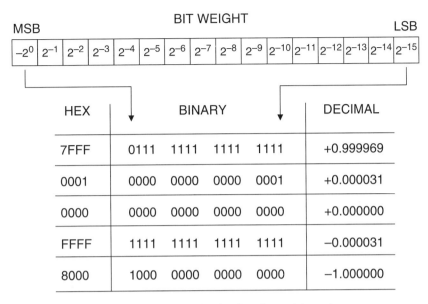

Figure 7-24: 16-Bit Fixed-Point Arithmetic
Fractional 1.15 Format

This convention may be generalized as "I.Q," in which I is the number of bits to the left of the radix point, and Q is the number of bits to the right. For example, full unsigned integer number representation is 16.0 format. For most signal processing applications, however, fractional numbers (1.15) are assumed. Fractional numbers have the advantage that the product of two fractional numbers is smaller than either of the numbers.

By comparison, floating-point DSPs typically use a minimum of 32 bits to represent each number. This results in many more possible numbers than for 16-bit fixed-point, $2^{32} = 4,294,967,296$ to be exact. More importantly, floating-point greatly increases the range of values that can be expressed. The most common floating-point standard is ANSI/IEEE Standard 754-1985, wherein the largest and smallest numbers allowed by the standard are $\pm 3.4 \times 10^{38}$ and $\pm 1.2 \times 10^{-38}$, respectively. Note, for example, that the 754 standard reserves some of the possible range to free up bit patterns which allow other special classes of numbers such as ± 0 and $\pm \infty$.

The IEEE-754 floating-point standard is described in more detail in Figure 7-25. The 32-bit word is divided into a sign bit, S, an 8-bit exponent, E, and a 23-bit mantissa, M. The relationship between the decimal and binary IEEE-754 floating-point equivalent is given by the equation:

$$\text{NUMBER}_{10} = (-1)^S \times 1.M \times 2^{(E-127)}$$

Notice that the "1" is assumed to precede the "M," and that a "bias" of 127 is subtracted from the exponent "E" so that "E" is always a positive number.

Section Seven

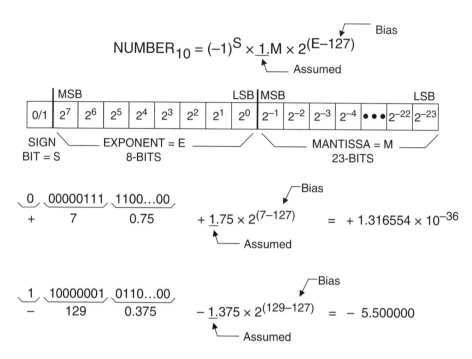

Figure 7-25: Single-Precision IEEE-754 32-Bit Floating-Point Format

In the case of *extended-precision* floating-point arithmetic, there is one sign bit, the mantissa is 31 bits, the exponent is 11 bits, and the total word length is 43 bits.

Extended precision thus adds 8 bits of dynamic range to the mantissa and can execute almost as fast as single-precision, since accumulators can readily be extended beyond 32 bits. On the other hand, true 64-bit *double precision* (52-bit mantissa, 11-bit exponent, and 1 sign bit) requires extra processor cycles. Requirements for double precision are rare in most DSP applications.

Many DSP applications can benefit from the extra dynamic range provided by 32-bit floating-point arithmetic. In addition, programming floating-point processors is generally easier, because fixed-point problems such as overflow, underflow, data scaling, and round-off error are minimized, if not completely eliminated. Although the floating-point DSP may cost slightly more than the fixed-point DSP, development time may well be shorter with floating-point.

While all floating-point DSPs can also handle fixed-point numbers (required to implement counters, loops, and ADC/DAC signals), this doesn't necessarily mean that fixed-point math will be carried out as quickly as the floating-point operations; it depends on the internal DSP architectures. For instance, the SHARC DSPs from Analog Devices are optimized for both floating-point and fixed-point operations, and execute them with equal efficiency. For this reason, the SHARC devices are often referred to as "32-bit DSPs" rather than just "floating-point."

DSP Hardware

- 16-Bit Fixed-Point:
  - 2^{16} = 65,536 Possible Numbers
- 32-Bit Floating-Point:
  - Biggest Number: $\pm 6.8 \times 10^{38}$ 754 Std: $\pm 3.4 \times 10^{38}$
  - Smallest Number: $\pm 5.9 \times 10^{-39}$ 754 Std: $\pm 1.2 \times 10^{-38}$
- Extended-Precision (40-Bit: Sign + 8-Bit Exponent + 31-Bit Mantissa)
- Double-Precision (64-Bit: Sign + 11-Bit Exponent + 52-Bit Mantissa)
- 32-Bit Floating-Point
  - More Precision
  - Much Larger Dynamic Range
  - Easier to Program

Figure 7-26: Fixed-Point vs. Floating-Point Arithmetic

ADI SHARC Floating-Point DSPs

The ADSP-2106x Super Harvard Architecture SHARC is a high performance 32-bit DSP. The SHARC builds on the ADSP-21000 family DSP core to form a complete system-on-a-chip, adding dual-ported on-chip SRAM and integrated I/O peripherals supported by a dedicated I/O bus. With its on-chip instruction cache, the processor can execute every instruction in a single cycle. Four independent buses for dual data, instructions, and I/O plus crossbar switch memory connections, comprise the Super Harvard Architecture of the ADSP-2106x shown in Figure 7-27.

A general-purpose data register file is used for transferring data between the computation units and the data buses, and for storing intermediate results. The register file has two sets (primary and alternate) of 16 registers each, for fast context switching. All of the registers are 40 bits wide. The register file, combined with the core processor's super Harvard architecture, allows unconstrained data flow between computation units and internal memory.

The ADSP-2106x SHARC processors address the five central requirements for DSPs established in the ADSP-21xx family of 16-bit fixed-point DSPs: (1) fast, flexible, arithmetic computation units, (2) unconstrained data flow to and from the computation units, (3) extended precision and dynamic range in the computation units, (4) dual address generators, and (5) efficient program sequencing with zero-overhead looping.

The program sequencer includes a 32-word instruction cache that enables three-bus operation for fetching an instruction and two data values. The cache is selective; only instructions whose fetches conflict with program memory data accesses are cached. This allows full-speed multiply-accumulates and FFT butterfly processing.

Section Seven

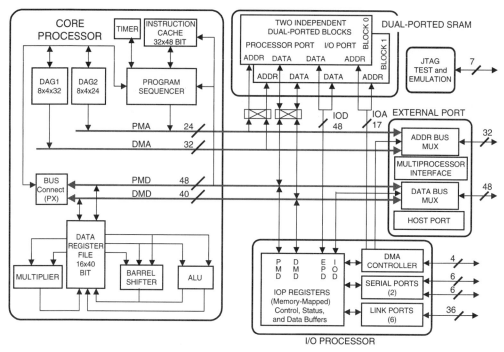

Figure 7-27: ADI Super Harvard Architecture (SHARC) 32-Bit DSP Architecture for ADSP-2106x Family

- 100MHz Core/300 MFLOPS Peak
- Parallel Operation of: Multiplier, ALU, 2 Address Generators and Sequencer
  - ◆ No Arithmetic Pipeline; All Computations Are Single-Cycle
- High Precision and Extended Dynamic Range
  - ◆ 32-/40-Bit IEEE Floating-Point Math
  - ◆ 32-Bit Fixed-Point MACs with 64-Bit Product and 80-Bit Accumulation
- Single-Cycle Transfers with Dual-Ported Memory Structures
  - ◆ Supported by Cache Memory and Enhanced Harvard Architecture
- Glueless Multiprocessing Features
- JTAG Test and Emulation Port
- DMA Controller, Serial Ports, Link Ports, External Bus, SDRAM Controller, Timers

Figure 7-28: SHARC Key Features

DSP Hardware

- SHARC is the de facto standard in multiprocessing
- ADSP-21160 continues SHARC leadership in multiprocessing
- ADSP-21065L is the right choice for low cost floating-point

SUPER HARVARD ARCHITECTURE:

Balancing Memory, I/O and Computational Power...

- High Performance Computation Unit
- 4-Bus Performance
  - Fetch Next Instruction
  - Access Two Data Values
  - Perform DMA for I/O
- Memory Architecture
- Nonintrusive DMA

Figure 7-29: SHARC, the Leader in Floating-Point DSP

The ADSP-2106x family executes all instructions in a single cycle. This family handles 32-bit IEEE floating-point format, 32-bit integer and fractional fixed-point formats (twos complement and unsigned), and extended-precision 40-bit IEEE floating-point format. The processors carry extended precision throughout their computation units, minimizing intermediate data truncation errors. When working with data on-chip, the extended precision 32-bit mantissa can be transferred to and from all computation units. The 40-bit data bus may be extended off-chip if desired. The fixed-point formats have an 80-bit accumulator for true 32-bit fixed-point computations.

The ADSP-2106x has a super Harvard architecture combined with a 10-port data register file. In every cycle, (1) two operands can be read or written to or from the register file, (2) two operands can be supplied to the ALU, (3) two operands can be supplied to the multiplier, and (4) two results can be received from the ALU and multiplier.

The ADSP-2106x family instruction set provides a wide variety of programming capabilities. Multifunction instructions enable computations in parallel with data transfers, as well as simultaneous multiplier and ALU operations.

The ADSP-21060 contains 4 Mbits of on-chip SRAM, organized as two blocks of 2 Mbits each, which can be configured for different combinations of code and data storage. The ADSP-21062, ADSP-21061, and ADSP-21065 each include 2 Mbits, 1 Mbit, and 544 Kbits of on-chip SRAM, respectively. Each memory block is dual-ported for single-cycle, independent accesses by the core processor and I/O processor or DMA controller. The dual-ported memory and separate on-chip buses allow two data transfers from the core and one from I/O, all in a single cycle.

While each memory block can store combinations of code and data, accesses are most efficient when one block stores instructions and data, using the DM bus for transfers, and the other block stores instructions and data, using the PM bus for transfers. Using the DM bus and PM bus in this way, with one dedicated to each memory block, assures single-cycle execution with two data transfers. In this case, the instruction must be available in the cache. Single-cycle execution is also maintained when one of the data operands is transferred to or from off-chip, via the ADSP-2106x's external port.

The ADSP-2106x's external port provides the processor's interface to off-chip memory and peripherals. The 4 Gword off-chip address space is included in the ADSP-2106x's unified address space. The separate on-chip buses—for PM addresses, PM data, and DM addresses, DM data, I/O addresses, and I/O data—are multiplexed at the external port to create an external system bus with a single 32-bit address bus and a single 48-bit data bus. The ADSP-2106x provides programmable memory wait states and external memory acknowledge controls to allow interfacing to DRAM and peripherals with variable access, hold, and disable time requirements.

The ADSP-2106x's host interface allows easy connection to standard microprocessor buses, both 16-bit and 32-bit, with little additional hardware required. Four channels of DMA are available for the host interface; code and data transfers are accomplished with low software overhead. The host can directly read and write the internal memory of the ADSP-2106x, and can access the DMA channel setup and mailbox registers. Vector interrupt support is provided for efficient execution of host commands.

The ADS-2106x offers powerful features tailored to multiprocessing DSP systems. The unified address space allows direct interprocessor accesses of each ADSP-2106x's internal memory. Distributed bus arbitration logic is included on-chip for simple, glueless connection of systems containing up to six ADSP-2106xs and a host processor. Master processor changeover incurs only one cycle of overhead. Bus arbitration is selectable as either fixed or rotating priority. Maximum throughput for interprocessor data transfer is 240 Mbytes/second (with a 40 MHz clock) over the link ports or external port.

The ADSP-2106x's I/O Processor (IOP) includes two serial ports, six 4-bit link ports, and a DMA controller. The ADSP-2106x features two synchronous serial ports that provide an inexpensive interface to a wide variety of digital and mixed-signal peripheral devices. The serial ports can operate at the full external clock rate of the processor, providing each with a maximum data rate of 50 Mbit/second. Independent transmit and receive functions provide greater flexibility for serial communications. Serial port data can be automatically

DSP Hardware

transferred to and from on-chip memory via DMA. Each of the serial ports offers a TDM multichannel mode. They offer optional μ-law or A-law companding. Serial port clocks and frame syncs can be internally or externally generated.

The ADSP-21060 and ADSP-21062 feature six 4-bit link ports that provide additional I/O capabilities. The link ports can be clocked twice per cycle, allowing each to transfer 8 bits per cycle. Link port I/O is especially useful for point-to-point interprocessor communication in multiprocessing systems. The link ports can operate independently and simultaneously, with a maximum data throughput of 240 Mbytes/second. Link port data is packed into 32-bit or 48-bit words, and can be directly read by the core processor or DMA-transferred to on-chip memory. Each link port has its own double-buffered input and output registers. Clock/acknowledge handshaking controls link port transfers. Transfers are programmable as either transmit or receive. There are no link ports on the ADSP-21061 or ADSP-21065 devices.

The ADSP-2106x's on-chip DMA controller allows zero-overhead data transfers without processor intervention. The DMA controller operates independently and invisibly to the processor core, allowing DMA operations to occur while the core is simultaneously executing its program. Both code and data can be downloaded to the ADSP-2106x using DMA transfers. DMA transfers can occur between the ADSP-2106x's internal memory and external memory, external peripherals, or a host processor. DMA transfers can also occur between the ADSP-2106x's internal memory and its serial ports or link ports. DMA transfers between external memory and external peripheral devices are another option.

The internal memory of the ADSP-2106x can be booted at system power-up from an 8-bit EPROM or a host processor. Additionally, the ADSP-21060 and the ADSP-21062 can also be booted through one of the link ports. Both 32-bit and 16-bit host processors can be used for booting.

The ADSP-2106x supports the IEEE standard P1149.1 Joint Test Action Group (JTAG) standard for system test. This standard defines a method for serially scanning the I/O status of each component in a system. In-circuit emulators also use the JTAG serial port to access the processor's on-chip emulation features. EZ-ICE Emulators use the JTAG test access port to monitor and control the target board processor during emulation. The EZ-ICE in-circuit emulator provides full-speed emulation to enable inspection and modification of memory, registers, and processor stacks. Use of the processor's JTAG interface assures nonintrusive in-circuit emulation—the emulator does not affect target system loading or timing.

The SHARC architecture avoids processor bottlenecks by balancing core, memory, I/O processor, and peripherals as shown in Figure 7-30. The core supports 32-bit fixed- and floating-point data. The memory contributes to the balance by offering large size and dual ports. The core can access data from one port, and the other port is used to move data to and from the I/O processor. The I/O processor moves data to and from the peripherals to the internal memory using zero-overhead DMAs. These operations run simultaneously to the core operation.

Section Seven

ADSP-2116x Single-Instruction, Multiple-Data (SIMD) Core Architecture

The ADSP-21160 is the first member of the second-generation ADI 32-bit DSPs. Its core architecture is shown in Figure 7-30. Notice that the core is similar to the ADSP-2106x core except for the width of the buses and the addition of a second computational unit complete with its own multiplier, ALU, shifter, and register file. This architecture is called *single-instruction, multiple-data* (SIMD) as opposed to *single-instruction, single-data* (SISD). The second computational unit allows the DSP to process multiple data streams in parallel. The core operates at up to 100 MIPS. At 100 MHz clock operation the core is capable of 400 MFLOPS (millions of floating-point operations per second) sustained and 600 MFLOPS peak operation. SIMD is a natural next step in increasing performance for ADI DSPs. Because their basic architecture already allows single-instruction and multiple-data access, adding another computational unit lets the architecture process the multiple data. The SIMD architectural extension allows code compatible higher performance parts.

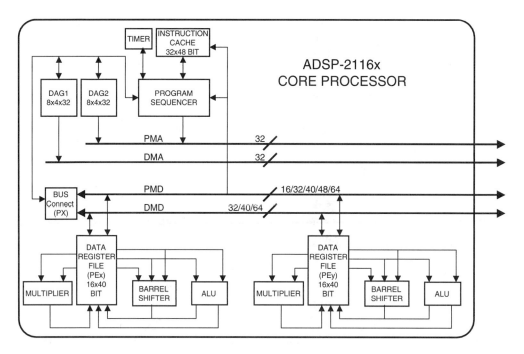

Figure 7-30: ADSP-2116x Core Processor Featuring Single-Instruction, Multiple-Data (SIMD)

The SIMD features of the ADSP-2116x include two computational units (PEx, PEy) and double data-word size buses (DMD and PMD). The primary processing element, PEx, is always enabled. The secondary processing element, PEy, is mode control enabled. The double-wide data buses provide each computational unit with its own data set in each cycle. With SIMD enabled, both processing elements execute the same instruction each cycle (that's the single-instruction), but they execute that instruction using different data (that's the multiple-data). The SIMD-based performance increase appears in algorithms that can be optimized by splitting the processing of data between the two computational units. By taking advantage of the second computational unit the cycle time can be cut in half compared to the SISD approach for many algorithms.

The ADSP-21160 has a complete set of integrated peripherals: I/O processor, 4 Mbyte on-chip dual-ported SRAM, glueless multiprocessing features, and ports (serial, link, external bus, host, and JTAG). Power dissipation is approximately 2 W at 100 MHz in a 400-ball 27 mm × 27 mm PBGA package. The complete SHARC family road map is shown in Figure 7-35.

- SIMD (Single-Instruction, Multiple-Data) Architecture
- Code Compatible with ADSP-2106x Family
- 100 MHz Core / 600 MFLOPS Peak
- On-Chip Peripherals Similar to ADSP-2106x Family
- Dual-Ported 4 Mbit SRAM
- Glueless Multiprocessing Features
- 400-Ball, PBGA 27 mm × 27 mm Package

Figure 7-31: ADSP-21160 32-Bit SHARC Key Features

Figure 7-36 shows some typical coding using the SHARC family of DSPs. Note the algebraic syntax of the assembly language, which facilitates coding of algorithms and reading the code after it is written. In a single cycle, the SHARC performs multiplication, addition, subtraction, memory read, memory write, and address pointer updates. In the same cycle, the I/O processor can transfer data to and from the serial port, the link ports, memory or DMA, and update the DMA pointer.

Section Seven

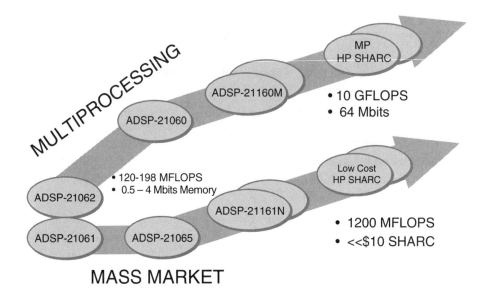

Figure 7-32: SHARC Road Map
Commitment to Code Compatibility into Tomorrow

---

f11 = f1 * f7, f3 = f9 + f14, f9 = f9 − f14, dm (i2, m0) = f13, f7 = pm (i8, m8);

■ In this Single-Cycle Instruction the SHARC Performs:

   ◆ 1 (2) Multiply
   ◆ 1 (2) Addition
   ◆ 1 (2) Subtraction   ( ) = ADSP-2116x SIMD DSP
   ◆ 1 (2) Memory Read
   ◆ 1 (2) Memory Write
   ◆ 2 Address Pointer Updates

■ Plus the I/O Processor Performs:

   ◆ Active Serial Port Channels: Transmit and Receive on all Ports
   ◆ 6 Active Link Ports if Present
   ◆ Memory DMA
   ◆ 2 (4) DMA Pointer Updates

The Algebraic Syntax of the Assembly Language Facilitates Coding of DSP Algorithms

Figure 7-33: Example: SHARC Multifunction Instruction

Multiprocessing Using SHARCs

Analog Devices' SHARC DSPs such as the ADSP-21160 are optimized for multiprocessing applications such as telephony, medical imaging, radar/sonar, communications, 3D graphics, and imaging. Figure 7-34 shows SHARC benchmark performance on common DSP algorithms.

	ADSP-21065L SHARC	ADSP-21160 SISD	ADSP-21160 SIMD/ Multiple Channels
Clock Cycle	66 MHz	100 MHz	100 MHz
Instruction Cycle Time	15 ns	10 ns	10 ns
MFLOPS Sustained	132 MFLOPS	200 MFLOPS	400 MFLOPS
MFLOPS Peak	198 MFLOPS	300 MFLOPS	600 MFLOPS
1024-Point Complex FFT (Radix 4, with reversal)	274 µs	180 µs	90 µs
FIR Filter (per tap)	15 ns	10 ns	5 ns
IIR Filter (per biquad)	60 ns	40 ns	20 ns
Matrix Multiply (pipelined) [3x3] * [3x1] [4x4] * [4x1]	135 ns 240 ns	90 ns 160 ns	45 ns 80 ns
Divide (y/x)	90 ns	60 ns	30 ns
Square Root	135 ns	90 ns	45 ns

Figure 7-34: DSP Benchmarks for SHARC Family

Multiprocessor systems typically use one or both of two methods to communicate between processor nodes. One method uses dedicated point-to-point communication channels. This method is often called *data flow multiprocessing*. In the other method, nodes communicate through a single shared global memory via a parallel bus. The SHARC family supports the implementation of point-to-point communication through its six link ports, called *link port multiprocessing*. It also supports an enhanced version of shared parallel bus communication called *cluster multiprocessing*.

For applications that require high computational bandwidth, but only limited flexibility, data flow multiprocessing is the best solution. The DSP algorithm is partitioned sequentially across several processors and data is passed directly across them as shown on the right side of Figure 7-35. The SHARC is ideally suited for data flow multiprocessing applications because it eliminates the need for interprocessor data FIFOs and external memory. Each SHARC has six link ports, allowing 2D and 3D arrays as well as traditional data flow. The internal memory of the SHARC is usually large enough to contain both code and data for most applications using this topology. All a data flow system requires are a number of SHARC processors and point-to-point signals connecting them.

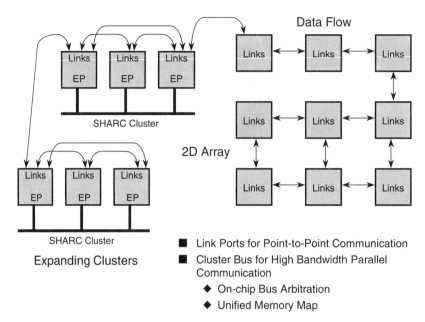

Figure 7-35: Multiprocessor Communication Examples for SHARC

- Advantages of External Ports (EP)
  - Communications through the EP have the Highest Bandwidth between Two SHARCs (400 Mbytes/s)
  - Allows up to Six SHARCs and a Host to Share the EP
  - The EP Offers Flexible Communication of Data and Control Information
  - The Shared Memory Model Allows a Simple Software Structure
- Advantage of Link Ports
  - Each Link Port Provides Independent, 100 MBytes/s communication between two SHARCs
  - Up to Six Link Ports (600 MBytes/s)
  - Easily scalable to any number of SHARCs
- Both Link Port and EP Communications Can Be Used Simultaneously

Figure 7-36: External Port vs. Link Port Communications

Cluster multiprocessing is best suited for applications in which a fair amount of flexibility is required. This is especially true when a system must be able to support a variety of different tasks, some of which may be running concurrently. SHARC processors also have an on-chip host interface that allows a cluster to be easily interfaced to a host processor or even to another cluster.

Cluster multiprocessing systems include multiple SHARC processors connected by a parallel bus that allows interprocessor access to on-chip memory as well as access to shared global memory. In a typical cluster of SHARCs, up to six ADSP-21160 processors and a host can arbitrate for the bus. The on-chip bus arbitration logic allows these processors to share the common bus. The SHARC's other on-chip features help eliminate the need for any extra glue hardware in the cluster multiprocessor configuration. External memory, both local and global, can frequently be eliminated in this type of system.

TigerSHARC: The ADSP-TS001 Static Superscalar DSP

The ADSP-TS001 is the first DSP from Analog Devices to use the new TigerSHARC static superscalar architecture. The TigerSHARC targets telecommunications infrastructure equipment with a new level of integration and the unique ability to process 8-, 16-, and 32-bit fixed- and floating-point data types on a single chip. Each of these data types is critical to the next generation of telecommunications protocols currently under development, including IMT-2000 (also known as 3G wireless) and xDSL (digital subscriber line). Unlike any other DSP, the ADSP-TS001 has the unique ability to accelerate processing speed based on the data type. Moreover, the chip delivers the highest performance floating-point processing.

In telecommunications infrastructure equipment, voice coder and channel coder protocols are developed around 16-bit data types. To improve signal quality, many telecom applications employ line equalization and echo cancellation techniques that boost overall signal quality and system performance. These algorithms benefit from the added precision of 32-bit and floating-point data processing. The 8-bit native support is well-suited to the commonly used Viterbi channel decoder algorithm, as well as image processing where it is more straightforward and cost-effective to represent red, green, and blue components of the signal with 8-bit data types. Many of these applications require high levels of performance and may require algorithms to be executed consecutively or even concurrently. The end application determines the exact requirements. The flexibility of the TigerSHARC architecture enables the software engineer to match the application precision requirements without any loss of system performance. In the TigerSHARC, performance is traded directly against numerical precision.

The TigerSHARC architecture uses key elements from multiple microprocessor types—RISC (reduced instruction set computer), VLIW (very long instruction word), and DSP in order to provide the highest performance digital signal processing engine. The new architecture leverages existing DSP product attributes such as fast and deterministic execution cycles, highly responsive interrupts, and an excellent peripheral interface to support large core computation rates and a high data rate I/O. To achieve

Section Seven

excellent core performance, RISC-like features such as load/store operations, a deeply pipelined sequencer with branch prediction, and large interlocked register files are introduced. Additionally, the VLIW (very long instruction word) attributes offer more efficient use of code space, especially for control code.

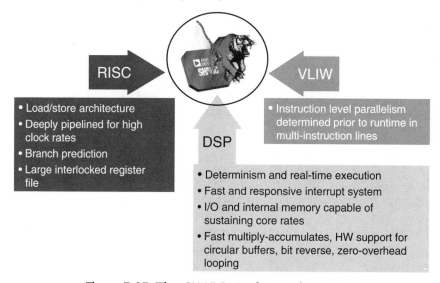

Figure 7-37: TigerSHARC: Analog Devices' New Static Superscaler DSP Architecture

Core
- 1200 MMACs/s @ 150 MHz, 16-Bit Fixed-Point
- 300 MMACs/s @150 MHz, 32-Bit Floating-Point
- 900 MFLOPS, 32-Bit Floating-Point

Memory
- 6 Mbits of On-Chip SRAM Organized in a Unified Memory Map as Opposed to the Traditional Harvard Architecture

I/O, Peripherals, and Package
- 600 Mbytes/s Transfer Rate through External Bus
- 600 Mbytes/s Aggregate Transfer Rate through 4 Link Ports
- Glueless Multiprocessor Cluster Support for up to 8 ADSP-TS001s
- 4 General-Purpose I/O Ports
- SDRAM Controller
- 360-Ball, SBGA Package 35 mm × 35 mm

Figure 7-38: TigerSHARC Key Architectural Features

DSP Hardware

Finally, to supply all functional blocks with instructions, clever management of the instruction word is necessary. Specifically, multiple instructions must be dispatched to processing units simultaneously, and functional parallelism must be calculated prior to runtime.

By incorporating the best of all worlds, the TigerSHARC architecture will provide a state-of-the-art platform for the most demanding signal processing applications.

The TigerSHARC core shown in Figure 7-39 consists of multiple functional blocks: computation blocks, memory, integer ALUs, and a sequencer. There are two computational blocks (X and Y) in the TigerSHARC architecture, each containing a multiplier, ALU, and 64-bit shifter. With the resources in these blocks, it is possible to execute eight 40-bit MACs on 16-bit data, two 40-bit MACs on 16-bit complex data, or two 80-bit MACs on 32-bit data, all in a single cycle. TigerSHARC is a register-based load/store architecture, where each computational block has access to a fully orthogonal 32-word register file.

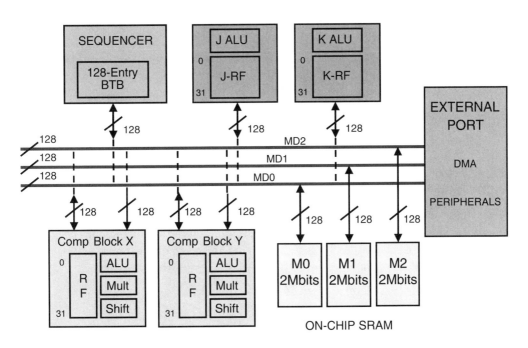

Figure 7-39: ADSP-TS001 TigerSHARC Architecture

Section Seven

The TigerSHARC DSP features a short-vector memory architecture organized in three 128-bit-wide banks. Quad, long, and normal word accesses move data from the memory banks to the *register* files for operations. In a given cycle, four 32-bit instruction words can be fetched, and 256 bits of data can be loaded to the register files or stored into memory. Data in 8-, 16-, or 32-bit words can be stored in contiguous, packed memory. Internal and external memories are organized in a unified memory map, which leaves specific partitioning to the programmer. The internal memory bandwidth for data and instructions is 7.2 Gbytes/second when operating on a 150 MHz clock.

Two integer ALUs are available for data addressing and pointer updates. They support circular buffering and bit reversal, and each has its own 32-word register file. More than simple data address generations units, both integer ALUs support general-purpose integer computations. The general-purpose nature of the integer ALUs benefits the compiler efficiency and increases programming flexibility.

The TigerSHARC architecture is designated *static superscalar*, as it executes up to four 32-bit instructions per clock cycle, and the programmer has the flexibility to issue individual instructions to each of the computational units. The sequencer supports predicted execution, where any individual instruction executes according to the result of a previously defined condition. The same instruction can be executed by the two computation blocks concurrently using different data values (this is called SIMD, single-instruction, multiple-data operation).

The TigerSHARC architecture enables native operation using 8-, 16-, or 32-bit data values. The overall processor performance increases as the level of data precision decreases.

The inclusion of a *branch target buffer* (BTB) and *static branch prediction* logic eliminates the programming task of filling the instruction pipeline after branch instructions. If seen before, the branch is taken in a single cycle.

Three internal 128-bit-wide buses ensure a large data bandwidth between internal functional blocks and external peripherals. The three-bus structure matches typical mathematical instructions that require two inputs and compute one output. The programming model is orthogonal and provides for deterministic interrupts.

The TigerSHARC architecture is free of hardware modes. This eliminates wasted cycles and simplifies compiler operation. The instruction set directly supports all DSP, image, and video processing arithmetic types including signed, unsigned, fractional, and integer data types. There is optional saturation (clipping) arithmetic for all cases.

At 150 MHz, the ADSP-TS001 offers the highest integer and floating-point performance of any SHARC product. Additionally, at 6 Mbits of on-chip SRAM, Analog Devices has increased its level of memory integration by 50% over previous SHARC family members. The migration to smaller process geometries will enable ADI to increase clock frequencies and integrate additional memory for future product derivatives.

DSP Hardware

- Execution of One to Four 32-Bit Instructions Per Clock Cycle
- Single-Instruction, Multiple-Data (SIMD) Operations Supported by Two Computation Blocks
- Multiple-Data Type Computation Blocks
  - Each with Register File, MAC, ALU, Shifter
  - 32-/40-Bit Floating- or 32-Bit Fixed-Point Operations (Six Per Cycle)
  - 16-Bit (24 Per Cycle) or 8-Bit (32 Per Cycle) Operations
- Static Branch Prediction Mechanism, with 128-Entry Branch Target Buffer (BTB)
- Internal Bandwidth of 7.2 Gbytes/second
- Simple and Fully Interruptible Programming Model

Figure 7-40: TigerSHARC Key Features

The ADSP-TS001 reduces total material costs by integrating multiple I/O and peripheral functions that reduce or eliminate the need for external glue logic and support chips. Specifically, the ADSP-TS001 at 150 MHz integrates four glueless link ports with an aggregate transfer rate of 600 Mbytes/s, glueless multiprocessor cluster interface support for up to 8 ADSP-TS001s, an SDRAM controller, and a JTAG controller. This unprecedented functionality is packaged in a 35 mm × 35 mm 360-ball SBGA package.

Typical computation rates and coding details of the TigerSHARC are shown in Figure 7-41. Four 32-bit instructions are executed in parallel, forming one 128-bit instruction line. The entire instruction line is executed in one cycle. This example assembly code is for a single line and is performing the following:

xR3:0=Q[j0+=4];//	load four registers (xR0,xR1,xR2,xR3) in the X register file from memory
yR3:0=Q[k0+=4];//	load four registers in the Y register file from memory
FR5=R4*R4; //	multiply two 32-bit floats in X computational block and two more in Y (two multiplies)
FR9:8=R6+/-R7;;//	add and subtract in both X and Y computational blocks (four ALU operations)

A single semicolon separates each 32-bit instruction, and a double semicolon indicates the end of an instruction line. This particular example shows the syntax for 32-bit floating-point multiplies and ALU operations. Parallel 16-bit operands can easily be specified by using the "S" for "short" prefix instead of the "F" for the "float" prefix. J0 and K0 are IALU registers being used as indirect address pointers for the memory reads.

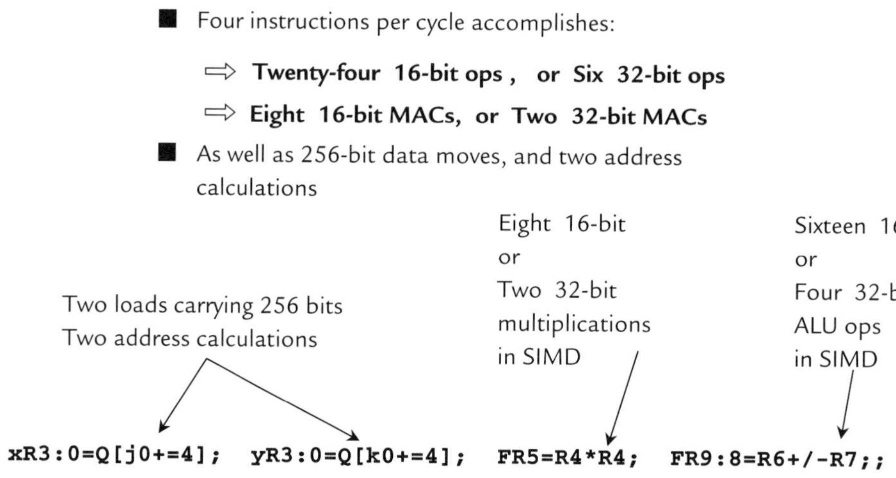

- Four instructions per cycle accomplishes:
  ⇒ Twenty-four 16-bit ops, or Six 32-bit ops
  ⇒ Eight 16-bit MACs, or Two 32-bit MACs
- As well as 256-bit data moves, and two address calculations

Two loads carrying 256 bits
Two address calculations

Eight 16-bit
or
Two 32-bit
multiplications
in SIMD

Sixteen 16-bit
or
Four 32-bit
ALU ops
in SIMD

```
xR3:0=Q[j0+=4];   yR3:0=Q[k0+=4];   FR5=R4*R4;   FR9:8=R6+/-R7;;
```

Figure 7-41: TigerSHARC Peak Computation Rates

DSP programmers demand and require the capability to program in both high level languages and low level assembly language. The determination of programming language is dependent upon a number of factors including speed performance, memory size, and time-to-market considerations. Ultimately, however, the whole DSP product should incorporate features that enable user-friendly coding in both high level and low level languages. The TigerSHARC architecture does indeed meet these requirements.

Specifically, the TigerSHARC core includes 128 each 32-bit general-purpose registers. This large number of registers allows C compilers sufficient flexibility to capitalize on the full potential performance of the architecture. In order to ensure data integrity, all registers are completely interlocked, meaning that the programmer does not have to be cognizant of architecture delays. The hardware ensures valid data is used in computations. Additionally, all registers can be accessed via all addressing modes (orthogonal) and a deterministic delay (two clock cycles) is achieved for all computational instructions. Lastly, the TigerSHARC architecture includes a branch target buffer which holds the effective address of the last 128 branches or jumps. This feature alleviates the programming task of filling the instruction pipeline after branch instructions. If seen before, the architecture jumps to the next instruction in a single clock cycle.

DSP Hardware

- 128 General-Purpose Registers
- All Registers Fully Interlocked
- General-Purpose Integer ALUs for Addressing
- Branch Prediction
- No Hardware Modes
- Orthogonal Addressing Modes
- Assembly Language Support

Figure 7-42: Architectural Features for High Level Language Support

Figure 7-43 depicts one possible configuration of a TigerSHARC design in a multiprocessing implementation. Up to eight ADSP-TS001 processors can communicate directly via the high speed 64-bit-wide external bus interface. In this type of communication, a commonly used master-slave protocol is implemented that enables any two processors to communicate directly at any one time.

In addition to the primary external bus, a limitless number of processors can be connected via the ADSP-TS001 link ports. While offering more flexibility, link port connectivity provides lower per-port bandwidth than the primary external bus interface. Again, all data transfers via the link ports are managed by a dedicated I/O processor and require no CPU intervention.

To summarize, the data I/O bandwidth of the link port (600 MBytes/s) and external port (600 MBytes/s) can be aggregated, yielding an overall individual processor data bandwidth of 1200 Mbytes/s with 150 MHz clock operation. Additionally, both the link port interface and the multiprocessor cluster interface are both completely glueless.

The ADSP-TS001 is the first member of a planned family of TigerSHARC-based products. Specifically, future members of the TigerSHARC family will contain optimized mixes of memory and peripherals to meet the requirements of specific target markets. These markets include third generation cellular base stations and VOIP (Voice Over the Internet Protocol) servers/concentrators. Additionally, process and design improvements will double the baseline performance of the general-purpose TigerSHARC family members.

Section Seven

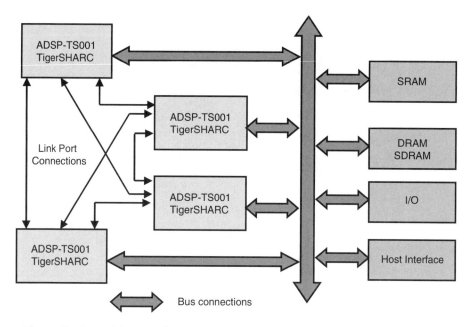

Figure 7-43: Multiprocessing Communication via Link Ports and Cluster Bus

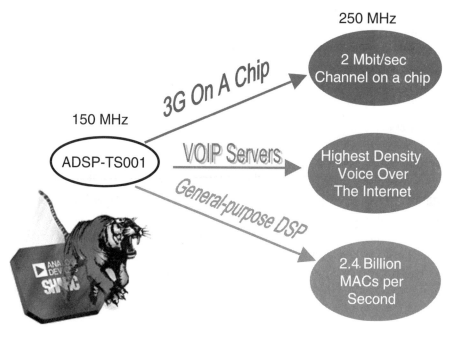

Figure 7-44: TigerSHARC Road Map

DSP Hardware

Comparing DSPs based solely on MIPS, MOPS, or MFLOPS does not tell the entire performance story. It is much more useful to compare DSPs based on their performance with respect to specific algorithms. The FFT and the FIR filter are popular benchmarks, as well as the IIR biquad filter, matrix multiplications, division, and square root.

Figure 7-45 shows the benchmark performance of the ADSP-TS001 TigerSHARC operating on 16-bit fixed-point data. Figure 7-46 shows its benchmark performance operating on 32-bit floating-point data.

- 16-Bit performance—1200 MMACs/s Peak Performance

Algorithm	Execution Time	Cycles to Execute
256-Point Complex FFT (Radix 2)	7.3 μs	1100
50 Tap FIR on 1024 inputs	48 μs	7200
Single FIR MAC	0.93 ns	0.14
Single Complex FIR MAC	3.80 ns	0.57
Single FFT Butterfly	6.7 ns	1.0

Figure 7-45: ADSP-TS001 TigerSHARC Benchmarks @ 150 MHz 16-Bit Performance

- 32-Bit performance—300 MMACs/s Peak Performance

Algorithm	Execution Time	Cycles to Execute
1024-Point Complex FFT (Radix 2)	69 μs	10300
50-Tap FIR on 1024 Input	184 μs	27500
Single FIR MAC	3.7 ns	0.55
Single FFT Butterfly	13.3 ns	2.0
Single Complex FIR MAC	13.3 ns	2.0
Divide	20 ns	3.0
Square Root	33.3 ns	5.0
Viterbi Decode (per Add/Compare/Select)	3.3 ns	0.5

Figure 7-46: ADSP-TS001 TigerSHARC Benchmarks @ 150 MHz 32-Bit Performance

DSP Evaluation and CROSSCORE™ Development Tools

The availability of a complete set of hardware and software development tools is essential to any DSP-based design. A typical DSP system design cycle is described as follows.

The first step in the process is to describe the system architecture. This would include such things as the type of processor, the peripherals (external memory, codec, host processor, links), and the configuration. This information is placed in a file known as the *link descriptor file (or LDF)*.

The next step in the process is to generate the actual DSP code. This can be done using a higher level language (usually C or C++), the DSP assembly language, or a combination of both. DSP code developed in C/C++ must be *compiled* and *assembled* in order to generate the assembly language code. The most effective way for producing application code is in assembly; however, C/C++ will expedite code creation. And mixing the languages provides the best combination of optimized code and gets designs to market faster. The Analog Devices DSP assembly language is based on algebraic syntax and is relatively easy to use directly. The *linker* then generates an executable file.

The software must then be debugged using the *software simulator* or in conjunction with an *evaluation board* such as the EZ-KIT Lite™ evaluation board or perhaps a third-party card that plugs into a slot in the PC. After the software is debugged using the simulator or evaluation board, it must be tested with the actual system target board (this is the board designed with the DSP in one's system). An in-circuit emulator interfaces with the target board, usually from a PCI to a JTAG interface port and connector on the EZ-KIT Lite or your target board.

The final step in the process is to generate the code required for booting the system using the *prom splitter*.

A summary of the tools available from Analog Devices is shown in Figure 7-47. Each one will be discussed in detail.

EZ-KIT Lites are basically DSP starter kit evaluation boards. In addition to the processor itself, these boards contain an ADC and a DAC (codec), interfaces to the PC over a serial port. All necessary analog and digital support circuitry is contained on the boards. The options on the board are controlled over a USB port connection to a PC as well as jumpers on the board. Windows 98/NT/2000/XP compatible software is supplied with the board. The software includes limited code-generation tools including a limited feature set compiler, assembler, linker, prom splitter (loader), and Visual DSP debugger. Application examples, such as a DTMF generator, echo cancellation, FFT, and simple digital filters are included as part of the software. The EZ-KIT Lite boards are primarily starter kit evaluation systems (and "lite" on the wallet).

DSP Hardware

- EZ-KIT Lite Evaluation Kits
- Emulators
- VisualDSP++™
  - ◆ Assembler, Linker, PROM Splitter, HIP Splitter, Simulator, C/C++ Compiler, Debugger, VisualDSP++ Kernel (VDK), Development Environment, Statistical Profiling, VisualDSP++ Component Software Engineering Capability
- Extensive Algorithm Libraries
- Factory, Field, and WWW Support
- Seminars
- ADI and Third-Party DSP Workshops
- ADI DSP Collaborative™ Third-Party Support

Figure 7-47: ADI CROSSCORE DSP Development Tools

- The EZ-KIT Lite is a Standalone (Desktop) System that Connects to a PC Running Windows
- The EZ-KIT Lites Provide:
  - ◆ A Cost-Effective Method for Initial Evaluation of the Capabilities of ADSP Series DSPs
  - ◆ A Powerful Development Environment for a Variety of General-Purpose Applications
- Target Market:
  - ◆ First-Time DSP Users
  - ◆ First-Time ADI DSP Users
  - ◆ Existing ADI DSP Users Implementing New Designs
  - ◆ Existing ADI DSP Users Upgrading to Faster Devices for Current Designs

Figure 7-48: EZ-Kit Lites for Analog Devices' DSPs

Section Seven

- **Hardware Features**
  - ADSP-2189M 75 MIPS Processor
  - AD73322L Stereo Codec
  - DSP-Programmable Codec Gain
  - 2 Mbit or Greater Boot-Protected Flash EPROM
  - RS-232 PC to EZ-Kit Lite Interface
  - Selectable Host vs. Full Memory Mode Implemented via Dip Switch
  - ADSP-218x EZ-ICE Emulator Port Connector
  - Expansion Connector Includes All Signal I/O plus 5 V, 3.3 V, 2.5 V, and GND Connections
  - LED Indicators for Master Power, RS-232 Interface, and One PF I/O

- **Software Features**
  - Windows 95/98/NT 4.0 PC Host Support
  - VisualDSP® ++ : Limited Feature Set Compiler, Assembler, Linker, Prom Splitter (Loader), VisualDSP Debugger Interface
  - Application Examples: DTMF Generator, Echo Cancellation, and FFT, (Similar to 2181 EZ-KIT Lite)
  - Email Support

Figure 7-49: ADSP-2189M EZ-Kit Lite

- **Hardware Features**
  - Single ADSP-2191
  - AD1885 48 kHz AC'97 SoundMAX® Codec
  - AD1803 Low Power Modem Codec
  - AD3338 and AD3339 Voltage Regulators
  - 4 Mbits of Flash Memory
  - Jumper Selectable Line-In or Mic-In 1/8" Stereo Jack and Line-Out 1/8" Stereo Jack
  - USB Version 1.1 Compliant Interface
  - 14-Pin Emulator Connector for JTAG Interface

- **Software Features**
  - Support for Win98 and Win2000
  - Evaluation Suite of VisualDSP++: Compiler, Assembler, Linker, Prom Splitter (Loader), VisualDSP++ Debugger Interface. VisualDSP++ Limited to Use With EZ-KIT Lite Hardware.

Figure 7-50: ADSP-2191 EZ-KIT Lite

DSP Hardware

- **Hardware Features**
    - ADSP-21161N SHARC DSP
    - 48 Mbits of SDRAM
    - AD1836 96 kHz Audio Codec
    - AD1852 96 kHz Auxiliary DAC
    - 4 Mbits of Flash Memory
    - USB Version 1.1 Compliant Interface
    - 14-Pin Emulator Connector for JTAG Interface

- **Software Features**
    - Support for Win98 or Win2000
    - Evaluation Suite of VisualDSP++: Compiler, Assembler, Linker, Prom Splitter (Loader), VisualDSP++ Debugger Interface. VisualDSP++ Limited to Use With EZ-KIT Lite Hardware.

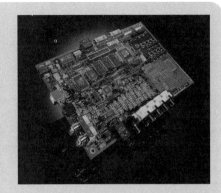

Figure 7-51: ADSP-21161N EZ-Kit Lite

- **Hardware Features**
    - ADSP-21535 DSP
    - 4 M × 32-Bit SDRAM
    - 272 K × 16-Bit FLASH Memory
    - AD1885 48 kHz AC'97 SoundMAX® Codec
    - ADP3088 Analog Devices Switching Regulator for Core Power Management
    - JTAG ICE 14 Pin Header
    - CE Compliant PCB and External Power Supply
    - Desktop Standalone Operation

- **Software Features**
    - Supports Win98 or Win2000
    - Evaluation Suite of VisualDSP++: Compiler, Assembler, Linker, Prom Splitter (Loader), VisualDSP++ Debugger Interface. VisualDSP++ Limited to Use With EZ-KIT Lite Hardware.

Figure 7-52: ADSP-21535 EZ-Kit Lite

- **Hardware Features**
    - ADSP-21065L SHARC DSP Running at 60 MHz
    - Full Duplex, 16-Bit Audio Codec
    - RS-232 Interface with UART
    - JTAG Emulation Connector
    - EMAFE Connector
    - SDRAM
    - Socketed EPROM

- **Software Features**
    - Support for Win9x, Win2000, and WinNT
    - Evaluation suite of VisualDSP++: Compiler, Assembler, Linker, Prom Splitter (Loader), VisualDSP Debugger Interface. VisualDSP Limited to Use with EZ-KIT Lite Hardware.
    - Demonstrations: Fast Fourier Transform (FFT), Discrete Fourier Transform (DFT), Band-Pass Filter, Pluck String Themes, Talk Through13

Figure 7-53: ADSP-21065L EZ-Kit Lite

The final step in the DSP system development is the debugging of the actual system, or "target" board. The Analog Devices in-circuit emulator interfaces with a JTAG connector on the target board for use in final system hardware and software debugging. Examples are shown in Figures 7-54 through 7-56. Figure 7-55 shows the Apex-ICE, which interfaces to the target board via a JTAG connector, which in turn interfaces to the JTAG DSP. A USB port connector is used to interface the emulator to a PC. Analog Devices currently offers two emulators supporting the JTAG compliant DSP processor families and one emulator specifically designed to accommodate the ADSP-218x series of DSPs (ADSP-218x is a non-JTAG part). The JTAG emulators currently support the ADSP-219X, ADSP-2116X, Blackfin™ DSP, TigerSHARC DSP, and SHARC DSP processors.

DSP Hardware

- Serial Port Interface, Printed Circuit Board, and 14-Pin Header
- Controls Equipment for Testing, Observing, and Debugging a Target System
- 6-Foot Cable
- Hardware Switch to Accommodate 2.5 V, 3.3 V, and 5 V
- Shielded Enclosure to Cover Bare Circuit Board
- Performance Increase via Faster Data Transfer

Figure 7-54: EZ-ICE For the ADSP-218x DSP Family

- Universal Serial Bus (USB)-Based Emulator for Analog Devices JTAG DSPs
- First Portable Solution for Analog Devices JTAG DSPs
- Small Hand-Held Unit
- Small Diameter Cable, 5 Meters in Length, for Hard to Reach Targets
- Power Provided Externally

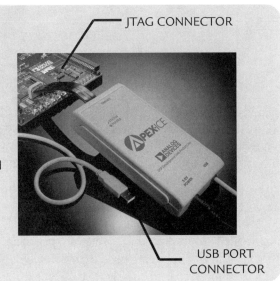

Figure 7-55: APEX-ICE™ USB Emulator

Section Seven

- 32-Bit PCI Interface Add-In Card
- 4-Inch, Flexible Shielded Target Board Cable for Easy Access to a 14-Pin JTAG Header
- Embedded ICEPAC Technology Provides a Rugged and Reliable Solution
- Remote 3 V/5 V JTAG Pod with Extended, Shielded Cable (1.5 m)
- Windows 95 and NT PNP

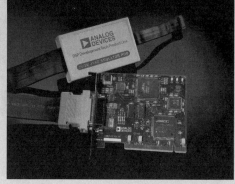

Figure 7-56: SUMMIT-ICE™ PCI Emulator

Platform and Processor Support

VisualDSP++ supports the Blackfin DSP SHARC DSP, TigerSHARC DSP, ADSP-218x (C only, does not support C++), and ADSP-219x DSP families on Windows® 98, Windows NT, Windows 2000, and Windows XP.

A "test drive" CD-ROM is available with a limited license for evaluation purposes. In addition to the tools and support functions described thus far, Analog Devices' DSP Collaborative consists of nearly 200 third-party companies that provide a range of products and services to make the DSP design task easier. DSP Collaborative members provide hardware products, software products, algorithms, and design services for a wide range of applications and markets. Our partners offer consulting services as well as commercial off-the-shelf (COTS) products for all of Analog Devices' DSP families.

A complete directory of our third-party developers can be found at:
    http://www.analog.com/dsp/3rdparty.

Further information about Analog Devices' DSP tools can be found at:
    http://www.analog.com/dsp/tools.

DSP Hardware

- VisualDSP++
- Integrated Development Environment for Analog Devices DSPs

  ◆ VisualDSP++ is an integrated development environment and debugger that delivers efficient project management, enabling programmers to move easily between editing, building, and debugging within a single interface. Key features include the C/C++ compiler, advanced plotting tools, statistical profiling, the VisualDSP++ Kernel (VDK), which allows the user's code to be implemented in a more structured and easier to scale manner. VisualDSP++ offers programmers a powerful DSP programming tool with flexibility that significantly reduces the time to market.

- Features
  ◆ Integrated Development and Debugger Environment
    - Develop within a single interface
    - Profile and trace instruction execution of C/C++ and assembly programs (simulator only)
    - Set watchpoints (conditional breakpoints) on processor registers and stacks, as well as program and data memory, including:
      - Statistical profiling
      - MP (multiprocessing)
      - Graphical plotting
  ◆ VisualDSP++ Component Software Engineering (VCSE)
  ◆ VisualDSP++ Kernel (VDK)
    - Scheduling and resource allocation
    - Supports threads, events, semaphores, and critical and unscheduled regions
  ◆ Code-Generation Features
    - Develop applications using an optimizing C/C++ compiler
    - Intersperse assembly statements within C/C++ source code
    - Create executables using a linker that supports multiprocessing, shared memory, and code overlays
    - Access numerous math, DSP, and C/C++ runtime library routines
  ◆ Expert Linker
  ◆ Assembler

Figure 7-57: Software Development Environment

Section Seven

- **Attributes**
  - Dual ADSP-TS101 Digital Signal Processors in 484-lead (19 mm × 19 mm) PBGA Packages
  - 90 MHz Oscillator and Buffer Logic ST Microelectronics DSM2150F5V Combination FLASH (512 K × 8) and Programmable Logic
  - Cypress EZ USB FX USB 1.1 Interface Controller Chip for High Speed Data Transfers with Host
  - USB Port Connector
  - MT4LSDT464A (4 M x 64) 32 MB Synchronous Dynamic RAM (SDRAM) DIMM Expandable up to 128 MB
  - Three 90-Pin Connectors for Analyzing and Interfacing with the Expansion Port
  - USB Version 1.1 Compliant Interface
  - JTAG ICE 14-Pin Header
  - Evaluation Suite of VisualDSP++
  - Desktop Standalone Operation
- **Software Features**
  - Support for Win98 and Win2000
  - Evaluation Suite of VisualDSP++: Compiler, Assembler, Linker, Prom Splitter (Loader), VisualDSP++ Debugger Interface, VisualDSP++ Limited to Use With EZ-KIT Lite Hardware

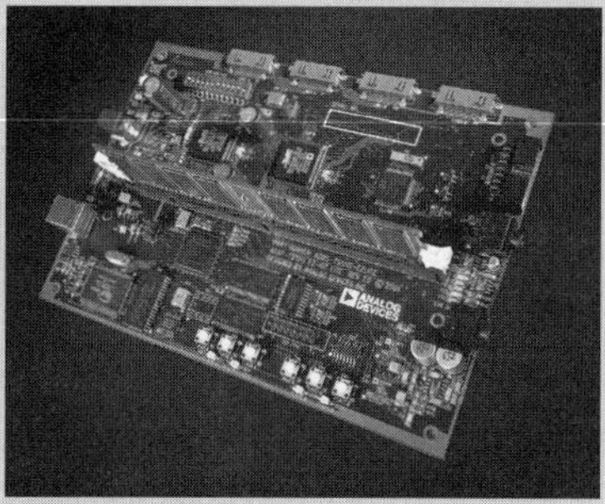

Figure 7-58: TigerSHARC Development Tools

- The test drive is a free 30-day trial of VisualDSP++. The test drive is a full version of VisualDSP++, with no limitations other than the time limit, and contains PDFs of the VisualDSP++ manuals.
- Test drive CDs may be ordered from the website or can be downloaded from the test drives from the DSP Tools website at: http://www.analog.com/dsp/tools/.
- Test drives require registration online to receive a serial number that activates the software: http://forms.analog.com/Form_Pages/DSP/tools/visualDSPTestDrive.asp. The test drive expires 30 days from the installation date.
- VisualDSP++ Demo
  - The VisualDSP ++ demonstration allows you to investigate the capabilities of Analog Devices' DSP software development environment and some of the many features of release 3.0. The demo is available on the ADI DSP Tools website at: http://www.analog.com/technology/dsp/training/tutorials/v_dsp_tutorial.html.

Figure 7-59: VisualDSP Test Drive

DSP Hardware

- The DSP Collaborative partners (Analog Devices' Third-Party Network) offer tools, services and solutions for a wide range of applications/markets
  - Communications
  - Audio
  - Medical Imaging
  - Speech Processing
  - Motor Control
  - Industrial Automation
  - Optical Networking
  - Voice Over IP

- When you select Analog Devices as your DSP vendor, you're broadening your design team to include the industry-leading resources of the DSP Collaborative. The DSP Collaborative is comprises more than 180 partners who offer more than 700 commercial products, in addition to hundreds of custom solutions that build on more than 35 years of signal processing experience found in every one of our DSPs. These partners offer consulting services as well as a wide range of commercial off-the-shelf (COTS) products. Their development tools are specifically designed to work with Analog Devices DSP-based systems.

- With the DSP Collaborative, you are supported by highly reputable brands, patented technologies, and the pioneers in real-time system design and debug. The DSP Collaborative partners offer products and services that provide both system and application level expertise.

- Speed up your design process by leveraging the solutions our partners have to offer
  - Algorithms and Libraries
  - MATLAB® DSP Support
  - Real-Time Operating Systems
  - Development and Evaluation Boards
  - COTS Hardware Boards
  - DSP Systems
  - Emulators
  - Debuggers

- Design with Analog Devices' DSP Collaborative team approach with a proven strategy for maximizing your resources.

Figure 7-60: ADI DSP Collaborative—What is it?

References

1. Steven W. Smith, **Digital Signal Processing: A Guide for Engineers and Scientists**, Newnes, 2002.
2. C. Britton Rorabaugh, **DSP Primer**, McGraw-Hill, 1999.
3. Richard J. Higgins, **Digital Signal Processing in VLSI,** Prentice-Hall, 1990.
4. Ethan Bordeaux, *Advanced DSP Performance Complicates Memory Architecures in Wireless Designs*, **Wireless Systems Design**, April 2000.
5. **DSP Designer's Reference (DSP Solutions)** CD-ROM, Analog Devices, 1999.
6. **DSP Navigators: Interactive Tutorials about Analog Devices' DSP Architectures** (Available for ADSP-218x family and SHARC family): http://www.analog.com/technology/dsp/index.html.
7. **General DSP Training and Workshops**: http://www.analog.com/technology/dsp/index.html.

The following DSP Reference Manuals and documentation are available for free download from: http://www.analog.com/technology/dsp/library.html.

8. **ADSP-2100 Family User's Manual, 3rd Edition**, Sept., 1995.
9. **ADSP-2100 Family EZ Tools Manual.**
10. **ADSP-2100 EZ-KIT Lite Reference Manual.**
11. **Using the ADSP-2100 Family, Vol. 1, Vol. 2.**
12. **ADSP-2106x SHARC User's Manual, 2nd Edition, July, 1996.**
13. **ADSP-2106x SHARC EZ-KIT Lite Manual.**
14. **ADSP-21065L SHARC User's Manual, Sept. 1, 1998.**
15. **ADSP-21065L SHARC EZ-LAB User's Manual.**
16. **ADSP-21160 SHARC DSP Hardware Reference.**

Section 8
Interfacing to DSPs

- Parallel Interfacing to DSP Processors: Reading Data From Memory-Mapped Peripheral ADCs

- Parallel Interfacing to DSP Processors: Writing Data to Memory-Mapped DACs

- Serial Interfacing to DSP Processors

- Interfacing I/O Ports, Analog Front Ends, and Codecs to DSPs

- DSP System Interface

Section 8
Interfacing to DSPs
Walt Kester and Dan King

Introduction

As technology in the rapidly growing field of mixed-signal processing evolves, more highly integrated DSP products (such as the ADSP-21ESP202) are being introduced that contain on-chip ADCs and DACs as well as the DSP, thereby eliminating most component level interface problems. Standalone ADCs and DACs are now available with interfaces especially designed for DSP chips, thereby minimizing or eliminating external interface support or *glue* logic. High performance sigma-delta ADCs and DACs are currently available in the same package (called a *codec* or COder/DECcoder) such as the AD73311 and AD73322. These products are also designed to require minimum glue logic when interfacing to the most common DSP chips. This section discusses the various data transfer and timing issues associated with the various interfaces.

Parallel Interfacing to DSP Processors: Reading Data from Memory-Mapped Peripheral ADCS

Interfacing an ADC or a DAC to a fast DSP parallel requires an understanding of how the DSP processor reads data from a memory-mapped peripheral (the ADC) and how the DSP processor writes data to a memory-mapped peripheral (the DAC). We will first consider some general timing requirements for reading and writing data. It should be noted that the same concepts presented here regarding ADCs and DACs apply equally when reading and writing from/to external memory.

A block diagram of a typical parallel DSP interface to an external ADC is shown in Figure 8-1. This diagram has been greatly simplified to show only those signals associated with *reading* data from an external memory-mapped peripheral device. The timing diagram for the ADSP-21xx read cycle is shown in Figure 8-2.

In this example it is assumed that the ADC is sampling at a continuous rate controlled by the external sampling clock, not the internal DSP clock. Using a separate clock for the ADC is the preferred method, since the DSP clock may be noisy and introduce jitter in the ADC sampling process, thereby increasing the noise level.

Assertion of the sampling clock at the ADC *convert start* input initiates the conversion process (step 1). The leading (or trailing) edge of this pulse causes the internal ADC sample-and-hold to switch from the sampling mode to the hold mode so that the conversion process can take place. When the conversion is complete, the *conversion complete* output of the ADC is asserted (step 2). The *read* process thus begins when this signal is applied to the *processor interrupt request line* ($\overline{\text{IRQ}}$) of the DSP. The

Section Eight

processor then places the address of the peripheral initiating the interrupt request (the ADC) on the *memory address bus* (A0–A13) (step 3). At the same time, the processor asserts a *memory select line* ($\overline{DMS}$ is shown here) (step 4). The two internal address buses of the ADSP-21xx (program memory address bus and data memory address bus) share a single external address bus, and the two internal data buses (program memory data bus and data memory data bus) share a single external data bus. The *boot memory select* ($\overline{BMS}$), *data memory select* ($\overline{DMS}$), *program memory select* ($\overline{PMS}$) and *input/output memory select* ($\overline{IOMS}$) signals indicate for which memory space the external buses are being used. These signals are typically used to enable an external *address decoder* as shown in Figure 8-1. The output of the address decoder drives the *chip select* input of the peripheral device (step 5).

The *memory read* ($\overline{RD}$) is asserted t_{ASR} ns after the $\overline{DMS}$ line is asserted (step 6). The sum of the address decode delay plus the peripheral chip select setup time should be less than t_{ASR} in order to take full advantage of the $\overline{RD}$ low time. The $\overline{RD}$ line remains low for t_{RP} ns. The *memory read* signal is used to enable the three-state parallel data outputs of the peripheral device (step 7). The $\overline{RD}$ line is connected to the appropriate pin on the peripheral device usually called *output enable* or *read*. The rising edge of the $\overline{RD}$ signal is used to clock the data on the data bus into the DSP processor (step 8). After the rising edge of the $\overline{RD}$ signal, the data on the data bus must remain valid for t_{RDH} ns, the data hold time. In the case of most members of the ADSP-21xx family, this specification value is 0 ns.

The key timing requirements for the peripheral device are shown in Figure 8-3. Values are given for the ADSP-2189M DSP operating at 75 MHz.

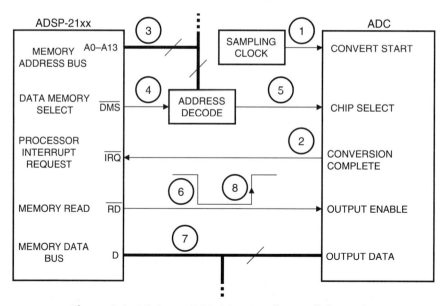

Figure 8-1: ADC to ADSP-21xx Family Parallel Interface

Interfacing to DSPs

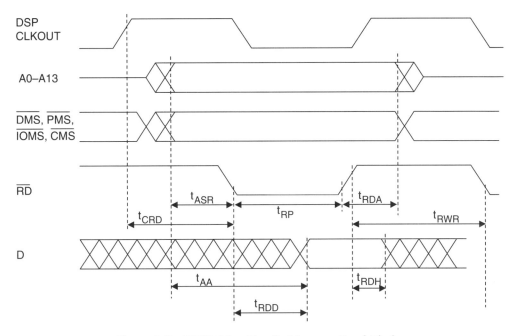

Figure 8-2: ADSP-21xx Family Memory Read Timing

- Peripheral Device Data Outputs Must Be Three-State Compatible

- Address Decode Delay + Peripheral Chip Select Setup Time Must Be Less Than Address and Memory Select Setup Time t_{ASR} (0.325 ns Min for ADSP-2189M)

- For Zero Wait State Access, the Time from a Negative-Going Edge of Read Signal ($\overline{RD}$) to Output Data Valid Must Be Less than t_{RDD} (1.65 ns Max for ADSP-2189M Operating at 75 MHz) or Software Wait States Must Be Added, or Processor Clock Frequency Reduced

- Output Data from Peripheral Must Remain Valid for t_{RDH} from the Rising Edge of Read Signal ($\overline{RD}$) (0 ns for ADSP-2189M)

- Peripheral Device Must Accept Minimum Output Enable Pulsewidth of t_{RP} (3.65 ns for ADSP-2189M Operating at 75 MHz) or Software Wait States Must Be Added, or Processor Clock Frequency Reduced

Figure 8-3: Parallel Peripheral Device Read Interface Key Requirements

The DSP t_{RDD} specification determines the peripheral device data access time requirement. In the case of the ADSP-2189M, t_{RDD} = 1.65 ns minimum at 75 MHz. If the access time of the peripheral is greater than this, wait states must be added or the processor speed reduced. This is a relatively common situation when interfacing external memory or ADCs to fast DSPs. The relationship between these timing parameters for the ADSP-2189M is given by the equations shown in Figure 8-4. Note that these specifications are dependent on the DSP clock frequency.

- t_{CK} = Processor Clock Period (13.3 ns)

- t_{ASR} = Address and Memory Select Setup Before Read Low
  = 0.25t_{CK} − 3 ns Minimum

- t_{RDD} = Read Low to Data Valid = 0.5t_{CK} − 5 ns + No. of Wait States × t_{CK} Maximum

- t_{RDH} = Data Hold from Read High = 0 ns Minimum

- t_{RP} = Read Pulsewidth = 0.5t_{CK} − 3 ns + No. of Wait States × t_{CK} Minimum

Figure 8-4: ADSP-2189M Parallel Read Timing at 75 MHz

The ADSP-2189M can easily be interfaced to slow peripheral devices using its programmable wait state generation capability. Three registers control wait state generation for boot, program, data, and I/O memory spaces. Zero to 15 wait states can be specified for each parallel memory interface. Each wait state added increases the allowable external data memory access time by an amount equal to the processor clock period (13.3 ns for the ADSP-2189M operating at 75 MHz). In this example, the *data memory address*, $\overline{DMS}$, and $\overline{RD}$ lines are all held stable for an additional amount of time equal to the duration of the wait states.

The AD7854/AD7854L is a 12-bit, 200 kSPS/100 kSPS ADC that operates in the parallel mode. It operates on a single 3 V to 5.5 V supply and dissipates only 5.5 mW (3 V supply, AD7854L). An automatic power-down after conversion feature reduces this to 650 μW.

Interfacing to DSPs

A functional block diagram of the AD7854/AD7854L is shown in Figure 8-5. The AD7854/AD7854L uses a successive-approximation architecture based on a charge redistribution (switched capacitor) DAC. A calibration mode removes offset and gain errors. The key interface timing specifications for the AD7854/AD7854L and the ADSP-2189M are compared in Figure 8-6. Specifications for the ADSP-2189M are given for a clock frequency of 75 MHz.

Examining the timing specifications shown in Figure 8-6 reveals that for the timing between the devices to be compatible, five software wait states must be programmed into the ADSP-2189M. This increases t_{RDD} to 68.15 ns, which is greater than the data access time of the AD7854/AD7854L (t_8 = 50 ns max.). The read pulse, t_{RP}, is likewise increased to 70.15 ns, which meets the ADC's read pulsewidth requirement (t_7 = 70 ns min.). Unless the memory-mapped peripheral has an extremely short access time, wait states are generally required, whether interfacing to ADCs, DACs, or external memory.

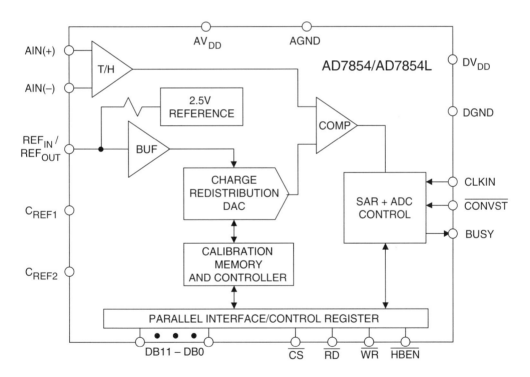

Figure 8-5: AD7854/AD7854L, 3 V Single Supply, 12-Bit, 200 kSPS/100 kSPS Parallel Output ADC

Section Eight

ADSP-2189M Processor (75 MHz)	AD7854/AD7854L ADC
t_{ASR} (Data Address, Memory Select Setup Time before $\overline{RD}$ Low) = 0.325 ns Min	t_5 ($\overline{CS}$ to $\overline{RD}$ Setup Time = 0 ns Min (Must Add Address Decode Time to this Value)
t_{RP} ($\overline{RD}$ Pulsewidth) = 3.65 ns + No. of Wait States × 13.3 ns Min = 70.15 ns Min	t_7 ($\overline{RD}$ Pulsewidth) = 70 ns Min
t_{RDD} ($\overline{RD}$ Low to Data Valid) = 1.65 ns + No. of Wait States × 13.3 ns Min = 68.15 ns Min	t_8 (Data Access Time after $\overline{RD}$) = 50 ns Max
t_{RDH} (Data Hold from $\overline{RD}$ High) = 0 ns Min	t_9 (Bus Relinquish Time After $\overline{RD}$) = 5 ns Min/40 ns Max

NOTES:
(1) Adding five wait states to the ADSP-2189M increases t_{RP} to 70.15 ns, which is greater than t_7 (70 ns) and meets the t_8 (50 ns) requirement.
(2) t_9 max (40 ns) may cause bus contention if a write cycle immediately follows the read cycle.

Figure 8-6: ADSP-2189M and AD7854/AD7854L Parallel Read Interface Timing Specification Comparison

A simplified interface diagram for the two devices is shown in Figure 8-7. The conversion complete signal from the AD7854/AD7854L corresponds to the BUSY output pin. Notice that the configuration allows the DSP to write data to the AD7854/AD7854L parallel interface control register. This is needed in order to set various options in the AD7854/AD7854L and perform the calibration routines. In normal operation, however, data is read from the AD7854/AD7854L as described above. Writing to external parallel memory-mapped peripherals is discussed later in this section.

Parallel interfaces between other DSP processors and external peripherals can be designed in a similar manner by carefully examining the timing specifications for all appropriate signals for each device. The data sheets for most ADCs contain sufficient information in the application section to interface them to the DSPs.

Interfacing to DSPs

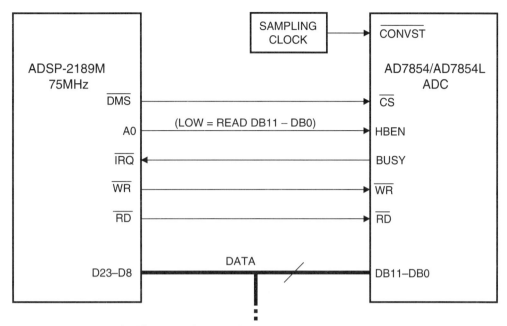

Figure 8-7: AD7854/AD7854L ADC Parallel Interface to ADSP-2189M

Parallel Interfacing to DSP Processors: Writing Data to Memory-Mapped DACS

A simplified block diagram of a typical DSP interface to a parallel peripheral device (such as a DAC) is shown in Figure 8-8. The memory write cycle timing diagram for the ADSP-21xx family is shown in Figure 8-9.

In most real-time applications, the DAC is operated continuously from a stable sampling clock. Most DACs for these applications have double buffering: an input latch to handle the asynchronous DSP interface, followed by a second latch (called the DAC latch) that drives the DAC current switches. The DAC latch strobe is derived from an external stable sampling clock. In addition to clocking the DAC latch, the DAC latch strobe is also used to generate a processor interrupt to the DSP that indicates that the DAC is ready for a new input data-word.

Section Eight

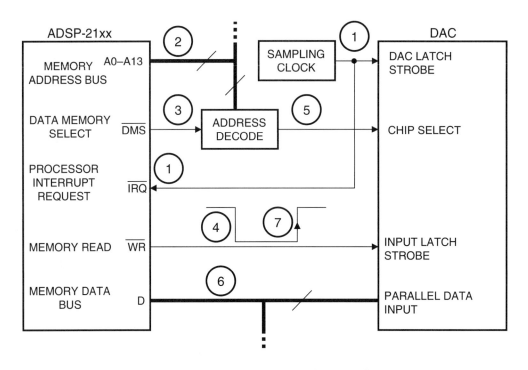

Figure 8-8: DAC to ADSP-21xx Family Parallel Interface

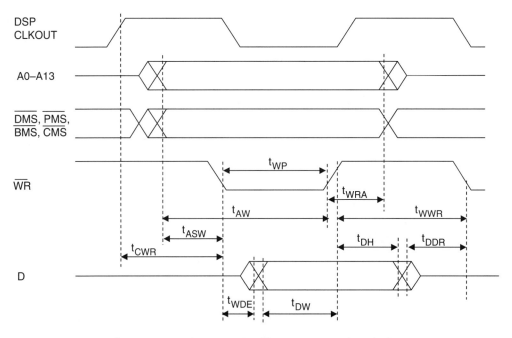

Figure 8-9: ADSP-21xx Family Memory Write Timing

Interfacing to DSPs

The write process is thus initiated by the peripheral device asserting the DSP *interrupt request* line indicating that the peripheral is ready to accept a new parallel data-word (step 1). The DSP then places the address of the peripheral device on the *address bus* (step 2) and asserts a *memory select* line ($\overline{\text{DMS}}$ is shown here) (step 3). This causes the output of the address decoder to assert the *chip select* input to the peripheral (step 5). The *write* (WR) output of the $\overline{\text{DSP}}$ is asserted t_{ASW} ns after the negative-going edge of the $\overline{\text{DMS}}$ signal (step 4). The width of the $\overline{\text{WR}}$ pulse is t_{WP} ns. Data is placed on the data bus (D) and is valid t_{DW} ns before the $\overline{\text{WR}}$ line goes high (step 6). The positive-going transition of the $\overline{\text{WR}}$ line is used to clock the data on the data bus (D) into the external parallel memory (step 7). The data on the data bus remains valid for t_{DH} ns after the positive-going edge of the $\overline{\text{WR}}$ signal.

The key timing requirements for writing to the peripheral device are shown in Figure 8-10. The key specification is t_{WP}, the write pulsewidth. All but the fastest peripheral devices will require wait states to be added due to their longer data access times. Figure 8-11 shows the key timing specifications for the ADSP-2189M. Note that they are all related to the processor clock frequency.

- Address Decode Delay + Peripheral Chip Select Setup Time Must Be Less Than Address and Memory Select Setup Time t_{ASW} (0.325 ns for ADSP-2189M Operating at 75 MHz)

- For Zero Wait State Access, Input Data *Setup* Time Must be Less Than t_{DW} (2.65 ns for ADSP-2189M Operating at 75 MHz) or Software Wait States Must be Added, or Processor Clock Frequency Reduced

- Input Data *Hold* Time Must be Less Than t_{DH} (2.325 ns for ADSP-2189M Operating at 75 MHz)

- Peripheral Device Must Accept Input Write Clock Pulsewidth t_{WP} (3.65 ns min for ADSP-2189M Operating at 75 MHz) or Software Wait States Must be Added, or Processor Clock Frequency Reduced

Figure 8-10: Parallel Peripheral Devices Write Interface Key Requirements

Section Eight

- t_{CK} = Processor Clock Period (13.3 ns)
- t_{ASW} = Address and Memory Select before $\overline{WR}$ Low
  = $0.25 t_{CK}$ − 3 ns Minimum
- t_{DW} = Data Setup Before $\overline{WR}$ High = $0.5 t_{CK}$ − 4 ns + No. of Wait States × t_{CK}
- t_{DH} = Data Hold After $\overline{WR}$ High = $0.25 t_{CK}$ − 1 ns
- t_{WP} = $\overline{WR}$ Pulsewidth = $0.5 t_{CK}$ − 3 ns + No. of Wait States × t_{CK} Minimum

Figure 8-11: ADSP-2189M Parallel Write Timing

The AD5340 is a 12-bit 100 kSPS DAC with a parallel data interface. It operates on a single 2.5 V to 5.5 V supply and dissipates only 345 µW (3 V supply). A power-down mode further reduces the power to 0.24 µW. The part incorporates an on-chip output buffer that can drive the output to both supply rails. The AD5340 allows the choice of a buffered or unbuffered reference input. The device has a power-on-reset circuit that ensures that the DAC output powers on at 0 V and remains there until valid data is written to the part. A block diagram is shown in Figure 8-12. The input is double-buffered. The key interface timing specifications for the two devices are compared in Figure 8-13. Specifications for the ADSP-2189M are given for a clock frequency of 75 MHz.

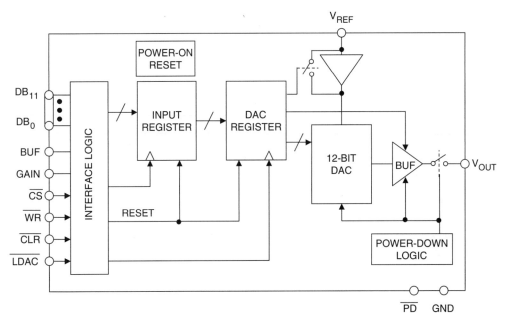

Figure 8-12: AD5340 12-Bit, 100 kSPS Parallel Input DAC

ADSP-2189M PROCESSOR (75 MHz)	AD5340 DAC
t_{ASW} (Address and Data Memory Select Setup before $\overline{WR}$ Low) = 0.325 ns Min	t_1 ($\overline{CS}$ to $\overline{WR}$ Setup Time) = 0 ns Min
t_{WP} ($\overline{WR}$ Pulsewidth) = 3.65 ns + No. of Wait States × 13.3 ns Min = 30.25 ns min	t_3 ($\overline{WR}$ Pulsewidth) = 20 ns Min
t_{DW} (Data Setup Before $\overline{WR}$ High) = 2.65 ns + No. of Wait States × 13.3 ns Min = 29.25 ns Min	t_4 (Data Valid to $\overline{WR}$ Setup Time) = 5 ns Min
t_{DH} (Data Hold after $\overline{WR}$ High) = 2.325 ns Min	t_5 (Data Valid to $\overline{WR}$ Hold Time) = 4.5 ns Min

NOTE: Adding two wait states to the ADSP-2189M increases t_{WP} to 30.25 ns and t_{DW} to 29.25 ns, which is greater than t_3 (20ns) and t_4 (5ns) respectively.

Figure 8-13: ADSP-2189M and AD5340 Parallel Write Interface Timing Specifications

Examining the timing specifications shown in Figure 8-13 reveals that for the timing between the devices to be compatible, two software wait states must be programmed into the ADSP-2189M. This increases the width of $\overline{WR}$ to 30.25 ns, which is greater than the minimum required AD5340 write pulsewidth (20 ns). The data setup time of 5 ns for the AD5340 is also met by adding two wait states. A simplified interface diagram for the two devices is shown in Figure 8-14.

Parallel interfaces with other DSP processors can be designed in a similar manner by carefully examining the timing specifications for all appropriate signals for each device.

Section Eight

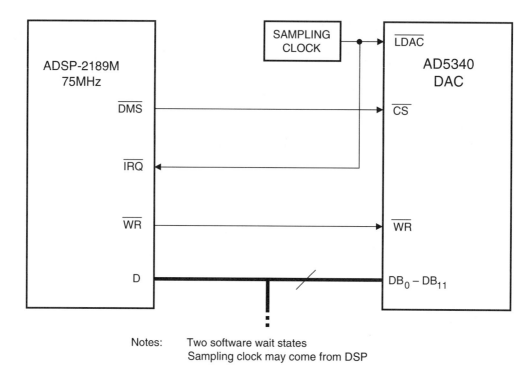

Notes: Two software wait states
Sampling clock may come from DSP

Figure 8-14: AD5340 DAC Parallel Interface to ADSP-2189M

Serial Interfacing to DSP Processors

DSP processors with serial ports (such as the ADSP-21xx family) provide a simple interface to peripheral ADCs and DACs. Use of the serial port eliminates the need for using large parallel buses to connect the ADCs and DACs to the DSP. In order to better understand serial data transfer, we will first examine the serial port operation of the ADSP-21xx series.

A block diagram of one of the two serial ports of the ADSP-21xx is shown in Figure 8-15. The *transmit* (Tx) and *receive* (Rx) registers are identified by name in the ADSP-21xx assembly language, and are not memory-mapped.

Interfacing to DSPs

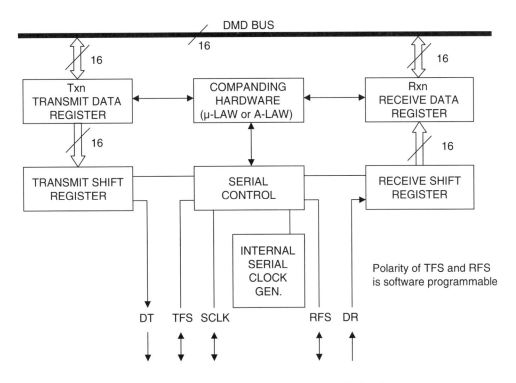

Figure 8-15: ADSP-21xx Family Serial Port Block Diagram

- Separate Transmit and Receive Sections for Each Port
- Double-Buffered Transmit and Receive Registers
- Serial Clock Can be Internally or Externally Generated
- Transmit and Receive Frame Sync Signals Can be Internally or Externally Generated
- Serial Data-Words of 3 Bits to 16 Bits Supported
- Automatically Generated Processor Interrupts
- Hardware Companding Requires No Software Overhead

Figure 8-16: ADSP-21xx Family Serial Port Features

In the receiving portion of the serial port, the *receive frame sync* (RFS) signal initiates reception. The serial *receive data* (DR) from the external device (ADC) is transferred into the *receive shift register* one bit at a time. The negative-going edge of the *serial clock* (SCLK) is used to clock the serial data from the external device into the *receive shift register*. When a complete word has been received, it is written to the *receive data register* (Rx), and the receive interrupt for that serial port is generated. The *receive data register* is then read by the processor.

Writing to the *transmit data register* readies the serial port for transmission. The *transmit frame sync* (TFS) signal initiates transmission. The value in *the transmit data register* (Tx) is then written to the internal *transmit shift register*. The data in the *transmit shift register* is sent to the peripheral device (DAC) one bit at a time, and the positive-going edge of the *serial clock* (SCLK) is used to clock the serial *transmit data* (DT) into the external device. When the first bit has been transferred, the serial port generates the transmit interrupt. The *transmit data register* can then be written with new data, even though the transmission of the previous data is not complete.

In the *normal* framing mode, the frame sync signal (RFS or TFS) is checked at the falling edge of SCLK. If the framing signal is asserted, data is available (transmit mode) or latched (receive mode) on the *next* falling edge of SCLK. The framing signal is not checked again until the word has been transmitted or received. In the *alternate* framing mode, the framing signal is asserted in the *same* SCLK cycle as the first bit of a word. The data bits are latched on the falling edge of SCLK, but the framing signal is checked only on the first bit. Internally generated framing signals remain asserted for the length of the serial word. The *alternate* framing mode of the serial port in the ADSP-21xx is normally used to receive data from ADCs and transmit data to DACs.

The serial ports of the ADSP-21xx family are extremely versatile. The TFS, RFS, or SCLK signals can be generated from the ADSP-21xx clock (master mode) or generated externally (slave mode). The polarity of these signals can be reversed with software, thereby allowing more interface flexibility. The port also contains μ-law and A-law companding hardware for voice-band telecommunications applications.

Serial ADC to DSP Interface

A timing diagram of the ADSP-2189M serial port operating in the receive mode (alternate framing) is shown in Figure 8-17. The first negative-going edge of the SCLK to occur after the negative-going edge of the $\overline{RFS}$, clocks the MSB data from the ADC into the serial input latch. The process continues until all serial bits have been transferred into the serial input latch. The key timing specifications of concern are the serial data setup (t_{SCS}) and hold times (t_{SCH}) with respect to the negative-going edge of the SCLK. In the case of the ADSP-2189M, these values are 4 ns and 7 ns, respectively. The latest generation ADCs with high speed serial clocks will have no trouble meeting these specifications, even at the maximum serial data transfer rate.

Interfacing to DSPs

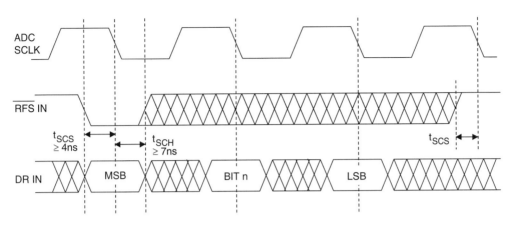

ALTERNATE FRAMING MODE, ADC IS MASTER

Figure 8-17: ADSP-2189M Serial Port Receive Timing

The AD7853/AD7853L is a 12-bit, 200 kSPS/100 kSPS ADC that operates on a single 3 V to 5.5 V supply and dissipates only 4.5 mW (3 V supply, AD7853L). After each conversion, the device automatically powers down to 25 µW. The AD7853/AD7853L is based on a successive-approximation architecture and uses a charge redistribution (switched capacitor) DAC. A calibration feature removes gain and offset errors. A block diagram of the device is shown in Figure 8-18.

The AD7853 operates on a 4 MHz maximum external clock frequency. The AD7853L operates on a 1.8 MHz maximum external clock frequency. The timing diagram for AD7853L is shown in Figure 8-19. The AD7853/AD7853L has modes that configure the $\overline{\text{SYNC}}$ and SCLK as inputs or outputs. In the example shown here they are generated by the AD7853L. The AD7853L serial clock operates at a maximum frequency of 1.8 MHz (556 ns period). The data bits are valid 330 ns after the positive-going edges of SCLK. This allows a setup time of approximately 330 ns minimum before the negative-going edges of SCLK, easily meeting the ADSP-2189M 4 ns t_{SCS} requirement. The hold time after the negative-going edge of SCLK is approximately 226 ns, again easily meeting the ADSP-2189M 7 ns t_{SCH} timing requirement. These simple calculations show that the data and RFS setup and hold requirements of the ADSP-2189M are met with considerable margin.

Section Eight

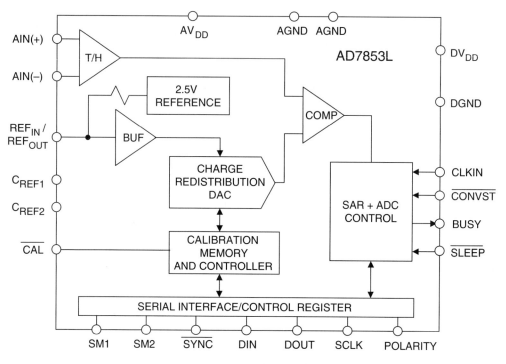

Figure 8-18: AD7853/AD7853L 3 V Single-Supply 12-Bit 200 kSPS/100 kSPS Serial Output ADC

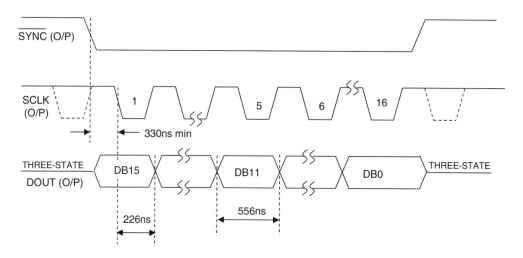

Figure 8-19: AD7853L Serial ADC Output Timing 3 V Supply, SCLK = 1.8 MHz

Figure 8-20 shows the AD7853L interfaced to the ADSP-2189M connected in a mode to transmit data from the ADC to the DSP (alternate/master mode). The AD7853/AD7853L contains internal registers that can be accessed by writing from the DSP to the ADC via the serial port. These registers are used to set various modes in the AD7853/AD7853L as well as to initiate the calibration routines. These connections are not shown in the diagram.

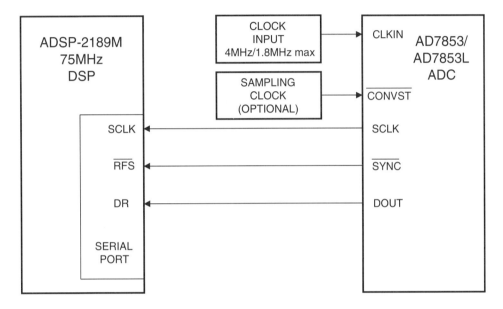

Figure 8-20: AD7853/AD7853L Serial ADC Interface to ADSP-2189M

Serial DAC to DSP Interface

Interfacing serial input DACs to the serial ports of DSPs such as the ADSP-21xx family is also relatively straightforward and similar to the previous discussion regarding serial output ADCs. The details will not be repeated here, but a simple interface example will be shown.

The AD5322 is a 12-bit, 100 kSPS dual DAC with a serial input interface. It operates on a single 2.5 V to 5.5 V supply, and a block diagram is shown in Figure 8-21. Power dissipation on a 3 V supply is 690 µW. A power-down feature reduces this to 0.15 µW. Total harmonic distortion is greater than 70 dB below full scale for a 10 kHz output. The references for the two DACs are derived from two reference pins (one per DAC). The reference inputs may be configured as buffered or unbuffered inputs. The outputs of both DACs may be updated simultaneously using the asynchronous LDAC input. The device contains a power-on-reset circuit that ensures that the DAC outputs power up to 0 V and remains there until a valid write takes place to the device.

Section Eight

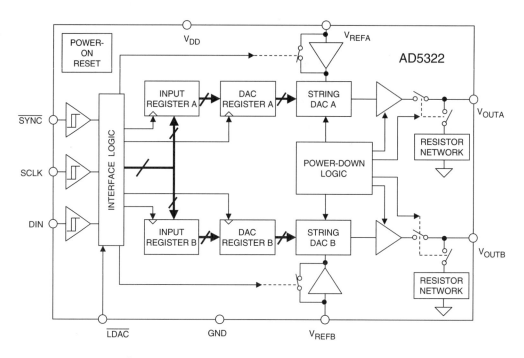

Figure 8-21: AD5322 12-Bit, 100 kSPS Dual DAC

Data is normally input to the AD5322 via the SCLK, DIN, and SYNC pins from the serial port of the DSP. When the SYNC signal goes low, the input shift register is enabled. Data is transferred into the AD5322 on the falling edges of the following 16 clocks. A typical interface between the ADSP-2189M and the AD5322 is shown in Figure 8-22. Notice that the clocks to the AD5322 are generated from the ADSP-2189M clock. It is also possible to generate the SCLK and SYNC signals externally to the AD5322 and use them to drive the ADSP-2189M. The serial interface of the AD5322 is not fast enough to handle the ADSP-2189M maximum master clock frequency. However, the serial interface clocks are programmable and can be set to generate the proper timing for fast or slow DACs.

The input shift register in the AD5322 is 16 bits wide. The 16-bit word consists of four control bits followed by 12 bits of DAC data. The first bit loaded determines whether the data is for DAC A or DAC B. The second bit determines if the reference input will be buffered or unbuffered. The next two bits control the operating modes of the DAC (normal, power-down with 1 kΩ to ground, power-down with 100 kΩ to ground, or power-down with a high impedance output).

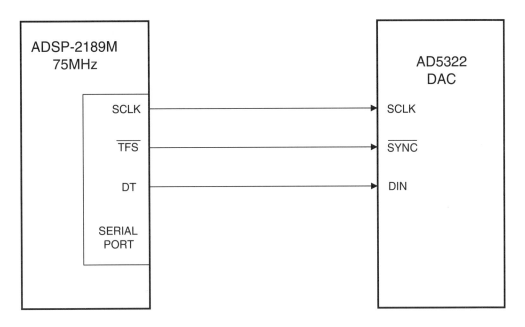

Figure 8-22: AD5322 DAC Serial Interface to ADSP-2189M

Interfacing I/O Ports, Analog Front Ends, and Codecs to DSPs

Since most DSP applications require both an ADC and a DAC, I/O ports and codecs have been developed that integrate the two functions on a single chip as well as provide easy-to-use interfaces to standard DSPs. These devices also go by the name of *analog front ends*.

A functional block diagram of the AD73322 is shown in Figure 8-23. This device is a dual analog front end (AFE) with two 16-bit ADCs and two 16-bit DACs capable of sampling at 64 kSPS. It is designed for general-purpose applications, including speech and telephony using sigma-delta ADCs and sigma-delta DACs. Each channel provides 77 dB signal-to-noise ratio over a voice-band signal bandwidth.

The ADC and DAC channels feature programmable input/output gains with ranges of 38 dB and 21 dB, respectively. An on-chip voltage reference is included to allow single-supply operation on 2.7 V to 5.5 V. Power dissipation is 73 mW with a 3 V supply.

Section Eight

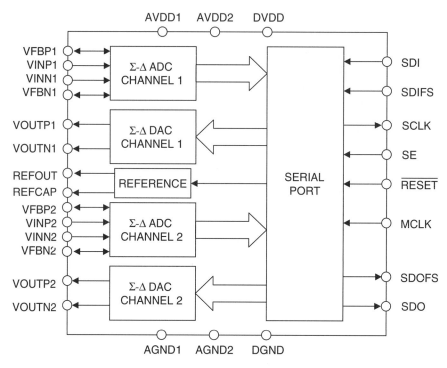

Figure 8-23: AD73322 Single-Supply 16-Bit, 64 kSPS Codec with Serial Interface

The sampling rate of the codecs is programmable, with four separate settings of 64 kHz, 32 kHz, 16 kHz, and 8 kHz when operating from a master clock of 16.384 MHz. The serial port allows easy interfacing of single or cascaded devices to industry-standard DSP engines, such as the ADSP-21xx family. The SPORT transfer rate is programmable to allow interfacing to both fast and slow DSP engines. The interface to the ADSP-218x family is shown in Figure 8-24. The SE pin (SPORT enable) may be controlled from a parallel output pin or a flag pin such as FL1, or where SPORT power-down is not required, it can be permanently strapped high using a suitable pull-up resistor. The RESET pin may be connected to the system hardware reset, or it may be controlled with another flag bit.

In the *program* mode, data is transferred from the DSP to the AD73322 control registers to set up the device for desired operation. Once the device has been configured by programming the correct settings to the various control registers, the device may exit the *program* mode and enter the *data* mode. The dual ADC data is transmitted to the DSP in two blocks of 16-bit words. Similarly, the dual DAC data is transmitted from the DSP to the AD73322 in two blocks of 16-bit words. Simplified interface timing is also shown in Figure 8-24.

Interfacing to DSPs

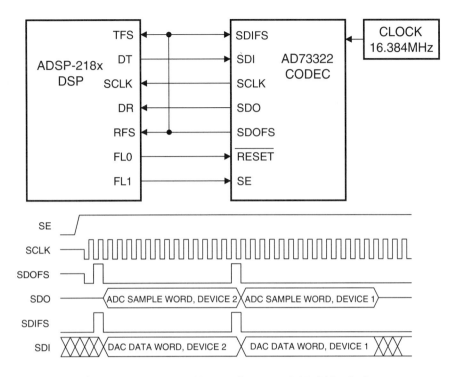

Figure 8-24: AD73322 Interface to ADSP-218x Series
(Data Transfer Mode)

The AD73422 is the first product in the dspConverter™ family of products to integrate a dual analog front end (AD73322) and a DSP (52 MIPS ADSP-2185L/ADSP-2186L). The entire functionality of the dual-channel codec and the DSP fits into a small, 119-ball, 14 mm × 22 mm plastic ball grid array (PBGA) package. The obvious advantage is the saving of circuit board real estate. ADC and DAC signal-to-noise ratios are approximately 77 dB over voice-band frequencies.

The AD74222-80 integrates 80 Kbytes of on-chip memory configured as 16 Kwords (24-bit) of program RAM, and 16 Kwords (16-bit) of data RAM. The AD73422-40 integrates 40 Kbytes of on-chip memory configured as 8 Kwords (24-bit) of program RAM, and 8 Kwords (16-bit) of data RAM. Power-down circuitry is also provided to meet the low power needs of battery-operated portable equipment. The AD73422 operates on a 3 V supply and dissipates approximately 120 mW with all functions operational.

Section Eight

- Complete Dual CODEC (AD73322) and DSP (ADSP-2185L/ADSP-2186L)
- 14 mm × 22 mm BGA Package
- 3 V Single-Supply Operation, 73 mW Power Dissipation
- Power-Down Mode
- Codec
  - Dual 16-Bit Sigma-Delta ADCs and DACs
  - Data Rates: 8, 16, 32, and 64 kSPS
  - 77 dB SNR
- DSP
  - 52 MIPS
  - ADSP-218x Code Compatible
  - 80 Kbyte and 40 Kbyte On-Chip Memory Options

Figure 8-25: AD73422 dspConverter

High Speed Interfacing

With the advent of ever faster DSP clock rates and newer architectures, it has become possible to acquire and process high speed signals. The programmability of DSPs makes it possible to run different algorithms on the same hardware, while providing different system functionality. Figure 8-26 shows a simplified ADSP-21065L system connected to a high speed ADC and high speed DAC. The ADC and DAC both have parallel interfaces connected to the external port of the DSP. With the SHARC family of DSPs there are several ways of connecting the converters to this port. The access to the converters can be done using the *direct memory access* (DMA) controller of the DSP, or it can be done under program control using the core of the DSP. Using the DMA places no load on the DSP core, so it can continue processing (executing program instructions) while the data is transferred to/from the on-chip memory.

The AD9201 is a dual-channel, 10-bit, 20 MSPS ADC that operates on a single 2.7 V to 5.5 V supply and dissipates only 215 mW (3 V supply). The AD9201 offers closely matched ADCs needed for many applications such as I/Q communications. Input buffers, an internal voltage reference, and multiplexed digital outputs buffers make interfacing to the AD9201 very simple.

The companion part to the AD9201 ADC is the AD9761 DAC. The AD9761 is a dual 10-bit, 20 MSPS per channel DAC operating on a single 2.7 V to 5.5 V supply and dissipating only 200 mW (3 V supply). A voltage reference, digital latches, and 2× interpolation make the AD9761 useful for I/Q transmitter applications.

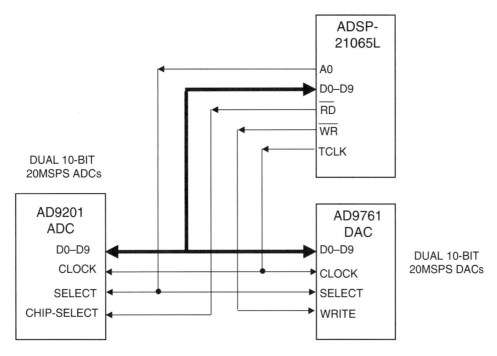

Figure 8-26: AD9201 ADC and AD9761 DAC
Interface to ADSP-21065L

DSP System Interface

Figure 8-26 shows a simplified ADSP-2189M system using the *full memory mode* configuration with two serial devices, a byte-wide EPROM, and optional external program and data overlay memories. Programmable wait state generation allows the fast processor to easily connect to slower peripheral devices. The ADSP-2189M also provides four external interrupts, seven general-purpose input/output pins, and two serial ports. One of the serial ports can be configured as two additional interrupts, a general-purpose input, and a general-purpose output pin for a total of six external interrupts, nine IOs, and one serial port. The ADSP-2189M can also be operated in the *host memory mode* which allows access to the full external data bus, but limits addressing to a single address bit. Additional system peripherals can be added in the *host memory mode* through the use of external hardware to generate and latch address signals.

Section Eight

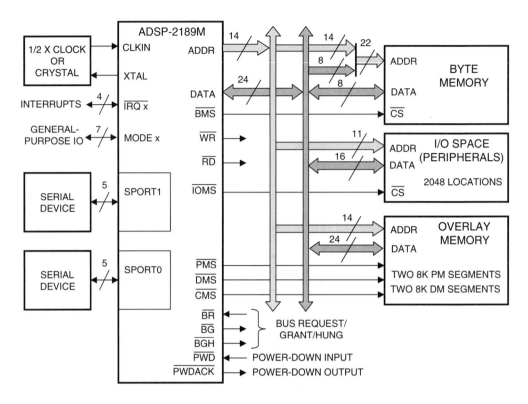

Figure 8-27: ADSP-2189M System Interface:
Full Memory Mode

References

1. Steven W. Smith, **Digital Signal Processing: A Guide for Engineers and Scientists**, Newnes, 2002.
2. C. Britton Rorabaugh, **DSP Primer**, McGraw-Hill, 1999.
3. Richard J. Higgins, **Digital Signal Processing in VLSI,** Prentice-Hall, 1990.
4. **DSP Designer's Reference (DSP Solutions)** CD-ROM, Analog Devices, 1999.
5. **DSP Navigators: Interactive Tutorials about Analog Devices' DSP Architectures** (Available for ADSP-218x family and SHARC family): http://www.analog.com/technology/dsp/index.html.
6. **General DSP Training and Workshops**: http://www.analog.com/technology/dsp/index.html.

The following DSP Reference Manuals and documentation are available for free download from: http://www.analog.com/technology/dsp/library.html.

7. **ADSP-2100 Family User's Manual, 3rd Edition**, Sept., 1995.
8. **ADSP-2100 Family EZ Tools Manual.**
9. **ADSP-2100 EZ-KIT Lite Reference Manual.**
10. **Using the ADSP-2100 Family, Vol. 1, Vol. 2.**
11. **ADSP-2106x SHARC User's Manual, 2nd Edition, July, 1996.**
12. **ADSP-2106x SHARC EZ-KIT Lite Manual.**
13. **ADSP-21065L SHARC User's Manual, Sept. 1, 1998.**
14. **ADSP-21065L SHARC EZ-LAB User's Manual.**
15. **ADSP-21160 SHARC DSP Hardware Reference.**

Section 9
DSP Applications

- High Performance Modems for Plain Old Telephone Service (POTS)
- Remote Access Server (RAS) Modems
- ADSL (Asymmetric Digital Subscriber Line)
- Digital Cellular Telephones
- GSM Handset Using SoftFone™ Baseband Processor and Othello™ Radio
- Analog Cellular Base Stations
- Digital Cellular Base Stations
- Motor Control
- Codecs and DSPs in Voice-Band and Audio Applications
- A Sigma-Delta ADC with Programmable Digital Filter

Section 9
DSP Applications
Walt Kester

High Performance Modems for Plain Old Telephone Service (POTS)

Modems (*mo*dulator/*dem*odulator) are widely used to transmit and receive digital data using analog modulation over the plain old telephone service (POTS) network as well as private lines. Although the data to be transmitted is digital, the telephone channel is designed to carry voice signals having a bandwidth of approximately 300 Hz to 3300 Hz. The telephone transmission channel suffers from delay distortion, noise, crosstalk, impedance mismatches, near-end and far-end echoes, and other imperfections. While certain levels of these signal degradations are perfectly acceptable for voice communication, they can cause high error rates in digital data transmission. The fundamental purpose of the transmitter portion of the modem is to prepare the digital data for transmission over the analog voice line. The purpose of the receiver portion of the modem is to receive the signal that contains the analog representation of the data, and reconstruct the original digital data at an acceptable error rate. High performance modems make use of digital techniques to perform such functions as modulation, demodulation, error detection and correction, equalization, and echo cancellation.

A block diagram of an ordinary telephone channel (often referred to as plain old telephone service—or POTS) is shown in Figure 9-1. Most voice-band telephone connections involve several connections through the telephone network. The 2-wire twisted pair subscriber line available at most sites is generally converted to a 4-wire signal at the telephone central office: two wires for transmit, and two wires for receive. The signal is converted back to a 2-wire signal at the far-end subscriber line. The 2- to 4-wire interface is implemented with a circuit called a *hybrid*. The hybrid intentionally inserts impedance mismatches to prevent oscillations on the 4-wire trunk line. The mismatch forces a portion of the transmitted signal to be reflected or echoed back to the transmitter. This echo can corrupt data the transmitter receives from the far-end modem.

Half-duplex modems are capable of passing signals in either direction on a 2-wire line, but not simultaneously. *Full-duplex* modems operate on a 2-wire line and can transmit and receive data simultaneously. Full-duplex operation requires the ability to separate a receive signal from the reflection (echo) of the transmitted signal. This is accomplished by assigning the signals in the two directions different frequency bands separated by filtering, or by echo cancelling, in which a locally synthesized replica of the reflected transmitted signal is subtracted from the composite receive signal.

Section Nine

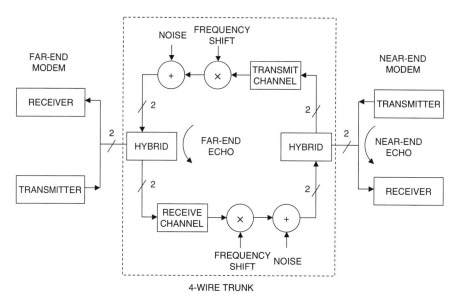

Figure 9-1: Analog Modem Using Plain Old Telephone Service (POTS) Analog Channel

There are two types of echo in a typical voice-band telephone connection. The first echo is the reflection from the near-end hybrid, and the second echo is from the far-end hybrid. In long-distance telephone transmissions, the transmitted signal is heterodyned to and from a carrier frequency. Since local oscillators in the network are not exactly matched, the carrier frequency of the far-end echo may be offset from the frequency of the transmitted carrier signal. In modern applications this shift can affect the degree to which the echo signal can be canceled. It is therefore desirable for the echo canceller to compensate for this frequency offset.

For transmission over the telephone voice network, the digital signal is modulated onto an audio sine wave carrier, producing a modulated tone signal. The frequency of the carrier is chosen to be well within the telephone band. The transmitting modem modulates the audio carrier with the transmit data signal, and the receiving modem demodulates the tone to recover the receive data signal.

The baseband data signal may be used to modulate the amplitude, the frequency, or the phase of the audio carrier, depending on the data rate required. These three types of modulation are known as amplitude shift keying (ASK), frequency shift keying (FSK), and phase shift keying (PSK). In its simplest form the modulated carrier takes on one of two states—that is, one of two amplitudes, one of two frequencies, or one of two phases. The two states represent a logic 0 or a logic 1.

Low to medium speed data links usually use FSK up to 1,200 bits/s. Multiphase PSK is used for 2,400 bits/s and 4,800 bits/s links. PSK utilizes bandwidth more efficiently than FSK but is more costly to implement. ASK is least efficient and is used only for very low speed links (less than 100 bits/s). For 9,600 bits/s up to 33,600 bits/s a combination of PSK and ASK is used, known as quadrature amplitude modulation (QAM).

The International Telegraph and Telephone Consultative Committee (CCITT in France) has established standards and recommendations for modems, which are given in Figure 9-2.

CCITT Rec.	Approximate Date	Speed (Bits/s) Maximum	Half-Duplex/ Full-Duplex/ Echo Cancel	Modulation Method
V.21	1964	300	FDX	FSK
V.22		1200	FDX	PSK
V.22bis		2400	FDX	16QAM
V.23		1200	HDX	FSK
V.26 bis		2400	HDX	PSK
V.26 ter		2400	FDX(EC)	PSK
V.27 ter		4800	HDX	8PSK
V.32		9600	FDX(EC)	32QAM
V.32 bis		14400	FDX(EC)	QAM
V.34		33600	FDX(EC)	QAM
V.90	1998	56000[1]	FDX(EC)	PCM
V.92	2001	56000[2]	FDX(EC)	PCM

[1] DOWNSTREAM ONLY, UPSTREAM IS V.34
[2] UPSTREAM AND DOWNSTREAM

Figure 9-2: Some Modem Standards

The goal in designing high performance modems is to achieve the highest data transfer rate possible over the POTS network and avoid the expense of using dedicated, conditioned private telephone lines. The V.90 recommendation describes a full-duplex (simultaneous transmission and reception) modem that operates on the POTS network. The V.90 modem communicates downstream from the central office to the subscriber modem at a rate of 56,000 bits/s using pulse code modulation (PCM). Upstream communication from the subscriber to the central office is at the V.34 rate of up to 33,600 bits/s (QAM).

A simplified block diagram for a V.90 analog modem is shown in Figure 9-3. The diagram shows that the bulk of the signal processing is done digitally. Both the transmit and receive portions of the modem subject the digital signals to a number of DSP algorithms that can be efficiently run on modern processors.

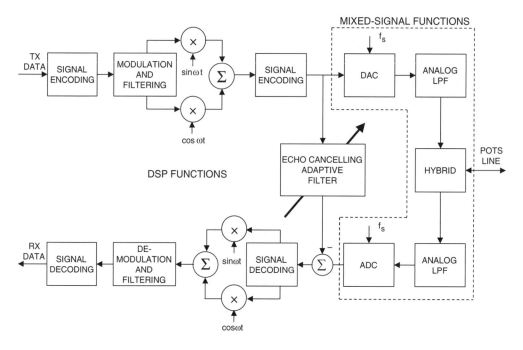

Figure 9-3: V.90 Analog Modem Simplified Block Diagram

The Tx input serial bit stream is first scrambled and encoded. Scrambling takes the input bit stream and produces a pseudorandom sequence. The purpose of the scrambler is to whiten the spectrum of the transmitted data. Without the scrambler, a long series of identical symbols could cause the receiver to lose carrier lock. Scrambling makes the transmitted spectrum resemble white noise (to utilize the bandwidth of the channel more efficiently), makes carrier recovery and timing synchronization easy, and makes adaptive equalization and echo cancellation possible.

The scrambled bit stream is divided into groups of bits, and the groups of bits are first differentially encoded and then convolutionally encoded.

The symbols are then mapped into the signal space using QAM as defined in the V.34 standard. The signal space mapping produces two coordinates, one for the real part of the QAM modulator and one for the imaginary part. As an example, a diagram of a 16-QAM signal constellation is shown in Figure 9-4. Larger constellations are used in V.90 modems, and the actual size of the constellation is adaptive and determined during the training, or "handshake" interval, when the modems synchronize with each other for upstream or downstream signaling.

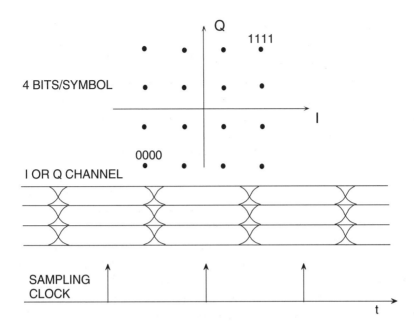

Figure 9-4: Quadrature Amplitude Modulated (QAM) Signal Transmits 4 Bits per Symbol (16-QAM)

Used prior to modulation, digital pulse-shaping filters attenuate frequencies above the Nyquist frequency that are generated in the signal mapping process. These filters are designed to have zero crossings at the appropriate frequencies to cancel inter-symbol interference.

QAM is easily implemented in modern DSP processors. The process of modulation requires the access of a sine or cosine value, the access of an input symbol (x or y coordinate) and a multiplication. The parallel architecture of the ADSP-21xx family permits all three operations to be performed in a single instruction cycle.

The output of the digital modulator drives a DAC. The output of the DAC is passed through an analog low-pass filter and to the 2-wire telephone line for transmission over the POTS network.

The receiver is made up of several functional blocks: the input antialiasing filter and ADC, a demodulator, an adaptive equalizer, a Viterbi decoder, an echo canceller, a differential decoder, and a descrambler. The receiver DSP algorithms are both memory intensive and computation intensive. The ADSP-218x family addresses both needs, providing sufficient program memory RAM (for both code and data) on-chip, data memory RAM on-chip, and an instruction execution rate of up to 75 MIPS.

The antialiasing filter and ADC in the receiver need to have a dynamic range from the largest echo signal to the smallest. The received signal can be as low as –40 dBm, while the near-end echo can be as high as –6 dBm. In order to ensure that the analog front end of the receiver does not contribute any significant impairment to the channel under these conditions, an instantaneous dynamic range of 84 dB and an SNR of 72 dB is required.

In order to compensate for amplitude and phase distortion in the telephone channel, equalization is required to recover the transmitted data at an acceptably low bit error rate. In order to respond to rapidly changing conditions on the telephone line, adaptive equalization is required for the V.90 modem receiver. An adaptive equalizer can be implemented digitally in an FIR filter whose coefficients are continuously updated based on current line conditions.

Separation between the transmit and receive signal in the V.90 modem is accomplished using echo cancellation. Both near-end and far-end echo must be cancelled in order to yield reliable communication. Echo cancellation is achieved by subtracting an estimate of the echo return signal from the actual received signal. The predicted echo is determined by feeding the transmitted signal into an adaptive filter with a transfer function that approximates the telephone channel. The adaptive filter commonly used in echo cancellers is the FIR filter (chosen for its stability and linear phase response). The taps are determined using the least-mean-square (LMS) algorithm during a training sequence executed prior to full-duplex communications.

The most common technique for decoding the received data is Viterbi decoding. Named after its inventor, the Viterbi algorithm is a general-purpose technique for making an error-corrected decision. Viterbi decoding provides a certain degree of error correction by examining the received bit pattern over time to deduce the value that was the most likely to have been transmitted at a particular time. Viterbi decoding is computation intensive. A history for each of the possible symbols sent at each symbol interval has to be maintained. At each symbol interval, the length of the path backward in time from each possible received symbol to a symbol sent some time ago is calculated. The symbol that has the shortest path back to the original signal is chosen to be the current decoded symbol. A complete description of Viterbi decoding and its implementation on the ADSP-21xx family of processors is given in documentation available from Analog Devices (Reference 2).

Figure 9-5 shows a comparison between V.34 and V.90 modems. Note that in the case of V.34 (Figure 9-5A), the communication is between two *analog* modems. This requires an ADC/DAC in both the transmit and receive path as shown in the diagram. The V.90 system requires an all-digital network and a V.90 digital modem as shown in Figure 9-5B. Note that the second ADC/DAC combination is eliminated, thereby allowing the faster downstream data rate of 56 Kbits/s. Downstream communication to the V.90 analog modem uses 64 Kbits/s PCM data, which is standard for all digital telephone networks. This serial data is converted into a pulse amplitude modulated (PAM) signal (8 bits, 8 kSPS) using an 8-bit DAC. The signal from the DAC to the analog modem is therefore a 256 K constellation with no imaginary component, i.e., the analog modem receiver must detect which one of the 256 levels is being sent during the symbol interval.

DSP Applications

The V.90 standard allows downstream data rates of up to slightly less than 56 Kbits/s, and upstream data rates of up to 33.6 Kbits/s (V.34). The V.92 standard will allow 56 Kbits/s transmission in both directions.

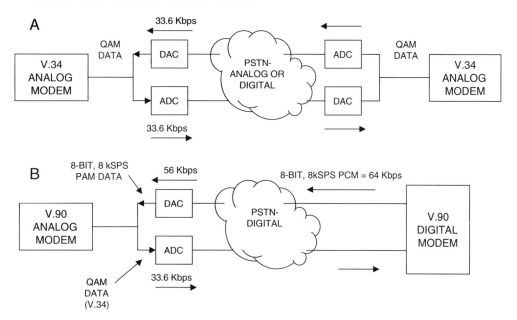

Figure 9-5: V.34 vs. V.90 Modems

Remote Access Server (RAS) Modems

Rapid growth and use of the Internet has created a problem in that there are more users trying to get on the Internet than there is equipment to accommodate all these users. internet service providers (ISPs) like America Online purchase modem equipment so their customers (referred to as subscribers) can remotely access a network (like the Internet from home). This application of accessing a network from a remote location is called remote network access. The equipment used in this application is called a remote access server (RAS) as shown in Figure 9-6. The remote access server is made up of many modem ports; each modem port can connect to a different user. The RAS can use analog modems, which connect to a POTS line, or digital modems, which are compatible with T1, E1, PRI, or BRI lines. Digital modems are used in most RAS systems since they are more efficient for eight ports or more.

Network access equipment enables individuals, small offices, and traveling employees to connect to corporate networks (intranets) and the Internet. Internet service providers use devices called *concentrators* to connect their telephone access lines to their networks. These concentrators are also referred to as remote access servers. The rapid growth in the use of the Internet and intranets has created a tremendous demand for modem equipment.

Section Nine

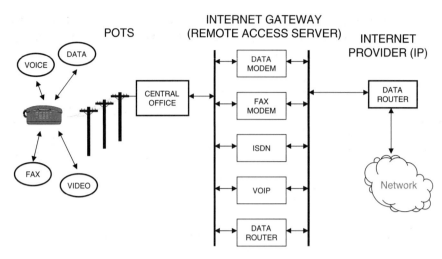

Figure 9-6: Internet Gateway Using Remote Access Server (RAS) Modem

Remote access servers accommodate employees and small offices/home offices (SOHO) wishing to connect individual computers to LANs (local area networks) or intranets. If a remote access server is installed in a corporate LAN, remote users can access the network in such a way that their computers appear to be directly connected to the LAN. This allows them to work from a remote location as if they were sitting at their desk in their office.

The ADSP-21mod870 acts as a bridge between the voice-based continuous connection switched network and the data-based IP network as shown in Figure 9-7. The high speed DMA interface and large on-chip RAM of the ADSP-21mod870 allow it to be configured to handle a large variety of tasks. The software with the ADSP-21mod870-110 can be configured for modem calls or HDLC (high bit-rate digital subscriber line) processing of digital ISDN-(integrated services digital network) based calls. Since the ADSP-21mod870 is an open platform, other functions can be loaded by users. Examples of other functions include voice-over-internet and FAX-over-internet. In these applications, the ADSP-21mod870 is a gateway for voice network users to save toll charges by routing their calls over the IP network. The ADSP-21mod870 DSP uses the ADSP-218x 16-bit fixed-point core and is code compatible with other members of the ADSP-21xx family.

As the number of remote network users has grown, capacity over the telephone central office switched network has often become strained. These bottlenecks often occur when thousands of calls from a metropolitan area are switched to a single point of presence (POP). To eliminate these bottlenecks, RAS equipment can be pushed outward from the POP toward the edges of the switched network as shown in Figure 9-8. When RAS equipment is located in the switching center for local exchanges, data calls can be separated from voice calls, eliminating the strain on the voice-based

DSP Applications

switched network. RAS equipment integrated into switching equipment is referred to as *on-switch*-based RAS. This differs for standalone RAS systems in that on-switch RAS can separate data calls before they are connected to trunk lines exiting the switch.

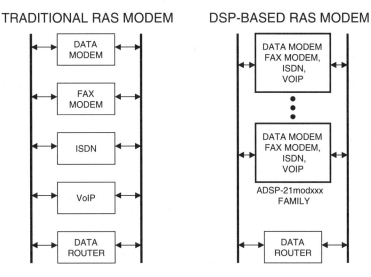

Figure 9-7: DSP-Based RAS Modem Using the ADSP-21modxxx Family

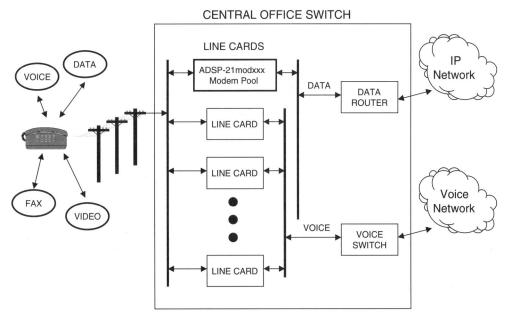

Figure 9-8: Expanding Central Office Capability with ADSP-21modxxx Family

Several types of RAS equipment have evolved to satisfy the needs of the different remote network access users. RAS equipment can take several forms. RAS concentrators combine a modem pool with a router in a standalone chassis. NT server RAS uses off-the-shelf Windows NT PCs to perform routing functions with a modem pool connected to its PCI or ISA expansion bus. On-switch RAS integrates modem pools into the line card chassis in switching systems. These various forms of RAS equipment serve the different needs of end users. Local exchange companies (LECs) or the new competing local exchange companies (CLECs) can take advantage of on-switch RAS equipment to eliminate bottlenecks in their switched networks. ISPs (internet service providers) use RAS concentrators to collect calls from a large area for connection to the Internet. Large businesses also use RAS concentrators to connect their employees to their LAN or intranets. Small office and home office (SOHO) users can use low cost NT server-based RAS to support all of their RAS, LAN, and other telecommunication needs.

The ADSP-21mod870 digital modem processor is the first complete digital RAS modem on a single chip. The device is V.34/56 K and V.42/V.42bis compatible, has a 16-bit DMA port for software downloads, provides TDM serial port interfaces directly to T1/E1, has 160 Kbytes on-chip SRAM, dissipates 140 mW at 3.3 V, and is packaged in a 16 mm TQFP package. The ADSP-21mod870's small footprint and power efficiency will enable internet service providers (ISPs) and central office access providers to quadruple the number of ports within the existing modem bank chassis. Moreover, the chip's unique capability to support any protocol on any port will improve ISP customer service and reduce business operating costs.

Part of a family of digital modem solutions from ADI, the ADSP-21mod870 includes silicon, software, and service. ADI is singular in its ability to provide the complete solution. Also available are the ADSP-21mod970 (six modem channels, 31 mm BGA) and the ADSP-21mod980N (16 modem channels, 35 mm BGA).

Multichannel Voice-Over-Internet-Provider (VOIP) Server

The ADSP-218x family of devices can efficiently implement multichannel telephony applications (such as RAS/VOIP servers and gateways) due to their high performance and large on-chip memories. A typical system is shown in Figure 9-9. The programmable nature of the DSP architecture gives the system architect the flexibility to implement various speech coding algorithms in addition to baseline telephony functionality.

The ADSP-2188M is the most highly integrated member of the ADSP-218x family (with over 2 MBits of on-chip SRAM). This high level of integration, combined with 75 MIPS performance, can support up to six voice channels per DSP (depending on the voice coder chosen).

DSP Applications

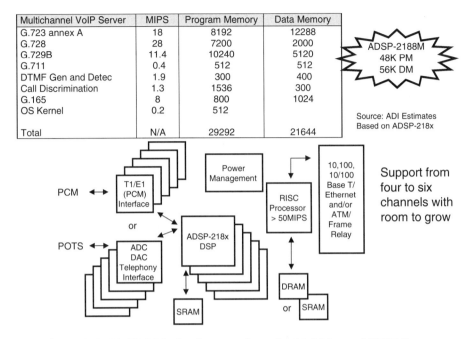

Figure 9-9: ADSP-218x Implementation of a Multichannel VOIP Server

ADSL (Asymmetric Digital Subscriber Line)

Thanks to the widespread popularity of the World Wide Web, Internet traffic is at an all-time high. A study recently conducted by the *Wall Street Journal* reported 58 million Internet users in the United States and Canada alone. Research firms all predict heavier Internet traffic as more and more people buy PCs and use the Internet for business, academic, and recreational purposes.

Unless something is done to improve the way we access the Internet, user traffic will ultimately burden the public switched telephone network (PSTN) beyond its original design limits. Internet users are frustrated by the amount of time it takes to view simple text-based Web sites, especially from 8:00 am to 6:00 pm on business days when congestion and bottlenecks are most prevalent. The problem will only get worse as users attempt to view graphically complex sites, download new video and audio clips, and access other types of multimedia services becoming available over the Internet.

Today's analog modem and telephone switch technology is simply inadequate. Assuming little or no network delays, a 10 MB data transmission—the equivalent of a four-minute audio/video clip—takes about 95 minutes to download when using a 14.4 Kbps analog modem, 45 minutes when using a 28.8 Kbps modem, and 25 minutes when using a 56 Kbps modem. Lengthy online calls are tying up telephone systems originally designed to handle short (three-minute) voice calls and switches built to accommodate approximately nine minutes per line during peak hours. How often do users browse the Internet for 10 minutes or less? It now appears that relief, in the form of ADSL (asymmetric digital subscriber line) technology, is on the way.

ADSL is a high speed digital switching/routing and signal processing technology. It promises to relieve network bottlenecks and provide enough user bandwidth for the Internet explosion. Originally conceived in 1994, ADSL delivers the huge amounts of bandwidth needed for interactive gaming, multimedia services, and video-on-demand. These applications, along with videoconferencing, remote schooling, and home shopping are still attractive applications. As more and more people use the Internet for electronic commerce in homes around the world, the need for high speed network access becomes ever more paramount.

ADSL can transfer data over ordinary telephone lines nearly 200 times faster than today's contemporary modems and 90 times faster than ISDN. Early tests and trials have shown promising results all over the world. And while GTE and other large telephone companies begin to deploy ADSL systems in select regions of the U.S. and abroad, others seek to deploy ADSL-based equipment as standards-based systems and modems become commercially available.

Due to ADSL's technical complexity, only a few semiconductor manufacturers are presently developing ADSL silicon chips. Analog Devices is one of them, and is considered a pioneer in ADSL after having produced the first complete ADSL chipset back in 1997. Early adopters of ADSL rallied behind the AD20msp910 chipset after having successfully evaluated its high speed, long reach capability. Soon thereafter, the world's most influential industry standards committees (ANSI, ETSI, and ITU) approved an advanced discrete multitone (DMT) signal processing technique used by the AD20msp910. Today, Analog Devices enjoys the advantages of having the industry's first standards-based solution, the largest installed customer base, and the most design wins of any semiconductor manufacturer to date.

ADSL is attractive for the following reasons:

- ADSL is fast. The same 10 MB video clip that takes 90 minutes to download using a conventional modem would be downloaded in just 10 seconds using an ADSL modem. Superfast ADSL modems can transmit data at rates as high as 8 Mb per second.

- ADSL is easy to install. It uses existing copper twisted-pair telephone lines from a local exchange carrier (central office) to the subscriber's home or office. Little or no rewiring is necessary.

- ADSL is cost-effective. It requires no major upgrades in the existing telephone network infrastructure.

- ADSL is viable. Issues that have slowed the deployment of high speed fiber networks to the home (e.g., prohibitive cost and installation) do not apply. ADSL works with existing POTS (plain old telephone service). High speed data transport can occur simultaneously with voice calls and fax transmissions.

Unlike other high speed data transmission technologies, ADSL requires no rewiring over the "last mile" of the network. Although commonly referred to as the last mile, transmission length is typically 12,000 to 18,000 feet. That "last mile" (the leg from the central office to a user's home or office) operates over existing copper loops of twisted-pair telephone wires. But ADSL does require the installation of new equipment at local exchange central offices and major switch offices. However, the technology developed for central office equipment is the same as that used in PC modems and home splitter boxes, thus ensuring interoperability throughout the entire network. A simplified block diagram of an ADSL system is shown in Figure 9-10.

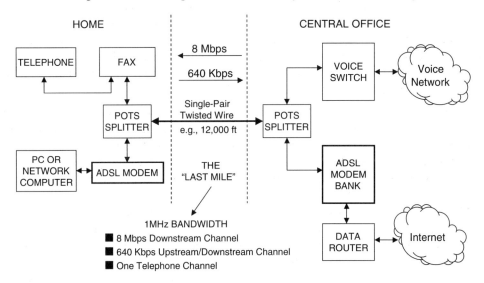

Figure 9-10: Asymmetric Digital Subscriber Line (ADSL) System Characteristics

Using sophisticated digital signal processing techniques, ADSL modems push as much data as possible across copper wires by making the best use of available bandwidth. Many perceived the telephone network as having only 4 kHz of bandwidth, but 4 kHz only represents the band used for the analog (voice) transmission. Using ADSL, the physical connection between the home and local exchange carrier (LEC) over conventional copper wire is capable of carrying 1 MHz. ADSL takes advantage of the portion of the bandwidth not used for voice call. Essentially, it splits the 1 MHz bandwidth into three information channels: one high speed downstream channel, one medium speed duplex (upstream/downstream) channel, and one conventional voice channel. (Downstream refers to data transmitted from the telephone network to the customer's premises; upstream is data routed from the customer to the network.)

In addition to providing high bandwidth data transmission, ADSL preserves "lifeline" services (i.e., the voice network that people depend upon for day-to-day communications and emergency situations). This 3-channel approach allows subscribers to send an email, download a video for viewing, and talk on the telephone at the same time. Telecommuters can access their corporate local area network, and simultaneously videoconference with a customer. In fact, using ADSL's full capability provides enough bandwidth for running four channels of MPEG compressed video with no interruption to normal telephone service.

Most Internet applications require uneven amounts of upstream and downstream bandwidth. In other words, users tend to move greater amounts of data in one direction than they do in the other direction. Generally speaking, Internet users access more information and receive more data than they transmit. That's the nature of the Internet. They read more electronic mail than they send, they download more video information than they produce. A user's upstream capacity is typically limited to sending commands or transmitting small data files to a server. Much more information travels downstream. ADSL was designed to take advantage of this trend in uneven bandwidth. It provides data rates in excess of 8 Mbps from the network to the subscriber (downstream), and up to 640 Kbps from the subscriber to the network (upstream).

For telephone companies, ADSL may be the key to overcoming the congestion created by explosive Internet growth. If telephone companies hope to capture the Internet data carrier market, they cannot use equipment designed for short voice traffic. The average voice call lasts about three minutes. According to a 1997 study by Bellcore, the research laboratory funded by U.S. local telephone companies (one of five submitted to the Federal Communications Commission), the average length of time for a typical Internet call is in excess of 20 minutes. Telephony magazine reported as early as 1994 that nearly 20 percent of all on-line data service connections lasted more than an hour. Cable companies like MediaOne, based in the Northeast, charge one flat rate for Internet use. Broadband subscribers can stay online 24 hours a day for roughly $40 a month.

ADSL was designed to take advantage of this natural unevenness in bandwidth requirements and make the most of available bandwidth, a limited resource. It provides for rates in excess of 8 Mbps from the network to the subscriber, and up to 640 Kbps from the subscriber to the network.

While some other telephony-based technologies offer relief for the PSTN, many carry high price tags and long-term implementation time frames. Replacing the copper network with fiber to the curb, for example, is a costly proposition. Industry experts estimate that creating a fiber-to-the-curb network will cost about $1,500 per telephone customer. Replacing the existing 560 million copper subscriber lines worldwide would cost more than three-quarters of a trillion dollars.

ADSL, by contrast, is easily installed. To complete an ADSL circuit, two modems are required, one at each end of a twisted-pair telephone line. One modem is located at the subscriber's premise; the other modem (usually a rack of modems with line cards) is located at the local telephone company's central office. Figure 9-10 shows a simplified connection.

At the customer premise, an ADSL modem is inside a PC or connected to the network computer, as well as the telephone and/or facsimile machine. A copper twisted pair telephone line links the POTS splitter to the central office. At the central office,

DSP Applications

where the switch and rack of modems (line cards) are located, a POTS splitter separates voice from data. Voice calls are forwarded to the central office switch for relay over the public telephone network. Data transmissions are sent through an Ethernet switch and a router and eventually over a high speed connection (for example, 155 Mbps OC-3) to the Internet service provider. The connection is made to the Internet backbone for access to the World Wide Web.

The key to ADSL's high speed data transmission is advanced digital signal processing (DSP). Drawing on its analog innovation and advanced digital signal processing capabilities, Analog Devices developed the first-generation AD20msp910 chipset (see Figure 9-11). The AD20msp910 offers three competitive advantages over modem manufacturers:

- It is the most functionally complete chipset solution on the market.
- It is fully compliant to ANSI, ETSI, and ITU industry standards.
- It is compatible with nearly every DLC (digital loop carrier) and modem manufacturer.

The AD20msp910 solution includes both hardware and software. It integrates a DSP host processor, line driver, and control software, plus DMT technology. Other supplier's chips offer only portions of the modem. The AD20msp910 speeds and simplifies development of ADSL modems for high speed Internet access and multimedia services and was released in 1997. The second-generation AD20msp918 adds ATM functionality, improved performance, and support for ADSL over ISDN for the European market. Both chipsets are fully compliant to all standards (ANSI T1.413 Issue 2, ETSI TR328, ITU G.dmt, and ITU G.lite for splitterless) and both are complete solutions, including all microcontroller hardware and software functions.

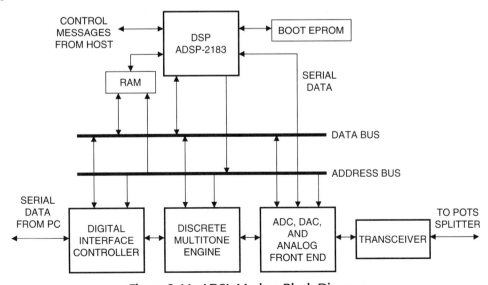

Figure 9-11: ADSL Modem Block Diagram

Now on our third-generation chipset, the AD20msp930, Analog Devices enables modem manufacturers to reduce their development time and costs. To further speed and streamline design efforts, Analog Devices will also provide a PC board layout, schematics, and a reference design. As a result, modem manufacturers can focus their development efforts on adding value to the core ADSL modem technology.

Digital Cellular Telephones

In the early 1990s, the GSM (Global System for Mobile Communications, a.k.a Groupe Speciale Mobile) was introduced in Europe, heralding the introduction of digital cellular radio. Out of necessity, other countries, including the U.S., have adopted various digital standards in order to relieve the congestion and other problems faced by analog cellular systems such as AMPS (advanced mobile phone service). The limitations of analog cellular radio are well known and include call blocking during busy hours, misconnects and disconnects due to rapidly fading signals, lack of privacy and security, and limited data transmission rates.

The basic cellular system is shown in Figure 9-12. A region is broken up into cells, with each cell having its own base station and its own group of assigned frequencies. Because the radius of each cell is small (10 miles, for example) low power transmitters and receivers can be used.

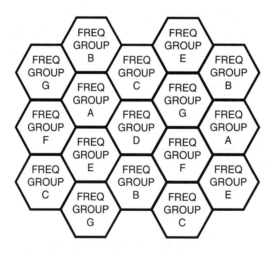

- Each Cell Covers a Radius of 5 to 10 Miles
- Each Cell Requires its Own Base Station for Reception and Retransmission
- Each Cell Must Handle Multiple Callers Simultaneously
- Callers May Be Handed Off from One Cell to Another

Figure 9-12: Cellular Phone Frequency Reuse

The cellular system lends itself to frequency reuse, since cells that are far enough apart can utilize the same band of frequencies without interference. The base stations must be linked together with an elaborate central control network so that a call may be handed off to another cell when the signal strength from the mobile unit becomes too low for the current cell to handle.

The frequency spectrum allocation for analog cellular radio (AMPS) in the United States is approximately 825 MHz to 850 MHz and 870 MHz to 895 MHz. Conventional architectures (both analog and digital) are channelized. The total spectrum is divided up into a large number of relatively narrow channels, defined by a carrier frequency. The carrier frequency is frequency modulated with the voice signal using analog techniques. Each full-duplex channel requires a pair of frequencies, each with a bandwidth of approximately 30 kHz. A user is assigned both frequencies for the duration of the call. The forward and reverse channel are widely separated, to help the radio keep the transmit and receive functions separated. The signal bandwidth for the "A" or "B" carriers serving a particular geographical area is 12.5 MHz (416 channels, each 30 kHz wide). There can be only one caller at a time per channel.

Time division multiple access (TDMA) allocates bandwidth on a time-slot basis. In the United States TDMA system, the entire 30 kHz channel is assigned to a particular transmission, but only for a short period of time. A 3:1 multiplexing scheme means that three conversations can take place with TDMA using the same amount of bandwidth as one analog cellular conversation does. Each transmit/receive sequence occurs on time slots lasting 6.7 ms. The TDMA system relies on an extensive amount of DSP technology to reduce the coded speech bit-rate as well as to prepare the digital data for transmission over the analog medium. The TDMA approach was chosen for the GSM system and will be discussed later in more detail.

The second digital approach used in the United States is called code division multiple access (CDMA). This technique has been used in secure military communications for a number of years under the name of *spread spectrum*. In spread spectrum, the transmitter transmits in a pseudorandom sequence of frequency hops over a relatively wide frequency range. The receiver has access to the same random sequence and can decode the transmission. The effect of adding additional users on the system is to decrease the overall signal-to-noise ratio for all the users. With this technique, the effect of allowing more calls than the normal capacity is to increase the bit-error rate for all users. New callers can keep coming in, and interference levels will rise gradually, until at some point the process will become self-regulating: the quality of the voice link will become so bad that users will cut short or refrain from making additional calls. No one is ever blocked in the conventional sense, as they are in FDMA or TDMA systems when all channels or slots are full.

Both TDMA and CDMA digital systems make extensive use of DSP algorithms in both speech encoding and in preparing the signal for transmission. In the receiver, DSP techniques are used for demodulation and decoding the speech signal.

At the present time, both the analog and digital systems are in use in the U.S. In many cases, analog and digital systems must coexist in the same geographical area and within the same service area. This implies that the cellular base stations must handle both analog and digital formats, and therefore there is extensive use of digital techniques in base station design to simplify the hardware.

The remainder of this section will concentrate on speech processing and channel coding as they relate to the GSM system. This will serve to illustrate the fundamental principles that are applicable to all digital mobile radio systems.

- Frequency Division Multiple Access (FDMA) – User Allocation Based on Frequency Slots (AMPS System)
- Time Division Multiple Access (TDMA) – User Allocation Based on Time Slots (At Least Three Times More Capacity than FDMA); GSM is an Example of TDMA
- Code Division Multiple Access (CDMA) – Based on Spread Spectrum Technology - More Users Cause Graceful Degradation in Bit-Error Rate
- Both TDMA and CDMA Make Extensive Use of DSP in Speech Encoding and Channel Coding for Transmission and Reception

Figure 9-13: Digital Mobile Radio Standards

The GSM System

Figure 9-14 shows a simplified block diagram of the GSM digital cellular telephone system. The *speech encoder and decoder* and *discontinuous transmission* function will be described in detail. Upconversion and downconversion portions of the system contain a digital modem similar to the ones previously discussed. Similar functions such as equalization, convolutional coding, Viterbi decoding, modulation, and demodulation are performed digitally.

The standard for encoding voice has been set in the T-carrier digital transmission system. In this system, speech is logarithmically encoded to 8 bits at a sampling rate of 8 kSPS. The logarithmic encoding and decoding to 8 bits is equivalent to linear encoding and decoding to 13 bits of resolution. This produces a bit rate of 104 Kbps. In most handsets, a 16-bit sigma-delta ADC is used, so the effective bit rate is 128 Kpbs. The speech encoder portion of the GSM system compresses the speech signal to 13 Kbps, and the decoder expands the compressed signal at the receiver. The speech encoder is based on an enhanced version of linear predictive coding (LPC). The LPC algorithm uses a model of the human vocal tract that represents the throat as a series of concentric cylinders of various diameters. An excitation (breath) is forced into the cylinders. This model can be mathematically represented by a series of simultaneous equations that describe the cylinders.

DSP Applications

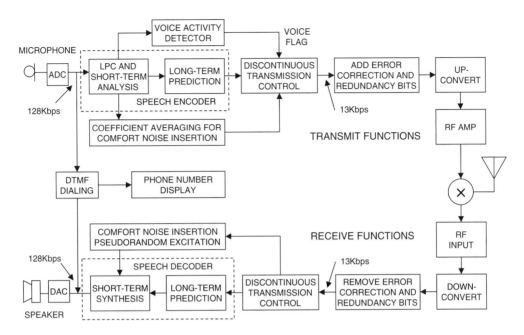

Figure 9-14: GSM Handset Block Diagram

The excitation signal is passed through the cylinders, producing an output signal. In the human body, the excitation signal is air moving over the vocal cords or through a constriction in the vocal tract. In a digital system, the excitation signal is a series of pulses for vocal excitation, or noise for a constriction. The signal is input to a digital lattice filter. Each filter coefficient represents the size of a cylinder.

An LPC system is characterized by the number of cylinders it uses in the model. Eight cylinders are used in the GSM system, and eight reflection coefficients must be generated.

Early LPC systems worked well enough to understand the encoded speech, but often the quality was too poor to recognize the voice of the speaker. The GSM LPC system employs two advanced techniques that improve the quality of the encoded speech. These techniques are *regular pulse excitation* (RPE) and *long-term prediction* (LTP). When these techniques are used, the resulting quality of encoded speech is nearly equal to that of logarithmic pulse code modulation (companded PCM as in the T-carrier system).

The actual input to the speech encoder is a series of 16-bit samples of uniform PCM speech data. The sampling rate is 8 kHz. The speech encoder operates on a 20 ms window (160 samples) and reduces it to 76 coefficients (260 bits total), resulting in an encoded data rate of 13 Kbps.

Discontinuous transmission (DTX) allows the system to shut off transmission during the pauses between words. This reduces transmitter power consumption and increases the overall GSM system's capacity.

Low power consumption prolongs battery life in the handset and is an important consideration for hand-held portable phones. Call capacity is increased by reducing the interference between channels, leading to better spectral efficiency. In a typical conversation each speaker talks for less than 40% of the time, and it has been estimated that DTX can approximately double the call capacity of the radio system.

The voice activity detector (VAD) is located at the transmitter. Its job is to distinguish between speech superimposed on the background noise and noise with no speech present. The input to the voice activity detector is a set of parameters computed by the speech encoder. The VAD uses this information to decide whether or not each 20 ms frame of the encoder contains speech.

Comfort noise insertion (CNI) is performed at the receiver. The comfort noise is generated when the DTX has switched off the transmitter; it is similar in amplitude and spectrum to the background noise at the transmitter. The purpose of the CNI is to eliminate the unpleasant effect of switching between speech with noise, and silence. If you were listening to a transmission without CNI, you would hear rapid alternating between speech in a high noise background (e.g., in a car), and silence. This effect greatly reduces the intelligibility of the conversation.

When DTX is in operation, each burst of speech is transmitted followed by a *silence descriptor* (SID) frame before the transmission is switched off. The SID serves as an end of speech marker for the receive side. It contains characteristic parameters of the background noise at the transmitter, such as spectrum information derived through the use of linear predictive coding.

The SID frame is used by the receiver's comfort noise generator to obtain a digital filter which, when excited by pseudorandom noise, will produce noise similar to the background noise at the transmitter. This comfort noise is inserted into the gaps between received speech bursts. The comfort noise characteristics are updated at regular intervals by the transmission of SID frames during speech pauses.

Redundant bits are then added by the processor for error detection and correction at the receiver, increasing the final encoded bit rate to 22.8 Kbps. The bits within one window, and their redundant bits, are interleaved and spread across several windows for robustness.

GSM Handset Using SoftFone Baseband Processor and Othello Radio

Analog Devices recently announced two chipsets that comprise the majority of a GSM handset. The SoftFone chipset performs the baseband and DSP functions, while the Othello radio chipset handles the RF functions.

The frequency bands originally allocated to GSM were 890 MHz to 915 MHz for mobile transmitting and 935 MHz to 960 MHz for mobile receiving. Another frequency allocation was made to further expand GSM capacity. This band, allocated to digital communications services (DCS), was 1710 MHz to 1785 MHz and 1805 MHz to 1880 MHz. All countries adopting GSM use one of these two pairs of frequencies. In the United States, these bands were already allocated by the FCC. In the mid-1990s, a set of bands were made available for GSM in the United States: 1850 MHz to 1910 MHz, and 1930 MHz to 1990 MHz.

Because of the frequency allocations in GSM countries (other than the U.S.), most GSM handsets must be dual-band: capable of handling both GSM and DCS frequencies. The SoftFone and Othello chipsets supply the main functions necessary for implementing dual- or triple-band radios for GSM cellular phones. The AD20msp430 SoftFone chipset comprises the baseband portion of the GSM handset. The AD20msp430 baseband processing chipset uses a combination of GSM system knowledge and advanced analog and digital signal processing technology to provide a new benchmark in GSM/GPRS terminal design.

The SoftFone architecture is entirely RAM-based. The software is loaded from FLASH memory and is executed from the on-chip RAM. This allows fast development cycles since no ROM code turns are required. Furthermore, the handset software can be updated in the field to enable new features. Combined with the Analog Devices Othello RF chipset, a complete multiband handset design contains fewer than 200 components, fits in a 20 cm^2 single-sided PCB layout, and has a total bill of materials cost 20% to 30% lower than previous solutions. A simplified block diagram of the handset is shown in Figure 9-15.

The AD20msp430 chipset comprises two chips, the AD6522 DSP-based baseband processor and the AD6521 voice-band/baseband mixed-signal codec. Together with the "Othello" radio chipset, the AD20msp430 allows a significant reduction in the component count and bill of materials (BOM) cost of GSM voice handsets and data terminals. The software and hardware foundations of the AD20msp430 chipset enjoy a long history of successful integration into GSM handsets.

Section Nine

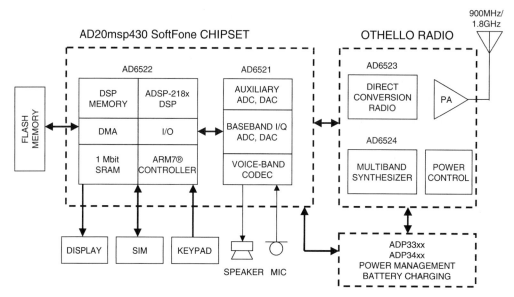

Figure 9-15: Othello Radio and SoftFone Chipsets Make Complete GSM/DCS Handset

This is Analog Devices' fourth generation of GSM chipsets, each of which has passed numerous type approvals and network operator approvals in OEM handsets. In each generation, additional features have been added, while cost and power have been reduced. Numerous power-saving features have been included in the AD20msp430 chipset to reduce the total power consumption. A programmable state machine allows events to be controlled with a resolution of one-quarter of a bit period. This reduces current in standby mode to 1 mA. This allows a handset to operate in standby mode for up to 1000 hours on a battery charge. The AD20msp430 chipset uses the SoftFone architecture, where all software resides in RAM or FLASH memory. Since ROM is not used, development time is reduced and additional features can easily be field-installed. A basic dual-band GSM terminal typically requires a single 8 Mb FLASH memory chip.

There are two processors in the AD20msp430 chipset. The DSP processor is the ADSP-218x core, proven in previous generations of GSM chipsets, and operated at 65 MIPS in the AD20msp430. This DSP performs the voice-band and channel coding functions previously discussed. The AD6521 voice-band/baseband codec chip contains all analog and mixed-signal functions. These include the I/Q channel ADCs and DACs, high performance multichannel voice-band codec, and several auxiliary ADCs and DACs for AGC (automatic gain control), AFC (automatic frequency control), and power amplifier ramp control. The microcontroller is an ARM7 TDMI, running at 39 MIPS. The ARM7 handles the protocol stack and the man/machine interface functions. Both processors are field-proven in digital wireless applications.

The AD20msp430 chipset is fully supported by a suite of development tools and software. The development tools allow easy customization of the DSP and/or ARM controller software to allow handset and terminal manufacturers to optimize the feature set and user interface of the end equipment. Software for all layers, including both voice and data applications, is available from Analog Devices' software partner TTPCom and is updated as new features become available. The system DMA and interrupt controllers are designed to allow easy upgrades to future generations of DSP and controller cores. The display interface can be used with either parallel or serial interface displays. System development can be shortened by the use of the debugging features in the AD20msp430. Most critical signals can be routed under software control to the universal system connector. This allows system debugging to take place in the final form factor. In addition, the architecture includes high speed logger and address trace functions in the DSP and single-wire trace/debug in the ARM controller.

Analog Devices recently announced the revolutionary Othello direct conversion radio for mobile applications. By eliminating intermediate frequency (IF) stages, this chipset will permit the mobile electronics industry to reduce the size and cost of radio sections and enable flexible, multistandard, multimode operation. The radio consists of two integrated circuits, the AD6523 Zero-IF Transceiver and the AD6524 Multiband Synthesizer.

The AD6523 contains the main functions necessary for both a direct conversion receiver and a direct VCO transmitter, known as the Virtual-IF™ transmitter. It also includes the local-oscillator generation block and a complete on-chip regulator that supplies power to all active circuitry for the radio. The AD6524 is a fractional-N synthesizer that features extremely fast lock times to enable advanced data services—such as high speed circuit-switched data (HSCSD) and general packet radio services (GPRS)—over cellular telephones.

Most digital cellular phones today include at least one "downconversion" in their signal chain. This frequency conversion shifts the desired signal from the allocated RF band for the standard (say, at 900 MHz) to some lower intermediate frequency (IF), where channel selection is performed with a narrow channel-select filter (usually a surface acoustic wave (SAW) or a ceramic type). The now filtered signal is then further downconverted to either a second IF or directly to baseband, where it is digitized and demodulated in a digital signal processor (DSP). Figure 9-16 shows the comparison between this *superheterodyne* architecture and the Superhomodyne™ architecture of the Othello radio receiver.

The idea of using direct conversion for receivers has long been of interest in RF design. The reason is obvious: in consumer equipment conversion stages add cost, bulk, and weight. Each conversion stage requires a local oscillator, (often including a frequency synthesizer to lock the LO onto a given frequency), a mixer, a filter, and (possibly) an amplifier. No wonder, then, that direct conversion receivers would be attractive. All intermediate stages are eliminated, reducing the cost, volume, and weight of the receiver.

Section Nine

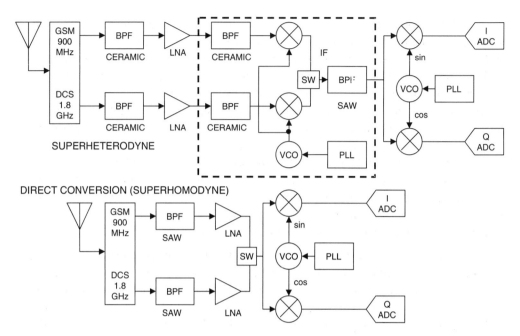

Figure 9-16: Direct Conversion Receiver Architecture Eliminates Components

The Othello radio reduces the component count even more by integrating the front-end GSM low noise amplifier (LNA). This eliminates an RF filter (the "image" filter) that is necessary to eliminate the image, or unwanted mixing product of a mixer and the off-chip LNA. This stage, normally implemented with a discrete transistor, plus biasing and matching networks, accounts for a total of about 12 components. Integrating the LNA saves a total of about 15 to 17 components, depending on the amount of matching called for by the (now eliminated) filter.

A functional block diagram of the Othello dual-band GSM radio's architecture is shown in Figure 9-17. The receive section is at the top of the figure. From the antenna connector, the desired signal enters the transmit/receive switch and exits on the appropriate path, either 925 MHz to 960 MHz for the GSM band or 1805 MHz to 1880 MHz for DCS. The signal then passes through an RF band filter (a so-called "roofing filter") that serves to pass the entire desired frequency band while attenuating all other out-of-band frequencies (blockers, including frequencies in the transmission band) to prevent them from saturating the active components in the radio front end. The roofing filter is followed by the low noise amplifier (LNA). This is the first gain element in the system, effectively reducing the contribution of all following stages to system noise. After the LNA, the direct conversion mixer translates the desired signal from radio frequency (RF) all the way to baseband by multiplying the desired signal with a local oscillator (LO) output at the same frequency.

DSP Applications

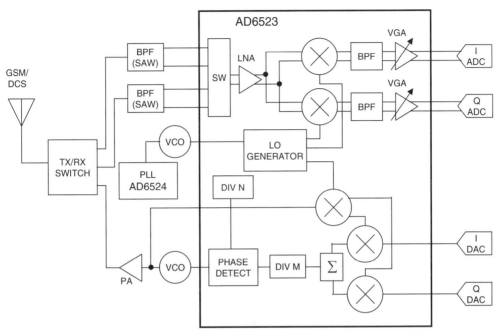

Figure 9-17: Superhomodyne Direct Conversion Dual-Band Transceiver Using the AD6523/AD6524 Chipset

The output of the mixer stage is then sent in quadrature (I and Q channels) to the variable-gain baseband amplifier stage. The VGA also provides some filtering of adjacent channels, and attenuation of in-band blockers. These blocking signals are other GSM channels that are some distance from the desired channel, say 3 MHz and beyond. The baseband amplifiers filter these signals so they will not saturate the receive ADCs. After the amplifier stage, the desired signal is digitized by the receive ADCs.

The transmit section begins on the right, at the multiplexed I and Q inputs/outputs. Because the GSM system is a time division duplex (TDD) system, the transmitter and receiver are never on at the same time. The Othello radio architecture takes advantage of this fact to save four pins on the transceiver IC's package. The quadrature transmit signals enter the transmitter through the multiplexed I/Os. These I and Q signals are then modulated onto a carrier at an intermediate frequency greater than 100 MHz.

The output of the modulator goes to a phase-frequency detector (PFD), where it is compared to a reference frequency that is generated from the external channel selecting LO. The output of the PFD is a charge pump, operating at above 100 MHz, whose output is filtered by a fairly wide (1 MHz) loop filter. The output of the loop filter drives the tuning port of a voltage-controlled oscillator (VCO), with frequency ranges that cover the GSM and DCS transmit bands.

The output of the transmit VCO is sent to two places. The main path is to the transmit power amplifier (PA), which amplifies the transmit signal from about +3 dBm to +35 dBm, sending it to the transmit/receive switch and low-pass filter (which attenuates power amplifier harmonics). The power amplifiers are dual-band, with a simple CMOS control voltage for the band switch. The VCO output also goes to the transmit feedback mixer by means of a coupler, which is either a printed circuit, built with discrete inductors and capacitors, or a monolithic (normally ceramic) coupling device. The feedback mixer downconverts the transmit signal to the transmit IF, and uses it as the local oscillator signal for the transmit modulator.

This type of modulator has several names, but the most descriptive is probably "translation loop." The translation loop modulator takes advantage of one key aspect of the GSM standard: the modulation scheme is Gaussian-filtered minimum-shift keying (GMSK). This type of modulation does not affect the envelope amplitude, which means that a power amplifier can be saturated and still not distort the GMSK signal sent through it.

GMSK can be generated in several different ways. In another European standard (for cordless telephones), GMSK is created by directly modulating a free-running VCO with the Gaussian filtered data stream. In GSM, the method of choice has been quadrature modulation. Quadrature modulation creates accurate phase GMSK, but imperfections in the modulator circuit (or upconversion stages) can produce envelope fluctuations, which can in turn degrade the phase trajectory when amplified by a saturated power amplifier. To avoid such degradations, GSM phone makers have been forced to use amplifiers with somewhat higher linearity, at the cost of reduced efficiency and talk time per battery charge cycle.

The translation loop modulator combines the advantages of directly modulating the VCO and the inherently more accurate quadrature modulation. In effect, the scheme creates a phase-locked loop (PLL), comprising the modulator, the LO signal, and the VCO output and feedback mixer. The result is a directly modulated VCO output with a perfectly constant envelope and almost perfect phase trajectory. Phase trajectory errors as low as 1.5 degrees have been measured in the AD6523 transceiver IC, using a signal generator as the LO signal to provide a reference for the loop.

Because Othello radios can be so compact, they enable GSM radio technology to be incorporated in many products from which it has been excluded, such as very compact phones or PCMCIA cards. However, the real power of direct conversion will be seen when versatile third-generation phones are designed to handle multiple standards. With direct conversion, hardware channel-selection filters will be unnecessary, because channel selection is performed in the digital signal processing section, which can be programmed to handle multiple standards. Contrast this with the superheterodyne architecture, where multiple radio circuits are required to handle the different standards (because each will require different channel-selection filters), and all the circuits will have to be crowded into a small space. With direct conversion, the same radio chain could in concept be used for several different standards, bandwidths, and modulation types. Thus, Web browsing and voice services could, in concept, occur over the GSM network using the same radio in the handset.

DSP Applications

Analog Cellular Base Stations

Consider the analog superheterodyne receiver invented in 1917 by Major Edwin H. Armstrong (see Figure 9-18). This architecture represented a significant improvement over single-stage direct conversion (homodyne) analog receivers, which had previously been constructed using tuned RF amplifiers, a single detector, and an audio gain stage. (Note that the homodyne technique has now gained favor in DSP-based receivers as explained above.) A significant advantage of the superhetrodyne analog receiver is that it is much easier and more economical to have the gain and selectivity of a receiver at fixed intermediate frequencies (IF) than to have the gain and frequency-selective circuits "tune" over a band of frequencies.

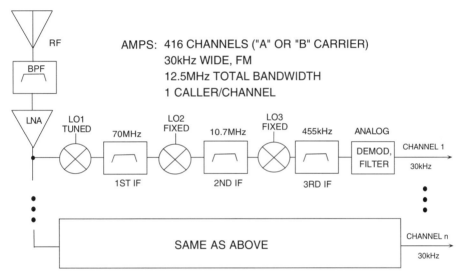

Figure 9-18: U.S. Advanced Mobile Phone Service (AMPS) Superheterodyne Analog Base Station Receiver

The frequencies shown in Figure 9-18 correspond to the AMPS (advanced mobile phone service) analog cellular phone system currently used in the U.S. The receiver is designed for AMPS signals at 900 MHz RF. The signal bandwidth for the "A" or "B" carriers serving a particular geographical area is 12.5 MHz (416 channels, each 30 kHz wide). The receiver shown uses triple conversion, with a first IF frequency of 70 MHz, a second IF of 10.7 MHz, and a third of 455 kHz. The image frequency at the receiver input is separated from the RF carrier frequency by an amount equal to twice the first IF frequency (illustrating the point that using relatively high first IF frequencies makes the design of the image rejection filter easier).

The output of the third IF stage is demodulated using analog techniques such as discriminators, envelope detectors, or synchronous detectors. In the case of AMPS the modulation is FM. An important point to notice about the above scheme is that there is *one receiver required per channel*, and only the antenna, prefilter, and LNA can be shared.

It should be noted that in order to make the receiver diagrams more manageable, the interstage amplifiers are not shown. They are, however, an important part of the receiver, and the reader should be aware that they must be present.

It should be seen by now that analog receiver design is a complicated art, and there are many trade-offs that can be made between IF frequencies, single-conversion versus double-conversion or triple-conversion, filter cost and complexity at each stage in the receiver, demodulation schemes, and so on. There are many excellent references on the subject, and the purpose of this discussion is to form an historical frame of reference for the following discussions on the application of digital techniques in the design of advanced communications and basestation receivers.

Digital Cellular Base Stations

Cellular telephone base stations form the backbone of the modern wireless cellular infrastructure. They must receive multiple calls, process the calls, and retransmit them. Hand-offs to base stations in adjacent cells must be done seamlessly with respect to the mobile customer. In addition, the base station must often handle multiple standards simultaneously. Many carriers in certain areas of the United States use multiple technologies in the same geographical area; AMPS and CDMA, for example.

Flexibility, performance, and low per-channel costs are the keys to modern base stations. Maximizing the use of DSP in the tranceiver allows multiple standards to be handled without the necessity for changing hardware. This has led to the widespread proliferation of so-called *software radios* which dominate the current base station market.

As in the case of the cell phone handset, direct conversion techniques are used in base stations. The signal is digitizied by a high performance wideband ADC after only one stage of downconversion. Figure 9-19 shows two fundamental approaches to a digital receiver: *narrow-band* and *wideband*.

By *narrow-band*, we mean that sufficient prefiltering has been done such that all undesired signals have been eliminated and that only the channel of interest is presented to the ADC input. *Wideband* simply means that a number of channels are presented to the input of the ADC and further filtering, tuning, and processing is performed digitally. Usually, a wideband receiver is designed to receive an entire band; cellular or other similar wireless service such as PCS or CDMA. In fact, one wideband digital receiver can be used to receive all channels within the band simultaneously, allowing almost all of the analog hardware (including the ADC) to be shared among all the channels.

DSP Applications

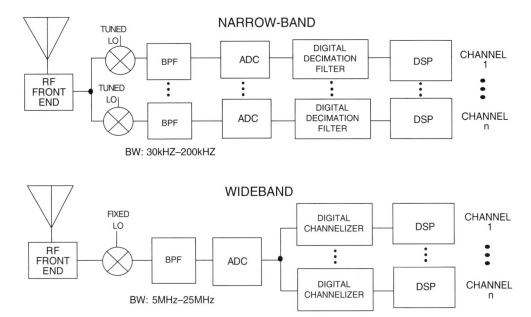

Figure 9-19: Narrow-Band and Wideband Digital Receivers for Cellular Basestations

The wideband approach places severe constraints on the ADC and requires high spurious-free dynamic range (SFDR) and signal-to-noise ratio (SNR), especially in cellular applications where signal strengths of channels can differ by more than 100 dB. This requires ADCs with bandwidths of greater than 100 MHz and sampling frequencies greater than 50 MSPS (required to handle a multicarrier bandwidth of 25 MHz, for example). On the other hand, the narrow-band approach provides more processing gain because each channel is highly oversampled, but this approach also requires more ADCs to process the same number of channels.

ADI's SoftCell chipset addresses key issues for wireless operators including the cost of coverage, flexibility and size, as well as quality of service. Base stations containing the SoftCell chipset are easy to modify in terms of adding services, additional channels, and changing wireless standards incrementally. In effect, operators will have the ability to use and move between any air interface standard (e.g., GSM, PHS, D-AMPS), deploy a higher number of channels, and offer frequency plans with greater efficiency. The new architecture also eliminates redundant channel radios for both transmitters and receivers.

The SoftCell chipset is optimized for four RF carrier channels and is easily expandable. This solution enables equipment manufacturers to create highly scalable multicarrier, multimode base stations at a fraction of the cost of traditional multichannel base stations using analog techniques. A block diagram of a system using the SoftCell chipset is shown in Figure 9-20.

Section Nine

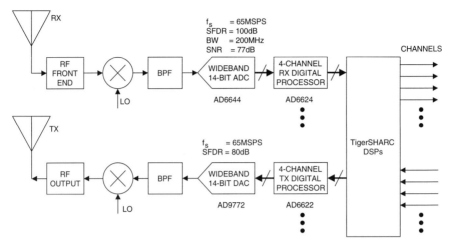

Figure 9-20: Base Station Block Diagram
Using SoftCell Multicarrier Wideband Chipset

The decrease in size and cost savings associated with SoftCell means base stations can be deployed in higher numbers and in tighter locations. The result is better coverage, better quality, and fewer busy signals for users. The deployment capability also makes the SoftCell chipset well-suited for wireless office applications. In addition, the software radio technology implemented in the chipset enables new applications such as Smart Antenna or phased-array antenna services that enable power efficiencies and cost savings, and small *picocell* installations for additional coverage capacity and in-building wireless systems.

The SoftCell chipset consists of ADI's AD6644 14-bit ADC, AD6624 quad digital receive signal processor (RSP), AD9772 TxDAC 14-bit DAC, and the AD6622 quad digital transmit signal processor (TSP). The solution leverages the benefits of digital signal processors to enable channel separation, equalization, error correction, and decoding with greater flexibility and efficiencies. This new chipset is optimized to work with ADI's TigerSHARC multiprocessor digital signal processor (DSP).

The TigerSHARC DSP is a ground-breaking processor optimized for communications applications capable of performing approximately 1 billion 16-bit multiply-accumulates per second at 150 MHz. Additionally, the TigerSHARC is unique in its ability to support 8-, 16-, and 32-bit word sizes on a single chip. Modulation/demodulation, channel encode/decoding, and other radio functions can be multiplexed to support multiple carriers in a single DSP.

SoftCell complements ADI's recently introduced AD6600 diversity receiver ADC. The AD6600 addresses narrow-band applications where a multicarrier architecture is still not feasible but allows direct IF sampling of signals at frequencies up to 250 MHz. Configured with the appropriate digital receiver signal processor, the AD6600 can address a variety of air interface standards, including GSM Macrocell.

Classic base station architectures require a complete transceiver for every RF carrier processed (from 4 to 80 channels for digital and analog systems, respectively). These radios must be duplicated for diversity and sectorized antennas as well. It is easy to see why base station electronics consume so much space, power, and cost. The advantage of a multicarrier software radio is the elimination of redundant radios in favor of a single, high performance radio per antenna, where each RF carrier is processed in the digital domain. Deployment of true software radios has been limited by the performance of analog-to-digital converters (ADC), which must digitize the enormous dynamic range demanded in a spectrum composed of multiple carriers, blockers, and adjacent channel interference.

Multicarrier transmitters have similar challenges to meet the performance demands of the newest air interface specifications. Digital-to-analog converters (DAC) and multicarrier power amplifiers (MCPAs) must preserve the spectrum of several digitally generated carriers without corruption or spurious signal generation in adjacent channels. The AD9772 is a 14-bit interpolating DAC optimized to accurately convert multiple carriers to a single IF frequency. The AD9744 is the latest in ADI's TxDAC® family of high performance converters.

The heart of ADI's SoftCell chipset is the AD6644, a 14-bit, 65 MSPS ADC that provides up to 100 dB of spurious-free dynamic range (SFDR) and 77 dB signal-to-noise ratio (SNR). This provides the receiver performance needed to implement a multicarrier digitizing radio for many applications. Shifting the channel tuning, filtering, and demodulation to the digital domain allows the flexibility to support different air interface standards, number of channels, and frequency plans with a single radio design.

Following the ADC, a digital receive signal processor (RSP) performs the channel tuning, filtering, and decimation required to provide baseband I and Q signals to a digital signal processor (DSP). The AD6624 is an 65 MSPS quad digital RSP developed to support GSM, IS136, and other narrow-band standards. The AD6624's four channels are independently programmable to change air interface characteristics on demand. It is a simple matter to add AD6624s in parallel for additional channel capacity. The AD6624 can also be configured to support EDGE extensions of GSM and IS136.

The AD6622 is a 4-channel digital transmit signal processor that takes baseband I and Q inputs from a DSP and provides all the signal processing functions required to drive the AD9772 DAC. Each independent channel can be programmed to provide the desired channel filtering for most air interface standards. The AD6622 supports IS95 and WCDMA standards and can be daisy-chained to combine an arbitrary number of channels onto a single 18-bit digital output.

Motor Control

Long known for its simplicity of construction, low cost, high efficiency, and long-term dependability, the ac induction motor has been limited by the inability to control its dynamic performance in all but the crudest fashion. This has severely restricted the application of ac induction motors where dynamic control of speed, torque, and response to changing load is required. However, recent advances in digital signal processing (DSP) and mixed-signal integrated circuit technology are providing the ac induction motor with performance never before thought possible. Manufacturers anxious to harness the power and economy of vector control can reduce R&D costs and time to market for applications ranging from industrial drives to electric automobiles and locomotives by utilizing a standard chipset/development system.

It is unlikely that Nikola Tesla (1856–1943), the inventor of the induction motor, could have envisaged that this workhorse of industry could be rejuvenated into a new class of motor that is competitive in most industrial applications.

Before discussing the advantages of vector control, it is necessary to have a basic understanding of the fundamental operation of the different types of electric motors in common use.

Until recently, motor applications requiring servo-control tasks such as tuned response to dynamic loads, constant torque, and speed control over a wide range were almost exclusively the domain of dc brush and dc permanent magnet synchronous motors. The fundamental reason for this preference was the availability of well understood and proven control schemes. Although easily controlled, dc brush motors suffer from several disadvantages; brushes wear and must be replaced at regular intervals, commutators wear and can be permanently damaged by inadequate brush maintenance, brush/commutator assemblies are a source of particulate contaminants, and the arcing of mechanical commutation can be a serious fire hazard is some environments.

The availability of power inverters capable of controlling high horsepower motors allowed practical implementation of alternate motor architectures such as the dc permanent magnet synchronous motor (PMSM) in servo-control applications. Although eliminating many of the mechanical problems associated with dc brush motors, these motors required more complex control schemes and suffered from several drawbacks of their own. Aside from being costly, dc PMSMs in larger, high horsepower configurations suffer from high rotor moment-of-inertia as well as limited use in high speed applications due to mechanical constraints of rotor construction and the need to implement field weakening to exceed baseplate speed.

In the 1960s, advances in control theory, in particular the development of *indirect field-oriented control*, provided the theoretical basis for dynamic control of ac induction motors. Indirect field-oriented control makes use of reference frame theory. Using these techniques, it is possible to transform the phase variable machine description of a motor to another reference frame. By judicious choice of the reference frame, it is possible to simplify considerably the complexity of the mathematical machine model. While these techniques were initially developed for the analysis and simulation of ac motors, they are now invaluable tools in the digital control of such machines. As digital control techniques are extended to the control of the currents, torque, and flux of such machines, the need for compact, accurate motor models is obvious.

Fortunately, the theory of reference frames is equally applicable to the synchronous machines, such as the permanent magnet synchronous machine (PMSM). This motor is sometimes known as the sinusoidal brushless motor or the brushless ac machine, and is very popular as a high performance servo drive.

Because of the intensive mathematical computations required by indirect field-oriented control, now commonly referred to as *vector control* or *reference frame theory*, practical implementation was not possible for many years. Available hardware could not perform the high speed precision sensing of rotor position and near real-time computation of dynamic flux vectors. The current availability of precision optical encoders, isolated gate bipolar transistors (IGBTs), high speed resolver-to-digital converters, and high speed digital signal processors (DSPs) has pushed vector control to the forefront of motor development due to the advantages inherent in the ac induction motor.

A simplified block diagram of an ac induction motor control system is shown in Figure 9-21. The inputs to the controller are the motor currents (normally three-phase) and the motor rotor position and velocity. Hall effect sensors are often used to monitor the currents, and a resolver and a resolver-to-digital converter (RDC) monitor the rotor position and velocity. The DSP is used to perform the real-time vector-type calculations necessary to generate the control outputs to the inverter processors. The transformations required for reference frame transformations and vector control are also accomplished with the DSP.

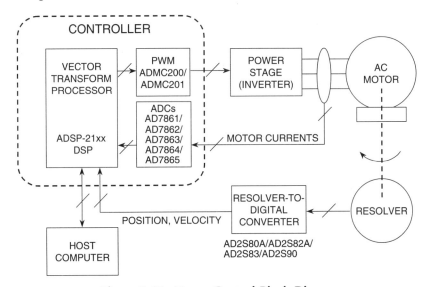

Figure 9-21: Motor Control Block Diagram

Section Nine

The functions in the controller block have been integrated into the Analog Devices ADMC300, ADMC331, ADMC401, and the ROM-based ADMC326 and ADMC328 DSP motor controller chips. These devices include the peripheral circuitry such as ADCs, voltage references, PWM controllers, and timers required to perform all the functions shown in Figure 9-21.

The latest members of the family are the ADMCF326 and ADMCF328 motor controllers, referred to as DashDSP™, meaning *digital* plus *analog* plus *flash* memory (see Figure 9-22). The use of flash memory allows the devices to be field programmable, thereby providing greater flexibility and shortening development time. These controllers include a 20 MIPS 16-bit fixed-point core based on the ADSP-217x family architecture. The memory consists of a 512 × 24-bit program memory RAM, a 512 × 16-bit data memory RAM, 4K × 24-bit program memory ROM, and a 4K × 24-bit programmable flash memory. A completely integrated ADC analog subsystem allows the three-phase motor currents to be monitored. A 16-bit, 3-phase PWM generates the control signals to the external inverter power stage. The parts are packaged in a 28-lead SOIC or PDIP package. A block diagram of the ADMCF328 is shown in Figure 9-23.

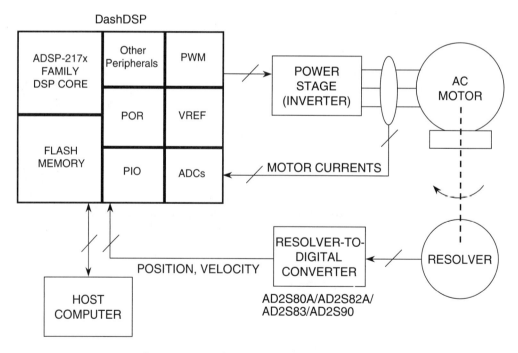

Figure 9-22: Fully Integrated Motor Control with DashDSP

DSP Applications

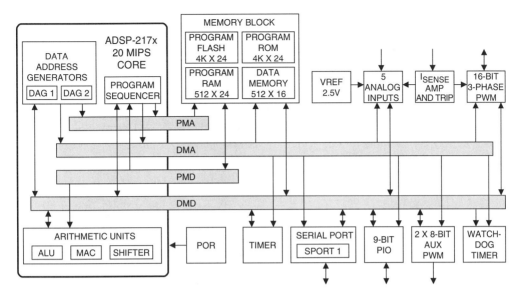

Figure 9-23: ADMCF328 DSP Motor Controller with Flash Memory

Third-party DSP software, reference designs, and evaluation systems are available to facilitate motor control system development using these chips.

To meet the future needs of embedded control applications, and to target a variety of new and emerging applications, ADI has announced a new family of mixed-signal DSPs, called the ADSP-2199x. The first three members of this family that have been announced are the ADSP-21990, ADSP-21991, and the ADSP-21992. These products combine an ADSP-219x DSP core, multichannel, high resolution ADC, and a selection of embedded control peripherals. As such, these products can be viewed as the merging of various technologies and competencies from within ADI—state of the art DSP cores (ADSP-219x core) with highest performance ADCs and embedded control intellectual property contained in our peripheral designs (leveraged and enhanced from the ADMC family of devices). These products form a new class of mixed-signal DSPs to address the needs of new and emerging markets as well as address the needs of the more established embedded control space.

In particular, the ADSP-21990 offers 160 MIPS, a 16-bit, 219x DSP core, 4 Kwords of on-chip program memory, 4 Kwords of on-chip data memory, and an external addressable memory space of 1 Mword. The mixed-signal integration consists of an 8-channel, 14-bit, 20 MSPS ADC core (with dual sample-and-hold amplifiers for simultaneous sampling needs), precision on-chip voltage reference, and an integrated power-on-reset (POR) circuit. The embedded control peripherals enhance the features of the released ADMC devices and include a three-phase PWM generator, a 32-bit incremental encoder, dual auxiliary PWM outputs, and a watchdog timer. In addition, a variety of general-purpose peripherals, leveraged from the ADSP-219x family of general-purpose DSP products, are also provided. These include three, 32-bit, general-purpose timers, a 16-bit general-purpose I/O port, a memory DMA controller, and a flexible, peripheral interrupt controller. Two high speed communications ports are included—a synchronous serial port (SPORT) and a standard SPI port.

The ADSP-21990 is offered in a 196-ball mini BGA package or a 176-lead LQFP package. The device is developed in 0.25 μm CMOS technology, with 2.5 V core voltage and 3.3 V I/O.

The ADSP-21991 device extends the available on-chip memory to 32 Kwords of program memory and 8 Kwords of data memory. The ADSP-21992 further extends the on-chip data memory to 16 Kwords and integrates a CAN (controller area network) interface. All products are available in pin-for-pin compatible versions.

Codecs and DSPs in Voice-Band and Audio Applications

In voice-band and audio applications such as modems, analog front ends (AFEs), also called codecs, make excellent system building blocks along with DSPs.

PC applications such as voice processing and modems require high performance codecs. Figure 9-24 shows a generalized audio/modem application based on the AD1819B SoundPort® codec.

This codec is fully compliant with the AC'97 specification (Audio Codec '97, Component Specification, Revision 1.03, © 1996, Intel Corporation). In addition, the AD1819 supports multiple codec configurations (up to three per AC Link), a DSP serial mode, variable sample rates, modem sample rates and filtering, and built-in Phat™ Stereo 3D enhancement.

DSP Applications

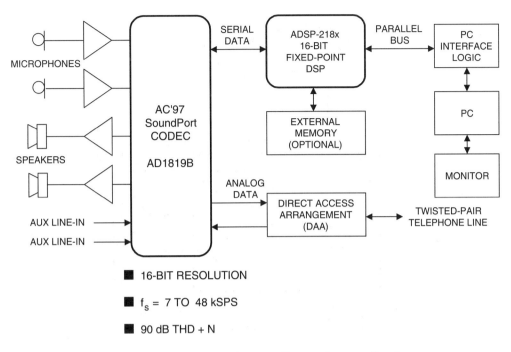

Figure 9-24: Generalized PC Audio, Modem Application

The AD1819B is an analog front end for high performance PC audio, modem, or DSP applications. The main architectural features of the AD1819B are the high quality analog mixer section, two channels of 16-bit sigma-delta A/D conversion, two channels of 16-bit sigma-delta D/A conversion, and a serial port. THD + N is 90 dB, and the sampling frequency is variable from 7 kSPS to 48 kSPS.

Analog Devices' 32-bit floating-point SHARC DSP has demonstrated the highest Dolby Digital AC-3 audio quality. The DSP architecture of the reference design shown in Figure 9-25 uses the ADSP-21065L SHARC and an integrated mixed-signal IC (AD1836) to provide low cost, high quality, multichannel audio. Key applications include A/V receivers for home theater and high end automotive audio. The AD1836 provides all the mixed-signal functions including four ADC input channels and six DAC output channels. The AD1836 codec provides 97 dB THD + N and 105 dB SNR for high quality audio. While a fixed-function DSP could be used, a programmable DSP provides greater flexibility. This reference design can be programmed for MP3, Dolby Digital AC-3, THX, or DTS decoding. Other audio processing algorithms can easily be added through additional programs.

Section Nine

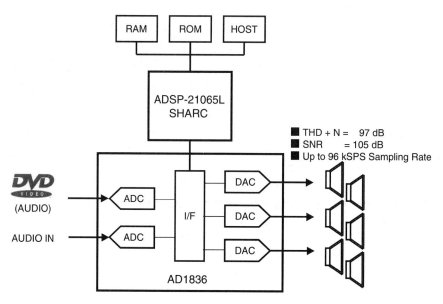

Figure 9-25: Automotive and Home Theater
Reference Design Using 32-Bit SHARC

In large digital audio systems it is often necessary to distribute the signal processing tasks between multiple processors. Figure 9-26 shows a 16-channel mixer that uses two ADSP-21160s. The data comes from sixteen 24-bit ADCs and is passed to the FPGA. This FPGA converts the serial ADC data into parallel data and sends it to the two ADSP-21160's external ports. The external port on each DSP has hardware support to enable the FPGA to write to both DSPs at once. DMA engines inside the DSP receive this data and move it to desired locations in internal memory. The hardware support and DMA engines reduce the complexity of the FPGA design, because the FPGA is simply required to drive the data onto the bus. No arbitration logic or address generation is necessary in the FPGA.

The DSPs perform various algorithms, such as mixing, left/right panning, equalization, and additional effects processing, such as reverberation or compression/expansion. The audio output samples are then transmitted to a 24-bit stereo DAC. These algorithms could be mapped in such a way that one of the DSPs does the mixing and effects, while the other does the equalization. Another possibility is to give each DSP the task of processing half of the data channels. The optimal algorithmic mapping depends on what type of processing is required.

For this example, it is estimated that the two ADSP-21160s have enough computation power to perform various algorithms on 16 channels of data sampled at 48 kSPS. In a 20 µs period each DSP is capable of executing 2000 instructions in the core. If each DSP is responsible for half of the channels of data (8 channels), that means that the DSP can allocate 250 instructions to each channel.

DSP Applications

With the ADSP-21160 SIMD architecture, there are roughly enough instructions for a 3-band (high, mid, and low) equalizer, mixing, delay effects, and compression for each channel. The movement of data into the memory does not enter into the processing loading calculation since it is a zero-overhead task.

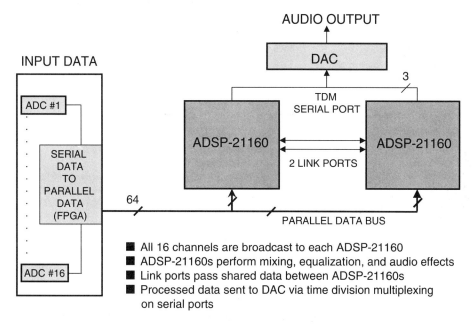

Figure 9-26: 16-Channel Audio Mixer Using ADSP-21160 SHARC Processors

A Sigma-Delta ADC with Programmable Digital Filter

Most sigma-delta ADCs have a fixed internal digital filter. The filter's cut-off frequency (and the ADC output data rate) scales with the master clock frequency. The AD7725 is a 16-bit sigma-delta ADC with a programmable internal digital filter. The block diagram in Figure 9-27 shows that the modulator operates at a maximum oversampling rate of 19.2 MSPS. The modulator is followed by a preset FIR filter that decimates the modulator output by a factor of eight, yielding an output data rate of 2.4 MSPS. The response of the preset FIR filter is shown in Figure 9-27. The output of the preset filter drives a programmable digital filter. The diagram shows typical response for a 300 kHz low-pass FIR filter.

Section Nine

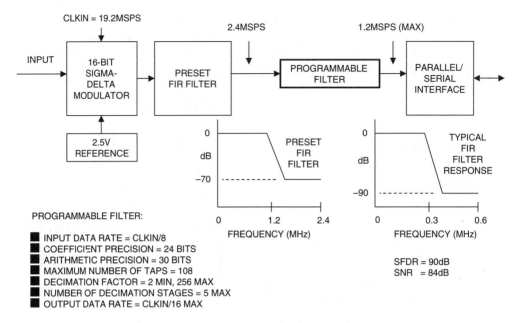

Figure 9-27: AD7725 16-Bit Sigma-Delta ADC
with Programmable Digital Filter

The programmable filter is flexible with respect to number of taps and decimation rate. The filter can have up to 108 taps, up to five decimation stages, and a decimation factor between two and 256. Coefficient precision is 24 bits, and arithmetic precision is 30 bits.

The AD7725 contains Systolix's PulseDSP post processor which permits the filter characteristics to be programmed through the parallel or serial microprocessor interface.

The postprocessor is a fully programmable core that provides processing power of up to 130 million multiply-accumulates (MAC) per second. To program the postprocessor, the user must produce a configuration file that contains the programming data for the filter function. This file is generated by the FilterWizard compiler, which is available from Analog Devices. The AD7725 compiler accepts filter coefficient data as an input and automatically generates the required device programming data.

The coefficient file for the filter response can be generated using a digital filter design package such as Systolix FilterExpress™ (http://www.systolix.co.uk) or QEDesign™ from Momentum Data Systems (http://www.mds.com). The response of the filter can be plotted so the user knows the response before generating the filter coefficients. The data is available to the processor at a 2.4 MSPS rate. When decimation is employed in a multistage filter, the first filter will be operated at 2.4 MSPS, and the user can then decimate between stages. The number of taps that can be contained in the processor is 108. Therefore, a single filter with 108 taps can be generated, or a multistage filter can be designed whereby the total number of taps adds up to 108. The filter characteristic can be low-pass, high-pass, band-stop, or band-pass and can be either FIR or IIR.

The AD7725 operates on a single 5 V supply, has an on-chip 2.5 V reference, and is packaged in a 44-lead PQFP. Power dissipation is approximately 350 mW when operating at full power. A half-power mode is available with a master clock frequency of 10 MSPS maximum. Power consumption in the standby mode is 200 mW maximum.

Summary

More examples of DSP applications are summarized in Figure 9-28. There are many other practical applications of DSP in today's rapidly expanding industrial, communication, medical, military, and consumer markets. A discussion of each would constitute a book in itself. It has been the purpose of this section to highlight just a few of the more common applications and give the reader a sense of how DSP affects practically every aspect of modern life.

- Hands-Free Car Telephone Kits
- Digital Answering Machines
- Voice Recognition Systems
- Cable Modems
- Computer Sound Cards
- Digital Audio: Professional and Consumer
- Digital Video Signal Processing
- High Definition Television (HDTV)
- Computer-Generated Graphics
- Digital Special Effects
- Direct Broadcast Satellite (DBS)
- Global Positioning Systems (GPS)
- Medical: Ultrasound, MRI, CT Scanners
- Military: Radar, Missile Guidance

Figure 9-28: Other DSP Applications

References

1. John Bingham, **The Theory and Practice of Modem Design**, John-Wiley, 1988.

2. **Digital Signal Processing Applications Using the ADSP-2100 Family, Vol. 1 and Vol. 2**, Analog Devices, Free download at: http://www.analog.com.

3. George Calhoun, **Digital Cellular Radio**, Artech House, Norwood, MA, 1988.

4. Nick Morley, *New Cellular Scheme Muscles In*, **EDN News**, November 15, 1990, p. 1.

5. P. Vary, K. Hellwig, R. Hofmann, R. J. Sluyter, C. Galand, M. Rosso, *Speech Codec for the European Mobile Radio System*, **ICASSP 1988 Proceedings**, p. 227.

6. D. K. Freeman, G. Cosier, C. B. Southcott, I. Boyd, *The Voice Activity Detector for the Pan-European Digital Cellular System*, **ICASSP 1989 Proceedings**, p. 369.

7. Brad Brannon, *Using Wide Dynamic Range Converters for Wide Band Radios*, **RF Design**, May 1995, pp.50–65.

8. Joe Mitola, *The Software Radio Architecture*, **IEEE Communications Magazine**, Vol. 33, No. 5, May 1995, pp. 26–38.

9. Jeffery Wepman, *Analog-to-Digital Converters and Their Applications in Radio Receivers*, **IEEE Communications Magazine**, Vol. 33, No. 5, May 1995, pp. 39–45.

10. Rupert Baines, *The DSP Bottleneck*, **IEEE Communications Magazine**, Vol. 33, No. 5, May 1995, pp. 46–54.

11. Brad Brannon, *Overcoming Converter Nonlinearities with Dither*, **Application Note AN-410**, Analog Devices, 1995.

12. P.J.M. Coussens, et al, *Three Phase Measurements with Vector Rotation Blocks in Mains and Motion Control*, **PCIM Conference, Intelligent Motion, April 1992 Proceedings**, available from Analog Devices.

13. Dennis Fu, *Digital to Synchro and Resolver Conversion with the AC Vector Processor AD2S100*, available from Analog Devices.

14. Dennis Fu, *Circuit Applications of the AD2S90 Resolver-to-Digital Converter*, AN-230, Analog Devices.

15. Aengus Murray and P. Kettle, *Towards a Single Chip DSP Based Motor Control Solution*, **Proceedings PCIM - Intelligent Motion**, May 1996, Nurnberg Germany, pp. 315–326. Also available at http://www.analog.com.

16. D. J. Lucey, P. J. Roche, M. B. Harrington, and J. R. Scannell, *Comparison of Various Space Vector Modulation Strategies*, **Proceedings Irish DSP and Control Colloquium**, July 1994, Dublin, Ireland, pp. 169–175.

17. Niall Lyne, *ADCs Lend Flexibility to Vector Motor Control Applications*, **Electronic Design**, May 1, 1998, pp. 93–100.

18. **Motor Control Products, Application Notes, and Tools**, CDROM, Analog Devices, Spring 2000.

19. **Application of Digital Signal Processing in Motion Control Seminar**, Analog Devices, Spring 2000.

20. Analog Devices' Motor Control website: http://www.analog.com/motorcontrol.

Section 10
Hardware Design Techniques

- Low Voltage Interfaces
- Grounding in Mixed-Signal Systems
- Digital Isolation Techniques
- Power Supply Noise Reduction and Filtering
- Dealing with High Speed Logic

Section 10
Hardware Design Techniques
Walt Kester

Low Voltage Interfaces
Ethan Bordeaux, Johannes Horvath, Walt Kester

For the past 30 years, the standard V_{DD} for digital circuits has been 5 V. This voltage level was used because bipolar transistor technology required 5 V to allow headroom for proper operation. However, in the late 1980s, complementary metal oxide semiconductor (CMOS) became the standard for digital IC design. This process did not necessarily require the same voltage levels as TTL circuits, but the industry adopted the 5 V TTL standard logic threshold levels to maintain backward compatibility with older systems (Reference 1).

The current revolution in supply voltage reduction has been driven by demand for faster and smaller products at lower costs. This push has caused silicon geometries to drop from 2 µm in the early 1980s to 0.25 µm, which is used in today's latest microprocessor and IC designs. As feature sizes have become increasingly smaller, the voltage for optimum device performance has also dropped below the 5 V level. This is illustrated in the current microprocessors for PCs, where the optimum core operating voltage is programmed externally using voltage identification (VID) pins, and can be as low as 1.3 V.

The strong interest in lower voltage DSPs is clearly visible in the shifting sales percentages for 5 V and 3.3 V parts. Sales growth for 3.3 V DSPs has increased at more than twice the rate of the rest of the DSP market (30% for all DSPs versus more than 70% for 3.3 V devices). This trend will continue as the high volume/high growth portable markets demand signal processors that contain all of the traits of the lower voltage DSPs.

On the one hand, the lower voltage ICs operate at lower power, allow smaller chip areas, and higher speeds. On the other hand, the lower voltage ICs must often interface to other ICs that operate at larger V_{DD} supply voltages, thereby causing interface compatibility problems. Although lower operating voltages mean smaller signal swings, and hence less switching noise, noise margins are lower for low supply voltage ICs.

The popularity of 2.5 V devices can be partially explained by their ability to operate from two AA alkaline cells. Figure 10-2 shows the typical discharge characteristics for a AA cell under various load conditions (Reference 2). Note that at a load current of 15 mA, the voltage remains above 1.25 V (2.5 V for two cells in series) for nearly 100 hours. Therefore, an IC that can operate effectively at low currents with a supply voltage of 2.5 V ±10% (2.25 V–2.75 V) is very useful in portable designs. Also, DSPs that have low mA/MIPS ratings and can integrate peripherals onto a single chip, such as the ADSP-218x L or M series, are useful in portable applications.

Section Ten

- Lower Power for Portable Applications
- 2.5 V ICs Can Operate on Two AA Alkaline Cells
- Faster CMOS Processes, Smaller Geometries, Lower Breakdown Voltages
- Multiple Voltages in System: 5 V, 3.3 V, 2.5 V, 1.8 V DSP Core Voltage (VID), Analog Supply Voltage
- Interfaces Required Between Multiple Logic Types
- Lower Voltage Swings Produce Less Switching Noise
- Lower Noise Margins
- Less Headroom in Analog Circuits Decreases Signal Swings and Increases Sensitivity to Noise

Figure 10-1: Low Voltage Mixed-Signal ICs

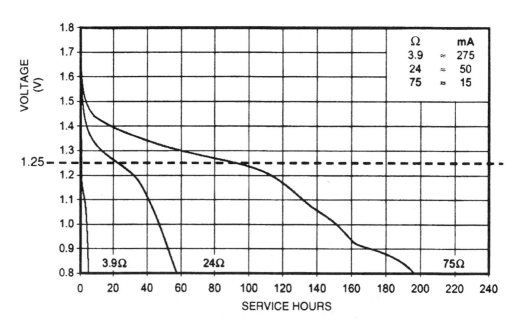

Courtesy: Duracell, Inc., Berkshire Corporate Park, Bethel, CT 06801
http://www.duracell.com

Figure 10-2: Duracell MN1500 AA Alkaline Battery Discharge Characteristics

Hardware Design Techniques

In order to understand the compatibility issues relating to interfacing ICs operated at different V_{DD} supplies, it is useful to first look at the structure of a typical CMOS logic stage as shown in Figure 10-3.

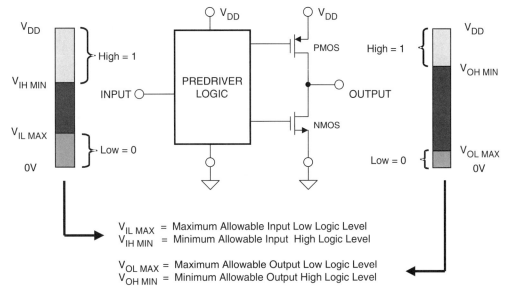

Figure 10-3: Typical CMOS IC Output Driver Configuration

Note that the output driver stage consists of a PMOS and an NMOS transistor. When the output is high, the PMOS transistor connects the output to the +V_{DD} supply through its low on resistance (R_{ON}), and the NMOS transistor is off. When the output is low, the NMOS transistor connects the output to ground through its on resistance, and the PMOS transistor is off. The R_{ON} of a CMOS output stage can vary between 5 Ω and 50 Ω depending on the size of the transistors, which in turn determines the output current drive capability.

A typical logic IC has its power supplies and grounds separated between the output drivers and the rest of the circuitry (including the predriver). This is done to maintain a clean power supply, which reduces the effect of noise and ground bounce on the I/O levels. This is increasingly important, since added tolerance and compliance are critical in I/O driver specifications, especially at low voltages.

Figure 10-3 also shows "bars" that define the minimum and maximum required input and output voltages to produce a valid high or low logic level. Note that for CMOS logic, the actual output logic levels are determined by the drive current and the R_{ON} of the transistors. For light loads, the output logic levels are very close to 0 V and +V_{DD}. The input logic thresholds, on the other hand, are determined by the input circuit of the IC.

There are three sections in the "input" bar. The bottom section shows the input range that is interpreted as a logic low. In the case of 5 V TTL, this range would be between

Section Ten

0 V and 0.8 V. The middle section shows the input voltage range where it is interpreted as neither a logic low nor a logic high. The upper section shows where an input is interpreted as a logic high. In the case of 5 V TTL, this would be between 2 V and 5 V.

Similarly, there are three sections in the "output" bar. The bottom range shows the allowable voltage for a logic low output. In the case of 5 V TTL, the IC must output a voltage between 0 V and 0.4 V. The middle section shows the voltage range that is not a valid high or low. The device should never transmit a voltage level in this region except when transitioning from one level to the other. The upper section shows the allowable voltage range for a logic high output signal. For 5 V TTL, this voltage is between 2.4 V and 5 V. The chart does not reflect a 10% overshoot/undershoot also allowed on the inputs of the logic standard.

A summary of the existing logic standards using these definitions is shown in Figure 10-4. Note that the input thresholds of classic CMOS logic (series 4000, for example) are defined as 0.3 V_{DD} and 0.7 V_{DD}. However, most CMOS logic circuits in use today are compatible with TTL and LVTTL levels, which are the dominant 5 V and 3.3 V operating standards for DSPs. Note that 5 V TTL and 3.3 V LVTTL input and output threshold voltages are identical. The difference is the upper range for the allowable high levels.

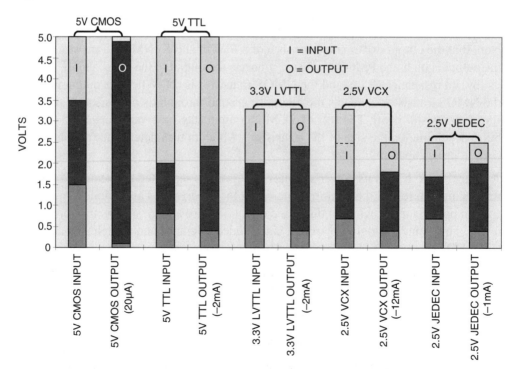

Figure 10-4: Low Voltage Logic Level Standards

Hardware Design Techniques

The international standards bureau JEDEC (Joint Electron Device Engineering Council) has created a 2.5 V standard (JEDEC standard 8-5) that will most likely become the minimum requirement for 2.5 V operation (Reference 3). However, there is no current (2000) dominant 2.5 V standard for IC transmission and reception, because few manufacturers are making products that operate at this voltage. There is one proposed 2.5 V standard created by a consortium of IC manufacturers, titled the Low Voltage Logic Alliance. Their specification provides a guideline for semiconductor operation between 1.8 V and 3.6 V. A standard covering this voltage range is useful because it ensures present and future compatibility. As an example, the 74VCX164245, a bus translator/transceiver from Fairchild Semiconductor, is designed to be operated anywhere between 1.8 V–3.6 V and has different input and output characteristics depending upon the supplied V_{DD}. This standard, named VCX, was formed by Motorola, Toshiba, and Fairchild Semiconductor. It currently consists primarily of bus transceivers, translators, FIFOs, and other building block logic. There is also a wide range of other low voltage standards, such as GTL (Gunning Transceiver Logic), BTL (Backplane Transceiver Logic), and PECL (PseudoECL Logic). However, most of these standards are aimed at application-specific markets and not for general-purpose semiconductor systems.

The VCX devices can be operated on a very wide range of voltage levels (1.8 V–3.6 V). The I/O characteristics of this standard are dependent upon the V_{DD} voltage and the load on each pin. In Figure 10-4, one voltage (2.5 V) was chosen to show the general I/O behavior of a VCX device. Each of the device's output voltages is listed for a specific current. As the current requirements increase, the output high voltage decreases while the output low voltage increases. Please refer to the appropriate data sheets for more specific I/O information.

From this chart, it is possible to visualize some of the possible problems in connecting two ICs operating on different standards. One example would be connecting a 5 V CMOS device to a 3.3 V LVTTL IC. The 5 V CMOS high level is too high for the LVTTL to handle (> 3.3 V). This could cause permanent damage to the LVTTL chip. Another possible problem would be a system with a 2.5 V JEDEC IC driving a 5 V CMOS device. The logic high level from the 2.5 V device is not high enough for it to register as a logic high on the 5 V CMOS input ($V_{IH\ MIN}$ = 3.5 V). These examples illustrate two possible types of logic level incompatibilities—either a device being driven with too high a voltage or a device not driving a voltage high enough for it to register a valid high logic level with the receiving IC. These interfacing problems introduce two important concepts: *voltage tolerance* and *voltage compliance*.

Voltage Tolerance and Voltage Compliance

A device that is *voltage tolerant* can withstand a voltage greater than its V_{DD} on its I/O pins. For example, if a device has a V_{DD} of 2.5 V and can accept inputs equal to 3.3 V and can withstand 3.3 V on its outputs, the 2.5 V device is called 3.3 V tolerant. The meaning of *input* voltage tolerance is fairly obvious, but the meaning of *output* voltage tolerance requires some explanation. The output of a 2.5 V CMOS driver in the high state appears like a small resistor (R_{ON} of the PMOS FET) connected to 2.5 V. Obviously, connecting its output directly to 3.3 V is likely to destroy the device due to excessive current. However, if the 2.5 V device has a three-state output connected to a bus that is also driven by a 3.3 V IC, the meaning becomes clearer. Even though the 2.5 V IC is in the off (third-state) condition, the 3.3 V IC can drive the bus voltage higher than 2.5 V, potentially causing damage to the 2.5 V IC output.

A device that is *voltage compliant* can receive signals from and transmit signals to a device that is operated at a voltage greater than its own V_{DD}. For example, if a device has a 2.5 V V_{DD} and can transmit and receive signals to and from a 3.3 V device, the 2.5 V device is said to be 3.3 V compliant.

The interface between the 5 V CMOS and 3.3 V LVTTL parts illustrates a lack of voltage tolerance; the LVTTL IC input is overdriven by the 5 V CMOS device output. The interface between the 2.5 V JEDEC and the 5 V CMOS part demonstrates a lack of voltage compliance; the output high level of the JEDEC IC does not comply to the input level requirement of the 5 V CMOS device.

■ Voltage Tolerance:
   ◆ A device that is *Voltage Tolerant* can *withstand* a voltage greater than its V_{DD} on its input and output pins. If a device has a V_{DD} of 2.5 V and can accept inputs of 3.3 V (±10%), the 2.5 V device is 3.3 V tolerant on its input. Input and output tolerance should be examined and specified separately.

■ Voltage Compliance:
   ◆ A device that is *Voltage Compliant* can *transmit and receive* signals to and from logic which is operated at a voltage greater than its own V_{DD}. If a device has a 2.5 V V_{DD} and can properly transmit signals to and from 3.3 V logic, the 2.5 V device is 3.3 V compliant. Input and output compliance should be examined and specified separately.

Figure 10-5: Logic Voltage Tolerance and Compatibility Definitions

Interfacing 5 V to 3.3 V Systems Using NMOS FET Bus Switches

When combining ICs that operate on different voltage standards, one is often forced to add additional discrete elements to ensure voltage tolerance and compliance. In order to achieve voltage tolerance between 5 V and 3.3 V logic, for instance, a bus switch voltage translator, or QuickSwitch™, can be used (References 4, 5). The bus switch limits the voltage applied to an IC. This is done to avoid applying a larger input high voltage than the receiving device can tolerate.

As an example, it is possible to place a bus switch between a 5 V CMOS and 3.3 V LVTTL IC, and the two devices can then properly transmit data as shown in Figure 10-6. The bus switch is basically an NMOS FET. If 4.3 V is placed on the gate of the FET, the maximum passable signal is 3.3 V (approximately 1 V less than the gate

Hardware Design Techniques

voltage). If both input and output are below 3.3 V, the NMOS FET acts as a low resistance ($R_{ON} \approx 5\ \Omega$). As the input approaches 3.3 V, the FET on resistance increases, thereby limiting the signal output. The QuickSwitch contains 10 bidirectional FETs with a gate drive enable as shown in Figure 10-6. The V_{CC} of the QuickSwitch sets the high level for the gate drive.

One way of creating a 4.3 V supply on a 5 V/3.3 V system board is to place a diode between the 5 V supply and V_{CC} on the QuickSwitch. In Figure 10-6, the 4.3 V is generated by a silicon diode in series with a Schottky diode connected to the 3.3 V supply. With 10% tolerances on both 5 V and 3.3 V supplies, this method produces a more stable gate bias voltage. Some bus switches are designed to operate on either 3.3 V or 5 V directly and generate the internal gate bias level internally.

A QuickSwitch removes voltage tolerance concerns in this mixed-logic design. One convenient feature of bus switches is that they are bidirectional; this allows the designer to place a bus translator between two ICs and not have to create additional routing logic for input and output signals.

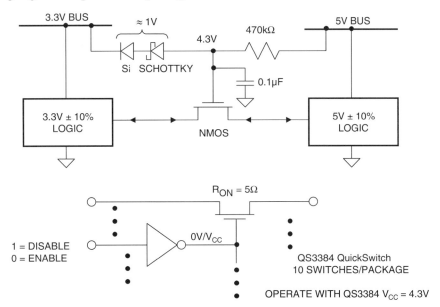

Figure 10-6: 3.3 V/5 V Bidirectional Interface Using NMOS FET Achieves Voltage Tolerance

A bus switch increases the total power dissipation along with the total area required to lay out a system. Since voltage bus switches are typically CMOS circuits, they have very low power dissipation ratings. An average value for added continuous power dissipation is 5 mW per package (10 switches), and this is independent of the frequency of signals that pass through the circuit. Bus switches typically have 8–20 I/O pins per package and take up approximately 25 mm² to 50 mm² of board space.

Section Ten

One concern when adding interface logic into a circuit is a possible increase in propagation delay. Added propagation delay can create many timing problems in a design. QuickSwitches have very low propagation delay values (< 0.25 ns) as shown in Figure 10-7.

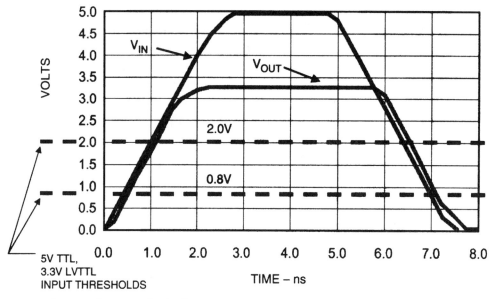

Courtesy: Integrated Device Technology (IDT), Inc., 2975 Stender Way, Santa Clara, CA 95054
http://www.idt.com

Figure 10-7: QS3384 QuickSwitch Transient Response with 4.3 V Supply

Internally Created Voltage Tolerance/Compliance

The requirement for low power, high performance ICs has triggered a race among manufacturers to design devices operating at and below 2.5 V that are also TTL/CMOS compatible. Figure 10-8 is a block diagram of a logic circuit that allows the logic core to operate at a reduced voltage, while the output driver operates at a standard supply voltage level of, for example, 3.3 V.

The technique followed by many IC manufacturers is to provide a secondary I/O ring, e.g., the I/O drivers in a 2.5 V IC are driven by a 3.3 V power supply; hence the device can be TTL compatible and meet the specification for V_{OH} and V_{OL}. The 3.3 V external power supply is *required* for the part to be 3.3 V tolerant. This causes the added complexity of two power supplies for the chip which have to be maintained in all future plug-in generations of the IC.

Hardware Design Techniques

A more flexible technique (used in the ADSP-218xM series DSPs) is to provide a separate I/O ring with an external voltage with the option of setting that voltage equal to the core's operating voltage, if desired. This design can provide tolerance to 3.3 V with the external voltage set to 2.5 V, or 3.3 V tolerance and compliance to 3.3 V with the external voltage set to 3.3 V. There are vendors today that use this option partially, i.e., the VCX devices are 3.3 V tolerant at 2.5 V internal and external voltages, but do not have the option of 3.3 V compliance. Other existing designs and patents that address this issue do not support complete tolerance and compliance and the low standby current specification. This approach is complicated, since the circuits must meet the noise and power requirements with the external voltages at 3.3 V or 2.5 V.

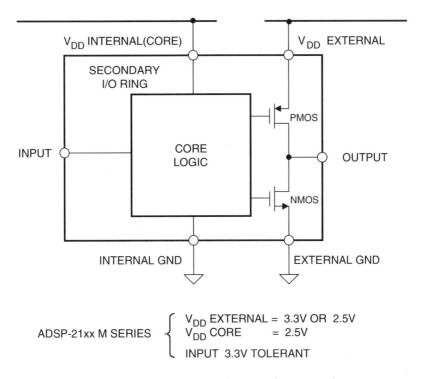

Figure 10-8: CMOS IC with Secondary I/O Ring

There are several issues to consider in a dual-supply logic IC design:
- *Power-Up Sequencing*: If two power supplies are required to give an IC additional tolerance/compliance, what is the power-up sequence? Is it a requirement that the power supplies are switched on simultaneously or can the device only have a voltage supplied on the core or only on the I/O ring?

Section Ten

- *Process Support and Electrostatic Discharge (ESD) Protection*: The transistors created in the IC's fabrication process must be able to both withstand and drive high voltages. The high voltage transistors create additional fabrication costs since they require more processing steps to build in high voltage tolerance. Designs with standard transistors require additional circuitry. The I/O drivers must also provide ESD protection for the device. Most current designs limit the overvoltage to below one diode drop (0.7 V) above the power supply. Protection for larger overvoltage requires more diodes in series.

- *Internal High Voltage Generation*: The PMOS transistors need to be placed in a substrate well that is tied to the highest on-chip voltage to prevent lateral diodes from turning on and drawing excessive current. This high voltage can either be generated on-chip using charge pumps, or from an external supply. This requirement can make the design complex, since one cannot efficiently use charge pumps to generate higher voltages and also achieve low standby current.

- *Chip Area*: Die size is a primary factor in reducing costs and increasing yields. Tolerance and compliance circuitry may require either more or larger I/O devices to achieve the desired performance levels.

- *Testing*: Since the core and the I/O can be at different voltages, testing the device for all possible combinations of voltages can be complicated, adding to the total cost of the IC.

3.3 V/2.5 V Interfaces

The Fairchild 74VCX164245 series are low voltage, 16-bit, dual-supply logic translators/transceivers with three-state outputs. A simplified block diagram is shown in Figure 10-9. These devices use the VCX low voltage standard previously discussed. The output driver circuit is supplied from the V_{DDB} power supply bus, ensuring V_{DDB} compliant and tolerant outputs. The input circuit is supplied from the V_{DDA} supply, and the input logic threshold adjust circuits optimize the input logic thresholds for the particular value of V_{DDA}. Figure 10-10 shows the VCX voltage standards for 3.3 V, 2.5 V, and 1.8 V supply voltages. Note that the input voltage is 3.3 V tolerant for all three supply voltages.

These devices dissipate about 2 mW per input/output and are packaged in a 48-lead TSSOP with a 2.5 V supply. Propagation delay is about 3.2 ns.

Figure 10-11 shows two possibilities for a 3.3 V to 2.5 V logic interface. The top diagram (A) shows a direct connection. This will work provided the 2.5 V IC is 3.3 V tolerant on its input. If the 2.5 V IC is not 3.3 V tolerant, the VCX translator can be used as shown in Figure 10-11B.

Hardware Design Techniques

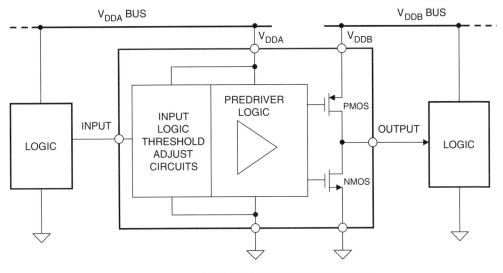

74VCX164245 CHARACTERISTICS:
- Power Dissipation = 2 mW/Input or Output
- 16 bits per 48-lead TSSOP 100 mm^2 Package
- 3.2 ns Propagation Delay at 2.5 V

Figure 10-9: Logic Translating Transceiver
(Fairchild 74VCX164245)

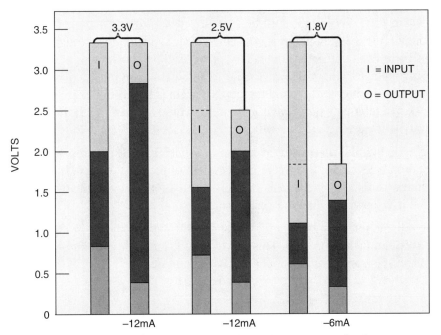

Figure 10-10: Voltage Compliance for VCX Standard
(Fairchild 74VCX164245 Translator)

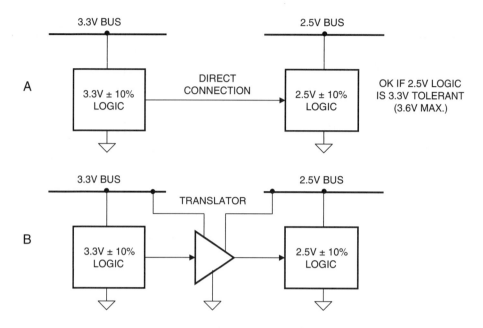

Figure 10-11: 3.3 V to 2.5 V Interface

Figure 10-12A shows a direct connection between 2.5 V and 3.3 V logic. In order for this to work, the 2.5 V output must be at least 2 V minimum. With no loading on the 2.5 V output, the 3.3 V IC input is connected directly to +2.5 V through the on resistance of the PMOS transistor driver. This provides 0.5 V noise margin for the nominal supply voltage of 2.5 V. However, the 10% tolerance on the 2.5 V bus allows it to drop to a minimum of 2.25 V, and the noise margin is reduced to 0.25 V. This may still work in a relatively quiet environment, but could be marginal if there is noise on the supply voltages.

Adding a 1.6 kΩ pull-up resistor as shown in Figure 10-12B ensures the 2.5 V output will not drop below 2.5 V due to the input current of the 3.3 V device, but the degraded noise margin still exists for a 2.25 V supply. With a 50% duty cycle, the resistor adds about 3.4 mW power dissipation per output.

A more reliable interface between 2.5 V and 3.3 V logic is shown in Figure 10-12C, where a VCX translator is used. This solves all noise margin problems associated with (A) and (B), and requires about 2 mW per output.

Hardware Design Techniques

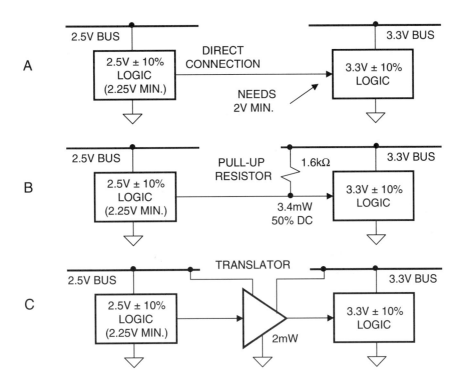

Figure 10-12: 2.5 V to 3.3 V Interface

References on Low Voltage Interfaces

1. P. Alfke, *Low-Voltage FPGAs Allow 3.3V/5V System Design*, **Electronic Design**, p. 70–76, August 18, 1997.

2. AA Alkaline Battery Discharge Characteristics, Duracell Inc., Berkshire Corporate Park, Bethel, CT 06801, http://www.duracell.com.

3. Joint Electron Device Engineering Council (JEDEC), Standard 8-5, October 1995.

4. QS3384 Data Sheet, Integrated Device Technology (IDT), Inc., 2975 Stender Way, Santa Clara, CA 95054, http://www.idt.com.

5. Pericom Semiconductor Corporation, 2380 Bering Drive, San Jose, CA 95131, http://www.pericom.com.

6. 74VCX164245 Data Sheet, Fairchild Semiconductor, 1997, http://www.fairchildsemi.com.

7. H. Johnson, M. Graham, **High-Speed Digital Design**, Prentice Hall, 1993.

Grounding in Mixed-Signal Systems

Walt Kester, James Bryant, Mike Byrne

Today's signal processing systems generally require mixed-signal devices such as analog-to-digital converters (ADCs) and digital-to-analog converters (DACs) as well as fast digital signal processors (DSPs). Requirements for processing analog signals having wide dynamic ranges increase the importance of high performance ADCs and DACs. Maintaining wide dynamic range with low noise in hostile digital environments is dependent upon using good high speed circuit design techniques including proper signal routing, decoupling, and grounding.

In the past, "high precision, low speed" circuits have generally been viewed differently than so-called "high speed" circuits. With respect to ADCs and DACs, the sampling (or update) frequency has generally been used as the distinguishing speed criteria. However, the following two examples show that, in practice, most of today's signal processing ICs are really high speed, and must be treated as such in order to maintain high performance. This is certainly true of DSPs, and also true of ADCs and DACs.

All sampling ADCs (ADCs with an internal sample-and-hold circuit) suitable for signal processing applications operate with relatively high speed clocks with fast rise and fall times (generally a few nanoseconds) and must be treated as high speed devices, even though throughput rates may appear low. For example, the 12-bit AD7892 successive-approximation (SAR) ADC operates on an 8 MHz internal clock, while the sampling rate is only 600 kSPS.

Sigma-delta (Σ-Δ) ADCs also require high speed clocks because of their high oversampling ratios. The AD7722 16-bit ADC has an output data rate (effective sampling rate) of 195 kSPS, but actually samples the input signal at 12.5 MSPS (64 times oversampling). Even high resolution, so-called "low frequency" Σ-Δ industrial measurement ADCs (having throughputs of 10 Hz to 7.5 kHz) operate on 5 MHz or higher clocks and offer resolution to 24 bits (for example, the Analog Devices AD7730 and AD7731).

To further complicate the issue, mixed-signal ICs have both analog and digital ports and, because of this, much confusion has resulted with respect to proper grounding techniques. In addition, some mixed-signal ICs have relatively low digital currents, while others have high digital currents. In many cases, these two types must be treated differently with respect to optimum grounding.

Digital and analog design engineers tend to view mixed-signal devices from different perspectives, and the purpose of this section is to develop a general grounding philosophy that will work for most mixed-signal devices, without having to know the specific details of their internal circuits.

Ground and Power Planes

The importance of maintaining a low impedance large area ground plane is critical to all analog circuits today. The ground plane not only acts as a low impedance return path for decoupling high frequency currents (caused by fast digital logic) but also minimizes EMI/RFI emissions. Because of the shielding action of the ground plane, the circuit's susceptibility to external EMI/RFI is also reduced.

Ground planes also allow the transmission of high speed digital or analog signals using transmission line techniques (microstrip or stripline) where controlled impedances are required.

The use of "buss wire" is totally unacceptable as a "ground" because of its impedance at the equivalent frequency of most logic transitions. For instance, #22 gauge wire has about 20 nH/inch inductance. A transient current having a slew rate of 10 mA/ns created by a logic signal would develop an unwanted voltage drop of 200 mV at this frequency flowing through 1 inch of this wire:

$$\Delta v = L \frac{\Delta i}{\Delta t} = 20_1$$

For a signal having a 2 V peak-to-peak range, this translates into an error of about 200 mV, or 10% (approximate 3.5-bit accuracy). Even in all-digital circuits, this error would result in considerable degradation of logic noise margins.

Figure 10-13 shows a situation where the digital return current modulates the analog return current (top figure). The ground return wire inductance and resistance is shared between the analog and digital circuits, and this is what causes the interaction and resulting error. A possible solution is to make the digital return current path flow directly to the GND REF as shown in the bottom figure. This is the fundamental concept of a "star," or single-point ground system. Implementing the true single-point ground in a system that contains multiple high frequency return paths is difficult because the physical length of the individual return current wires will introduce parasitic resistance and inductance that can make obtaining a low impedance, high frequency ground difficult. In practice, the current returns must consist of large area ground planes for low impedance to high frequency currents. Without a low impedance ground plane, it is therefore almost impossible to avoid these shared impedances, especially at high frequencies.

All integrated circuit ground pins should be soldered directly to the low impedance ground plane to minimize series inductance and resistance. The use of traditional IC sockets is not recommended with high speed devices. The extra inductance and capacitance of even "low profile" sockets may corrupt the device performance by introducing unwanted shared paths. If sockets must be used with DIP packages, as in prototyping, individual "pin sockets" or "cage jacks" may be acceptable. Both capped and uncapped versions of these pin sockets are available (AMP part numbers 5-330808-3, and 5-330808-6). They have spring-loaded gold contacts which make good electrical and mechanical connection to the IC pins. Multiple insertions, however, may degrade their performance.

Power supply pins should be decoupled directly to the ground plane using low inductance ceramic surface-mount capacitors. If through-hole mounted ceramic capacitors must be used, their leads should be less than 1 mm. The ceramic capacitors should be located as close as possible to the IC power pins. Ferrite beads may be also required for additional decoupling.

Hardware Design Techniques

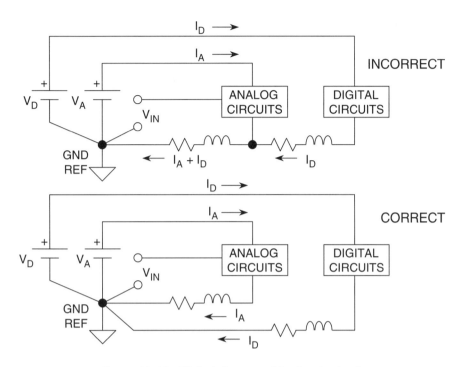

Figure 10-13: Digital Currents Flowing in Analog Return Path Create Error Voltages

Double-Sided Versus Multilayer Printed Circuit Boards

Each PCB in the system should have at least one complete layer dedicated to the ground plane. Ideally, a double-sided board should have one side completely dedicated to ground and the other side for interconnections. In practice, this is not possible, since some of the ground plane will certainly have to be removed to allow for signal and power crossovers, vias, and through-holes. Nevertheless, as much area as possible should be preserved, and at least 75% should remain. After completing an initial layout, the ground layer should be checked carefully to make sure there are no isolated ground "islands," because IC ground pins located in a ground "island" have no current return path to the ground plane. Also, the ground plane should be checked for "skinny" connections between adjacent large areas, which may significantly reduce the effectiveness of the ground plane. Needless to say, autorouting board layout techniques will generally lead to a layout disaster on a mixed-signal board, so manual intervention is highly recommended.

Systems that are densely packed with surface-mount ICs will have a large number of interconnections; therefore multilayer boards are mandatory. This allows at least one complete layer to be dedicated to ground. A simple four-layer board would have internal ground and power plane layers with the outer two layers used for interconnections between the surface-mount components. Placing the power and ground planes adjacent to each other provides additional interplane capacitance, which helps high frequency decoupling of the power supply. In most systems, four layers are not enough, and additional layers are required for routing signals as well as power.

- Use Large Area Ground (and Power) Planes for Low Impedance Current Return Paths (Must Use at Least a Double-Sided Board)
- Double-Sided Boards:
  - Avoid High Density Interconnection Crossovers and Vias, Which Reduce Ground Plane Area
  - Keep > 75% Board Area on One Side for Ground Plane
- Multilayer Boards: Mandatory for Dense Systems
  - Dedicate at Least One Layer for the Ground Plane
  - Dedicate at Least One Layer for the Power Plane
- Use at Least 30% to 40% of PCB Connector Pins for Ground
- Continue the Ground Plane on the Backplane Motherboard to Power Supply Return

Figure 10-14: Ground Planes are Mandatory

Multicard Mixed-Signal Systems

The best way of minimizing ground impedance in a multicard system is to use a "motherboard" PCB as a backplane for interconnections between cards, thus providing a continuous ground plane to the backplane. The PCB connector should have at least 30% to 40% of its pins devoted to ground, and these pins should be connected to the ground plane on the backplane mother card. To complete the overall system grounding scheme there are two possibilities:

1. The backplane ground plane can be connected to chassis ground at numerous points, thereby diffusing the various ground current return paths. This is commonly referred to as a "multipoint" grounding system and is shown in Figure 10-15.
2. The ground plane can be connected to a single system "star ground" point (generally at the power supply).

Hardware Design Techniques

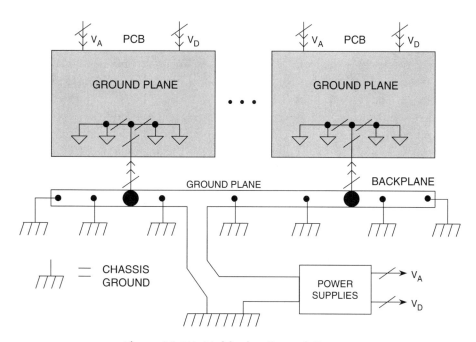

Figure 10-15: Multipoint Ground Concept

The first approach is most often used in all-digital systems, but can be used in mixed-signal systems provided the ground currents due to digital circuits are sufficiently low and diffused over a large area. The low ground impedance is maintained all the way through the PC boards, the backplane, and ultimately the chassis. However, it is critical that good electrical contact be made where the grounds are connected to the sheet metal chassis. This requires self-tapping sheet metal screws or "biting" washers. Special care must be taken where anodized aluminum is used for the chassis material, since its surface acts as an insulator.

The second approach ("star ground") is often used in high speed mixed-signal systems having separate analog and digital ground systems and warrants further discussion.

Separating Analog and Digital Grounds

In mixed-signal systems with large amounts of digital circuitry, it is highly desirable to *physically* separate sensitive analog components from noisy digital components. It may also be beneficial to use separate ground planes for the analog and the digital circuitry. These planes should not overlap in order to minimize capacitive coupling between the two. The separate analog and digital ground planes are continued on the backplane using either motherboard ground planes or "ground screens," which are made up of a series of wired interconnections between the connector ground pins.

The arrangement shown in Figure 10-16 illustrates that the two planes are kept separate all the way back to a common system "star" ground, generally located at the power supplies. The connections between the ground planes, the power supplies, and the "star" should be made up of multiple bus bars or wide copper braids for minimum resistance and inductance. The back-to-back Schottky diodes on each PCB are inserted to prevent accidental dc voltage from developing between the two ground systems when cards are plugged and unplugged. This voltage should be kept less than 300 mV to prevent damage to ICs that have connections to both the analog and digital ground planes. Schottky diodes are preferable because of their low capacitance and low forward voltage drop. The low capacitance prevents ac coupling between the analog and digital ground planes. Schottky diodes begin to conduct at about 300 mV, and several parallel diodes in parallel may be required if high currents are expected. In some cases, ferrite beads can be used instead of Schottky diodes; however, they introduce dc ground loops, which can be troublesome in precision systems.

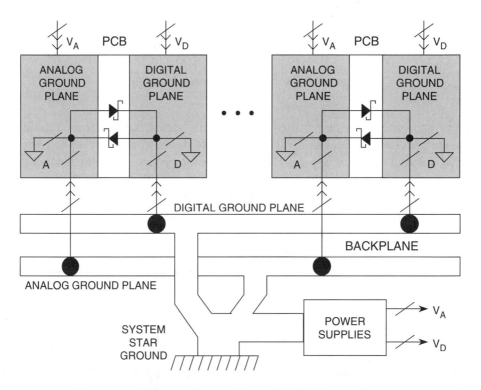

Figure 10-16: Separating Analog and Digital Ground Planes

It is mandatory that the impedance of the ground planes be kept as low as possible, all the way back to the system star ground. Not only can dc or ac voltages of more than 300 mV between the two ground planes damage ICs, but can also cause false triggering of logic gates and possible latch-up.

Hardware Design Techniques

Grounding and Decoupling Mixed-Signal ICs with Low Digital Currents

Sensitive analog components such as amplifiers and voltage references are always referenced and decoupled to the analog ground plane. *The ADCs and DACs (and other mixed-signal ICs) with low digital currents should generally be treated as analog components and also grounded and decoupled to the analog ground plane.* At first glance, this may seem somewhat contradictory, since a converter has an analog and digital interface and usually has pins designated as *analog ground* (AGND) and *digital ground* (DGND). The diagram shown in Figure 10-17 will help to explain this seeming dilemma.

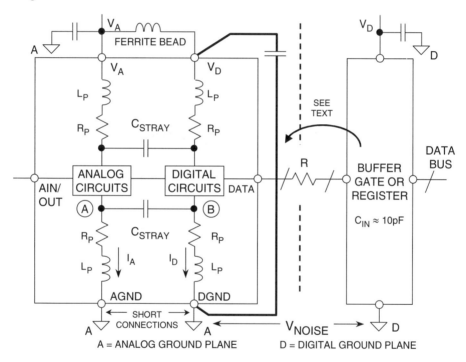

Figure 10-17: Proper Grounding of Mixed-Signal ICs with Low Internal Digital Currents

Inside an IC that has both analog and digital circuits, such as an ADC or a DAC, the grounds are usually kept separate to avoid coupling digital signals into the analog circuits. Figure 10-17 shows a simple model of a converter. There is nothing the IC designer can do about the wirebond inductance and resistance associated with connecting the bond pads on the chip to the package pins except to realize it's there. The rapidly changing digital currents produce a voltage at point B that will inevitably couple into point A of the analog circuits through the stray capacitance, C_{STRAY}. In addition, there is approximately 0.2 pF unavoidable stray capacitance between every pin of the IC package. It's the IC designer's job to make the chip work in spite of this. However, in order to prevent further coupling, the AGND and DGND pins should be joined together externally to the *analog* ground plane with minimum lead lengths. Any extra impedance in the DGND connection will cause more digital noise to be developed at point B; it will, in turn, couple more digital noise into the analog circuit through the stray capacitance. *Note that connecting DGND to the digital ground plane applies V_{NOISE} across the AGND and DGND pins and invites disaster.*

The name "DGND" on an IC tells us that this pin connects to the digital ground of the IC. It does not imply that this pin must be connected to the digital ground of the system.

It is true that this arrangement may inject a small amount of digital noise onto the analog ground plane. These currents should be quite small, and can be minimized by ensuring that the converter output does not drive a large fanout (they normally cannot, by design). Minimizing the fanout on the converter's digital port will also keep the converter logic transitions relatively free from ringing and minimize digital switching currents, thereby reducing any potential coupling into the analog port of the converter. The logic supply pin (V_D) can be further isolated from the analog supply by the insertion of a small lossy ferrite bead as shown in Figure 10-17. The internal transient digital currents of the converter will flow in the small loop from V_D through the decoupling capacitor and to DGND (this path is shown with a heavy line on the diagram). The transient digital currents will therefore not appear on the external analog ground plane, but are confined to the loop. The V_D pin decoupling capacitor should be mounted as close to the converter as possible to minimize parasitic inductance. These decoupling capacitors should be low inductance ceramic types, typically between 0.01 µF and 0.1 µF.

Treat the ADC Digital Outputs with Care

It is always a good idea (as shown in Figure 10-17) to place a buffer register adjacent to the converter to isolate the converter's digital lines from noise on the data bus. The register also serves to minimize loading on the digital outputs of the converter and acts as a Faraday shield between the digital outputs and the data bus. Even though many converters have three-state outputs/inputs, this isolation register still represents good design practice. In some cases it may be desirable to add an additional buffer

Hardware Design Techniques

register on the analog ground plane next to the converter output to provide greater isolation.

The series resistors (labeled "R" in Figure 10-17) between the ADC output and the buffer register input help to minimize the digital transient currents, which may affect converter performance. The resistors isolate the digital output drivers from the capacitance of the buffer register inputs. In addition, the RC network formed by the series resistor and the buffer register input capacitance acts as a low-pass filter to slow down the fast edges.

A typical CMOS gate combined with PCB trace and a through-hole will create a load of approximately 10 pF. A logic output slew rate of 1 V/ns will produce 10 mA of dynamic current if there is no isolation resistor:

$$\Delta I = C \frac{\Delta v}{\Delta t} = 10]$$

A 500 Ω series resistors will minimize this output current and result in a rise and fall time of approximately 11 ns when driving the 10 pF input capacitance of the register:

$$t_r = 2.2 \times \tau = 2.:$$

TTL registers should be avoided, since they can appreciably add to the dynamic switching currents because of their higher input capacitance.

The buffer register and other digital circuits should be grounded and decoupled to the *digital* ground plane of the PC board. Notice that any noise between the analog and digital ground plane reduces the noise margin at the converter digital interface. Since digital noise immunity is of the order of hundreds or thousands of millivolts, this is unlikely to matter. The analog ground plane will generally not be very noisy, but if the noise on the digital ground plane (relative to the analog ground plane) exceeds a few hundred millivolts, steps should be taken to reduce the digital ground plane impedance, thereby maintaining the digital noise margins at an acceptable level. Under no circumstances should the voltage between the two ground planes exceed 300 mV, or the ICs may be damaged.

Separate power supplies for analog and digital circuits are also highly desirable. The analog supply should be used to power the converter. If the converter has a pin designated as a digital supply pin (V_D), it should either be powered from a separate analog supply, or filtered as shown in the diagram. All converter power pins should be decoupled to the analog ground plane, and all logic circuit power pins should be decoupled to the digital ground plane as shown in Figure 10-18. If the digital power supply is relatively quiet, it may be possible to use it to supply analog circuits as well, but be very cautious.

In some cases it may not be possible to connect V_D to the analog supply. Some of the newer, high speed ICs may have their analog circuits powered by 5 V, but the digital interface powered by 3 V to interface to 3 V logic. In this case, the 3 V pin of the IC should be decoupled directly to the analog ground plane. It is also advisable to connect a ferrite bead in series with the power trace that connects the pin to the 3 V digital logic supply.

Section Ten

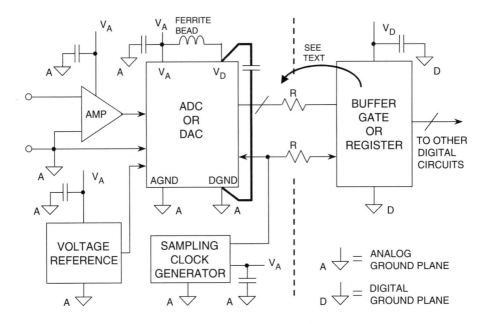

Figure 10-18: Grounding and Decoupling Points

The sampling clock generation circuitry should be treated like analog circuitry and also be grounded and heavily decoupled to the analog ground plane. Phase noise on the sampling clock produces degradation in system SNR, as will be discussed shortly.

Sampling Clock Considerations

In a high performance sampled data system a low phase-noise crystal oscillator should be used to generate the ADC (or DAC) sampling clock because sampling clock jitter modulates the analog input/output signal and raises the noise and distortion floor. The sampling clock generator should be isolated from noisy digital circuits and grounded and decoupled to the analog ground plane, as is true for the op amp and the ADC.

The effect of sampling clock jitter on ADC signal-to-noise ratio (SNR) is given approximately by the equation:

$$\mathrm{SNR} = 20\log_{10}\left[\frac{1}{2\pi f t_j}\right]$$

where *SNR* is the SNR of a perfect ADC of infinite resolution where the only source of noise is that caused by the rms sampling clock jitter, t_j. Note that f in the above equation is the analog input frequency. Just working through a simple example, if t_j = 50 ps rms, f = 100 kHz, then SNR = 90 dB, equivalent to about 15 bits of dynamic range.

It should be noted that t_j in the above example is the root-sum-square (RSS) value of the external clock jitter *and* the internal ADC clock jitter (called aperture jitter). However, in most high performance ADCs, the internal aperture jitter is negligible compared to the jitter on the sampling clock.

Since degradation in SNR is primarily due to external clock jitter, steps must be taken to ensure the sampling clock is as noise-free as possible and has the lowest possible phase jitter. This requires that a crystal oscillator be used. There are several manufacturers of small crystal oscillators with low jitter (less than 5 ps rms) CMOS compatible outputs. (For example, MF Electronics, 10 Commerce Dr., New Rochelle, NY 10801, Tel. 914-576-6570.)

Ideally, the sampling clock crystal oscillator should be referenced to the analog ground plane in a split-ground system. However, this is not always possible because of system constraints. In many cases, the sampling clock must be derived from a higher frequency multipurpose system clock generated on the digital ground plane. It must then pass from its origin on the digital ground plane to the ADC on the analog ground plane. Ground noise between the two planes adds directly to the clock signal and will produce excess jitter. The jitter can cause degradation in the signal-to-noise ratio and also produce unwanted harmonics. This can be somewhat remedied by transmitting the sampling clock signal as a differential signal using either a small RF transformer as shown in Figure 10-19 or a high speed differential driver and receiver IC. If an active differential driver and receiver are used, they should be ECL to minimize phase jitter. In a single 5 V supply system, ECL logic can be connected between ground and 5 V (PECL), and the outputs ac-coupled into the ADC sampling clock input. In either case, the original master system clock must be generated from a low phase-noise crystal oscillator.

Section Ten

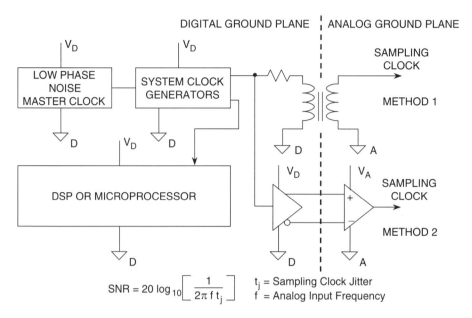

Figure 10-19: Sampling Clock Distribution from Digital to Analog Ground Planes

The Origins of the Confusion about Mixed-Signal Grounding: Applying Single-Card Grounding Concepts to Multicard Systems

Most ADC, DAC, and other mixed-signal device data sheets discuss grounding relative to a single PCB, usually the manufacturer's own evaluation board. This has been a source of confusion when trying to apply these principles to multicard or multi-ADC/DAC systems. The recommendation is usually to split the PCB ground plane into an analog plane and a digital plane. It is then further recommended that the AGND and DGND pins of a converter be tied together and that the analog ground plane and digital ground planes be connected at that same point, as shown in Figure 10-20. This essentially creates the system "star" ground at the mixed-signal device.

All noisy digital currents flow through the digital power supply to the digital ground plane and back to the digital supply; they are isolated from the sensitive analog portion of the board. The system star ground occurs where the analog and digital ground planes are joined together at the mixed-signal device. While this approach will generally work in a simple system with a single PCB and single ADC/DAC, it is not optimum for multicard mixed-signal systems. In systems having several ADCs or DACs on different PCBs (or on the same PCB), the analog and digital ground planes become connected at several points, creating the possibility of ground loops and making a single-point "star" ground system impossible. For these reasons, this grounding approach is not recommended for multicard systems,

Hardware Design Techniques

and the approach previously discussed should be used for mixed-signal ICs with low digital currents.

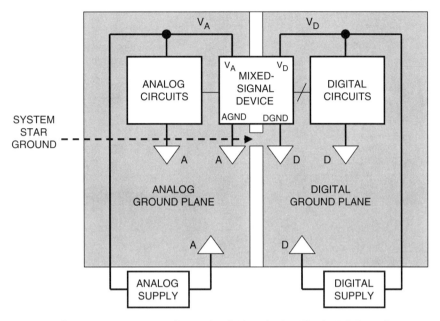

Figure 10-20: Grounding Mixed-Signal ICs: Single PC Board (Typical Evaluation/Test Board)

Summary: Grounding Mixed-Signal Devices with Low Digital Currents in a Multicard System

Figure 10-21 summarizes the approach previously described for grounding a mixed-signal device with low digital currents. The analog ground plane is not corrupted because the small digital transient currents flow in the small loop between V_D, the decoupling capacitor, and DGND (shown as a heavy line). The mixed-signal device is for all intents and purposes treated as an analog component. The noise V_N between the ground planes reduces the noise margin at the digital interface but is generally not harmful if kept less than 300 mV by using a low impedance digital ground plane all the way back to the system star ground.

However, mixed-signal devices such as sigma-delta ADCs, codecs, and DSPs with on-chip analog functions are becoming more and more digitally intensive. Along with the additional digital circuitry come larger digital currents and noise. For example, a sigma-delta ADC or DAC contains a complex digital filter that adds considerably to the digital current in the device. The method previously discussed depends on the decoupling capacitor between V_D and DGND to keep the digital transient currents isolated in a small loop. However, if the digital currents are significant enough and have components at dc or low frequencies, the decoupling capacitor may have to be so large that it is impractical. Any digital current that flows outside the loop between V_D and DGND must flow through the analog ground plane. This may degrade performance, especially in high resolution systems.

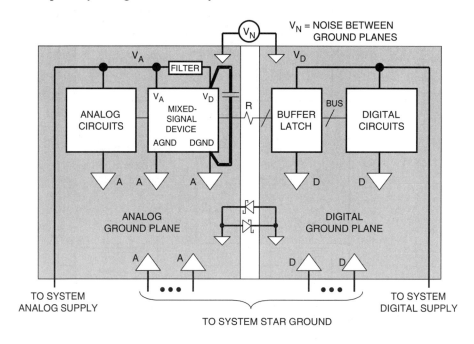

Figure 10-21: Grounding Mixed-Signal ICs with Low Internal Digital Currents: Multiple PC Boards

It is difficult to predict what level of digital current flowing into the analog ground plane will become unacceptable in a system. All we can do at this point is to suggest an alternative grounding method that may yield better performance.

Summary: Grounding Mixed-Signal Devices with High Digital Currents in a Multicard System

An alternative grounding method for a mixed-signal device with high levels of digital currents is shown in Figure 10-22. The AGND of the mixed-signal device is connected to the analog ground plane, and the DGND of the device is connected to the digital ground plane. The digital currents are isolated from the analog ground plane, but the noise between the two ground planes is applied directly between the AGND and DGND pins of the device. For this method to be successful, the analog and digital circuits within the mixed-signal device must be well isolated. The noise between AGND and DGND pins must not be large enough to reduce internal noise margins or cause corruption of the internal analog circuits.

Hardware Design Techniques

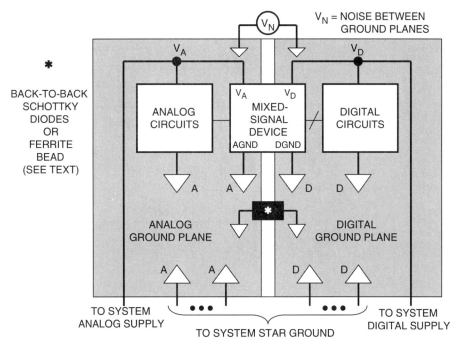

Figure 10-22: Gounding Alternative for Mixed-Signal ICs with High Digital Currents: Multiple PC Boards

Figure 10-22 shows optional Schottky diodes (back-to-back) or a ferrite bead connecting the analog and digital ground planes. The Schottky diodes prevent large dc voltages or low frequency voltage spikes from developing across the two planes. These voltages can potentially damage the mixed-signal IC if they exceed 300 mV because they appear directly between the AGND and DGND pins. As an alternative to the back-to-back Schottky diodes, a ferrite bead provides a dc connection between the two planes but isolates them at frequencies above a few MHz where the ferrite bead becomes resistive. This protects the IC from dc voltages between AGND and DGND, but the dc connection provided by the ferrite bead can introduce unwanted dc ground loops and may not be suitable for high resolution systems.

Grounding DSPs with Internal Phase-Locked Loops

As if dealing with mixed-signal ICs with AGND and DGNDs were not enough, newer DSPs such as the ADSP-21160 SHARC with internal phase-locked-loops (PLLs) raise issues with respect to proper grounding. The ADSP-21160 PLL allows the internal core clock (determines the instruction cycle time) to operate at a user-selectable ratio of 2, 3, or 4 times the external clock frequency, CLKIN. The CLKIN rate is the rate at which the synchronous external ports operates. Although this allows using a lower frequency external clock, care must be taken with the power and ground connections to the internal PLL, as shown in Figure 10-23.

Section Ten

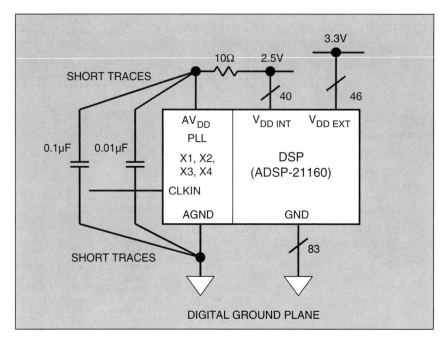

Figure 10-23: Grounding DSPs with Internal Phase-Locked Loops (PLLs)

In order to prevent internal coupling between digital currents and the PLL, the power and ground connections to the PLL are brought out separately on pins labeled AV_{DD} and AGND, respectively. The AV_{DD} 2.5 V supply should be derived from the $V_{DD\,INT}$ 2.5 V supply using the filter network as shown. This ensures a relatively noise-free supply for the internal PLL. The AGND pin of the PLL should be connected to the digital ground plane of the PC board using a short trace. The decoupling capacitors should be routed between the AV_{DD} pin and AGND pin using short traces.

Grounding Summary

There is no single grounding method that will guarantee optimum performance 100% of the time. This section has presented a number of possible options depending upon the characteristics of the particular mixed-signal devices in question. It is helpful, however, to provide for as many options as possible when laying out the initial PC board.

It is mandatory that at least one layer of the PC board be dedicated to the ground plane. The initial board layout should provide for nonoverlapping analog and digital ground planes, but pads and vias should be provided at several locations for the installation of back-to-back Schottky diodes or ferrite beads, if required. Pads and vias should also be provided so that the analog and digital ground planes can be connected together with jumpers if required.

Hardware Design Techniques

The AGND pins of mixed-signal devices should in general always be connected to the analog ground plane. An exception to this are DSPs that have internal phase-locked-loops (PLLs), such as the ADSP-21160 SHARC. The ground pin for the PLL is labeled AGND, but should be connected directly to the digital ground plane for the DSP.

- There is No Single Grounding Method Guaranteed to Work 100% of the Time
- Different Methods May or May Not Give the Same Levels of Performance
- At Least One Layer on Each PC Board MUST be Dedicated to the Ground Plane
- Do Initial Layout with Split Analog and Digital Ground Planes
- Provide Pads and Vias on each PC Board for Back-to-Back Schottky Diodes and Optional Ferrite Beads to Connect the Two Planes
- Provide "Jumpers" so that DGND Pins of Mixed-Signal Devices can be Connected to AGND Pins (Analog Ground Plane) or to Digital Ground Plane (AGND of PLLs in DSPs Should be Connected to Digital Ground Plane)
- Provide Pads and Vias for "Jumpers" so That Analog and Digital Ground Planes Can be Joined Together at Several Points on Each PC Board
- Follow Recommendations on Mixed-Signal Device Data Sheet

Figure 10-24: Grounding Philosophy Summary

Some General PC Board Layout Guidelines for Mixed-Signal Systems

It is evident that noise can be minimized by paying attention to the system layout and preventing different signals from interfering with each other. High level analog signals should be separated from low level analog signals, and both should be kept away from digital signals. We have seen elsewhere that in waveform sampling and reconstruction systems the sampling clock (which is a digital signal) is as vulnerable to noise as any analog signal, but is as liable to cause noise as any digital signal, and so must be kept isolated from both analog and digital systems. If clock driver packages are used in clock distribution, only one frequency clock should be passed through a single package. Sharing drivers between clocks of different frequencies in the same package will produce excess jitter and crosstalk and degrade performance.

The ground plane can act as a shield where sensitive signals cross. Figure 10-25 shows a good layout for a data acquisition board where all sensitive areas are isolated from each other and signal paths are kept as short as possible. While real life is rarely as tidy as this, the principle remains a valid one.

Section Ten

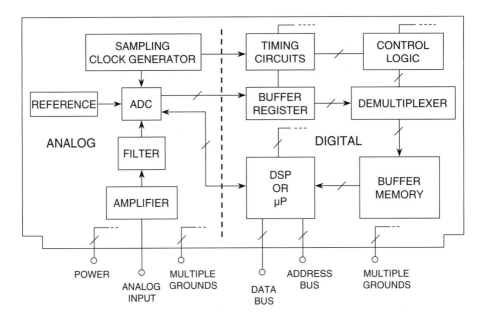

Figure 10-25: Analog and Digital Circuits Should be Partitioned on PCB Layout

There are a number of important points to be considered when making signal and power connections. First of all, a connector is one of the few places in the system where all signal conductors must run in parallel. It is therefore imperative to separate them with ground pins (creating a Faraday shield) to reduce coupling between them.

Multiple ground pins are important for another reason: they keep down the ground impedance at the junction between the board and the backplane. The contact resistance of a single pin of a PCB connector is quite low (of the order of 10 mW) when the board is new. As the board gets older, the contact resistance is likely to rise and the board's performance may be compromised. It is therefore well worthwhile to allocate extra PCB connector pins so that there are many ground connections (perhaps 30% to 40% of all the pins on the PCB connector should be ground pins). For similar reasons there should be several pins for each power connection, although there is no need to have as many as there are ground pins.

Manufacturers of high performance mixed-signal ICs like Analog Devices offer evaluation boards to assist customers in their initial evaluations and layout. ADC evaluation boards generally contain an on-board low jitter sampling clock oscillator, output registers, and appropriate power and signal connectors. They also may have additional support circuitry such as the ADC input buffer amplifier and external reference.

The layout of the evaluation board is optimized in terms of grounding, decoupling, and signal routing, and can be used as a model when laying out the ADC PC board in the system. The actual evaluation board layout is usually available from the ADC manufacturer in the form of computer CAD files (Gerber files). In many cases, the layout of the various layers appears on the data sheet for the device.

Section Ten

References on Grounding

1. William C. Rempfer, *Get All the Fast ADC Bits You Pay For*, **Electronic Design, Special Analog Issue**, June 24, 1996, p. 44.

2. Mark Sauerwald, *Keeping Analog Signals Pure in a Hostile Digital World*, **Electronic Design, Special Analog Issue**, June 24, 1996, p. 57.

3. Jerald Grame and Bonnie Baker, *Design Equations Help Optimize Supply Bypassing for Op Amps*, **Electronic Design, Special Analog Issue**, June 24, 1996, p. 9.

4. Jerald Grame and Bonnie Baker, *Fast Op Amps Demand More Than a Single-Capacitor Bypass*, **Electronic Design, Special Analog Issue**, November 18, 1996, p. 9.

5. Walt Kester and James Bryant, *Grounding in High Speed Systems*, **High Speed Design Techniques**, Analog Devices, 1996, Chapter 7, pp. 7–27.

6. Jeffrey S. Pattavina, *Bypassing PC Boards: Thumb Your Nose at Rules of Thumb*, **EDN**, Oct. 22, 1998, p. 149.

7. Henry Ott, **Noise Reduction Techniques in Electronic Systems, Second Edition**, New York, John Wiley and Sons, 1988.

8. Howard W. Johnson and Martin Graham, **High-Speed Digital Design**, PTR Prentice Hall, 1993.

9. Paul Brokaw, *An I.C. Amplifier User's Guide to Decoupling, Grounding and Making Things Go Right for a Change*, Application Note, Analog Devices, Inc., http://www.analog.com.

10. Walt Kester, *A Grounding Philosophy for Mixed-Signal Systems*, **Electronic Design Analog Applications Issue**, June 23, 1997, p. 29.

11. Ralph Morrison, **Grounding and Shielding Techniques**, Fourth Edition, John Wiley, 1998.

12. Ralph Morrison, **Solving Interference Problems in Electronics**, John Wiley, 1995.

13. C. D. Motchenbacher and J. A. Connelly, **Low Noise Electronic System Design**, John Wiley, 1993.

14. Crystal Oscillators: MF Electronics, 10 Commerce Drive, New Rochelle, NY, 10801, 914-576-6570.

15. Mark Montrose, **EMC and the Printed Circuit Board**, IEEE Press, 1999 (IEEE Order Number PC5756).

Hardware Design Techniques

Digital Isolation Techniques
Walt Kester

One way to break ground loops is to use isolation techniques. Analog isolation amplifiers find many applications in which a high degree of isolation is required, such as in medical instrumentation. Digital isolation techniques offer a reliable method of transmitting digital signals over interfaces without introducing ground noise.

Optocouplers (also called optoisolators) are useful and available in a wide variety of styles and packages. A typical optocoupler based on an LED and a phototransistor is shown in Figure 10-26. A current of approximately 10 mA is applied to an LED transmitter, and the light output is received by a phototransistor. The light produced by the LED is sufficient to saturate the phototransistor. Isolation of 5000 V rms to 7000 V rms is common. Although excellent for digital signals, optocouplers are too nonlinear for most analog applications. One should also realize that since the phototransistor is operated in a saturated mode, rise and fall times can range from 10 μs to 20 μs in slower devices, thereby limiting applications at high speeds.

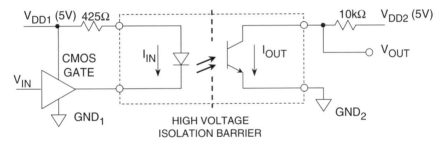

- Uses Light for Transmission over a High Voltage Barrier
- The LED is the Transmitter, and the Phototransistor is the Receiver
- High Voltage Isolation: 5000 V to 7000 V RMS
- Nonlinear—Best for Digital or Frequency Information
- Rise and Fall Times Can be 10 μs to 20 μs in Slower Devices
- Example: Siemens ILQ-1 Quad (http://www.siemens.com)

Figure 10-26: Digital Isolation Using LED/Phototransistor Optocouplers

A faster optocoupler architecture is shown in Figure 10-27 and is based on an LED and a photodiode. The LED is again driven with a current of approximately 10 mA. This produces a light output sufficient to generate enough current in the receiving photodiode to develop a valid high logic level at the output of the transimpedance amplifier. Speed can vary widely between optocouplers, and the fastest ones have propagation delays of 20 ns typical, and 40 ns maximum, and can handle data rates up to 25 MBd for NRZ data. This corresponds to a maximum square wave operating frequency of 12.5 MHz, and a minimum allowable passable pulsewidth of 40 ns.

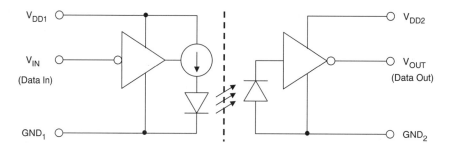

- 5 V Supply Voltage
- 2500 V RMS I/O Withstand Voltage
- Logic Signal Frequency: 12.5 MHz Maximum
- 25 MBd Maximum Data Rate
- 40 ns Maximum Propagation Delay
- 9 ns Typical Rise/Fall Time
- Example: Agilent HCPL-7720
  (http://www.semiconductor.agilent.com)

Figure 10-27: Digital Isolation Using LED/Photodiode Optocouplers

The ADuM1100A and ADuM1100B are digital isolators based on Analog Devices' μmIsolation™ (micromachined isolation) technology. Combining high speed CMOS and monolithic air core transformer technology, these isolation components provide outstanding performance characteristics superior to the traditional optocouplers previously described.

Configured as pin compatible replacements for existing high speed optocouplers, the ADuM1100A and ADuM1100B support data rates as high as 25 MBd and 100 MBb, respectively. A functional diagram of the devices is shown in Figure 10-28.

Both the ADuM1100A and ADuM1100B operate at either 3.3 V or 5 V supply voltages, have propagation delays < 10 ns, edge asymmetry of <2 ns, and rise and fall times < 2 ns. They operate at very low power, less than 600 µA of quiescent current (sum of both sides) and a dynamic current of less than 230 µA per MBd of data rate. Unlike common transformer implementations, the parts provide dc correctness with a patented refresh feature that continuously updates the output signal.

Hardware Design Techniques

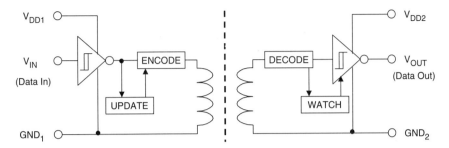

- 5 V/3.3 V Supply Voltage
- 2500 V RMS I/O Withstand Voltage
- 25 MBd Maximum Data Rate (ADuM1100A)
- 100 MBd Maximum Data Rate (ADuM1100B)
- 10 ns Maximum Propagation Delay
- 2 ns Typical Rise/Fall Time
- Pin Compatible with Popular Optocouplers

Figure 10-28: ADuM1100A/ADuM110B Digital Isolators

The AD260/AD261 family of digital isolators isolates five digital control signals to/from high speed DSPs, microcontrollers, or microprocessors. The AD260 also has a 1.5 W transformer for a 3.5 kV rms isolated external dc/dc power supply circuit.

Each line of the AD260 can handle digital signals up to 20 MHz (40 MBd) with a propagation delay of only 14 ns, which allows for extremely fast data transmission. Output waveform symmetry is maintained to within ±1 ns of the input so the AD260 can be used to accurately isolate time-based pulsewidth modulator (PWM) signals.

A simplified schematic of one channel of the AD260/AD261 is shown in Figure 10-29. The data input is passed through a Schmitt trigger circuit, through a latch, and a special transmitter circuit that differentiates the edges of the digital input signal and drives the primary winding of a proprietary transformer with a "set high/set low" signal. The secondary of the isolation transformer drives a receiver with the same "set high/set low" data that regenerates the original logic waveform. An internal circuit operates in the background that interrogates all inputs about every 5 μs and, in the absence of logic transitions, sends appropriate "set high/set low" data across the interface. Recovery time from a fault condition or at power-up is thus between 5 μs and 10 μs.

The power transformer (available on the AD260) is designed to operate between 150 kHz and 250 kHz and will easily deliver more than 1 W of isolated power when driven push-pull (5V) on the transmitter side. Different transformer taps, rectifier, and regu-

Section Ten

lator schemes will provide combinations of ±5 V, 15 V, 24 V, or even 30 V or higher. The output voltage, when driven with a low voltage-drop drive, will be 37 V p-p across the entire secondary with a 5 V push-pull drive.

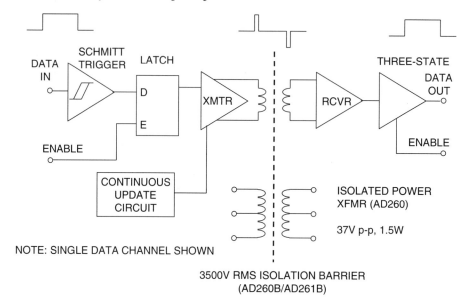

Figure 10-29: AD260/AD261 Digital Isolators

- Isolation Test Voltage to 3500 V RMS (AD260B/AD261B)
- Five Isolated Digital Lines Available in Six Input/Output Configurations
- Logic Signal Frequency: 20 MHz Max
- Data Rate: 40 MBd Max
- Isolated Power Transformer: 37 V p-p, 1.5 W (AD260)
- Waveform Edge Transmission Symmetry: ±1 ns
- Propagation Delay: 14 ns
- Rise and Fall-Times < 5 ns

Figure 10-30: AD260/AD261 Digital Isolator Key Specifications

Hardware Design Techniques

Power Supply Noise Reduction and Filtering
Walt Jung, Walt Kester, Bill Chestnut

Precision analog circuitry has traditionally been powered from well-regulated, low noise linear power supplies. During the last decade, however, switching power supplies have become much more common in electronic systems. As a consequence, they also are being used for analog supplies. Good reasons for the general popularity include their high efficiency, low temperature rise, small size, and light weight.

In spite of these benefits, switchers *do* have drawbacks, most notably high output noise. This noise generally extends over a broad band of frequencies, resulting in both conducted and radiated noise, as well as unwanted electric and magnetic fields. Voltage output noise of switching supplies are short-duration voltage transients, or spikes. Although the fundamental switching frequency can range from 20 kHz to 1 MHz, the spikes can contain frequency components extending to 100 MHz or more. While specifying switching supplies in terms of rms noise is common vendor practice, as a user you should also specify the *peak* (or p-p) amplitudes of the switching spikes, at the output loading of your system.

The following section discusses filter techniques for rendering a switching regulator output *analog ready*; that is, sufficiently quiet to power precision analog circuitry with relatively small loss of dc terminal voltage. The filter solutions presented are generally applicable to all power supply types incorporating switching element(s) in their energy path. This includes various dc-dc converters as well as popular 5 V (PC-type) supplies.

An understanding of the EMI process is necessary to understand the effects of supply noise on analog circuits and systems. Every interference problem has a *source*, a *path*, and a *receptor* (Reference 1). In general, there are three methods for dealing with interference. First, source emissions can be minimized by proper layout, pulse-edge rise time control/reduction, filtering, and proper grounding. Second, radiation and conduction paths should be reduced through shielding and physical separation. Third, receptor immunity to interference can be improved, via supply and signal line filtering, impedance level control, impedance balancing, and utilizing differential techniques to reject undesired common-mode signals. This section focuses on reducing switching power supply noise with external post filters.

Tools useful for combating high frequency switcher noise are shown in Figure 10-31. They differ in electrical characteristics as well as their practicality towards noise reduction, and are listed roughly in an order of priorities. Of these tools, L and C are the most powerful filter elements, and are the most cost-effective, as well as small in size.

Section Ten

- Capacitors
- Inductors
- Ferrites
- Resistors
- Linear Postregulation
- Proper Layout and Grounding Techniques
- PHYSICAL SEPARATION FROM SENSITIVE ANALOG CIRCUITS

Figure 10-31: Switching Regulator Noise Reduction Tools

Capacitors are probably the single most important filter component for switchers. There are many different types of capacitors, and an understanding of their individual characteristics is absolutely mandatory in the design of effective practical supply filters. There are generally four classes of capacitors useful in 10 kHz to 100 MHz filters, broadly distinguished as the generic dielectric types; *electrolytic, organic, film,* and *ceramic*. These can in turn can be further subdivided. A thumbnail sketch of capacitor characteristics is shown in the chart of Figure 10-32.

	Aluminum Electrolytic (General Purpose)	Aluminum Electrolytic (Switching Type)	Tantalum Electrolytic	OS-CON Electrolytic	Polyester (Stacked Film)	Ceramic (Multilayer)
Size	100 µF	120 µF	120 µF	100 µF	1 µF	0.1 µF
Rated Voltage	25 V	25 V	20 V	20 V	400 V	50 V
ESR	0.6 Ω @ 100 kHz	0.18 Ω @ 100 kHz	0.12 Ω @ 100 kHz	0.02 Ω @ 100 kHz	0.11 Ω @ 1 MHz	0.12 Ω @ 1 MHz
Operating Frequency (*)	≅ 100 kHz	≅ 500 kHz	≅ 1 MHz	≅ 1 MHz	≅ 10 MHz	≅ 1 GHz

(*) Upper frequency strongly size and package dependent

Figure 10-32: Types of Capacitors

With any dielectric, a major potential filter loss element is ESR (equivalent series resistance), the net parasitic resistance of the capacitor. ESR provides an ultimate limit to filter performance, and requires more than casual consideration, because it can vary both with frequency and temperature in some types. Another capacitor loss element is ESL (equivalent series inductance). ESL determines the frequency where the net impedance characteristic switches from capacitive to inductive. This varies from as low as 10 kHz in some electrolytics to as high as 100 MHz or more in chip ceramic types. Both ESR and ESL are minimized when a leadless package is used. All capacitor types mentioned are available in surface-mount packages, preferable for high speed uses.

The *electrolytic* family provides an excellent, cost-effective low frequency filter component, because of the wide range of values, a high capacitance-to-volume ratio, and a broad range of working voltages. It includes *general-purpose aluminum electrolytic* types, available in working voltages from below 10 V up to about 500 V, and in size from 1 to several thousand mF (with proportional case sizes). All electrolytic capacitors are polarized, and thus cannot withstand more than a volt or so of reverse bias without damage. They also have relatively high leakage currents (up to tens of mA, and are strongly dependent upon design specifics).

A subset of the general electrolytic family includes *tantalum* types, generally limited to voltages of 100 V or less, with capacitance of up to 500 mF (Reference 3). In a given size, tantalums exhibit a higher capacitance-to-volume ratios than do general-purpose electrolytics, and have both a higher frequency range and lower ESR. They are generally more expensive than standard electrolytics, and must be carefully applied with respect to surge and ripple currents.

A subset of aluminum electrolytic capacitors is the *switching* type, designed for handling high pulse currents at frequencies up to several hundred kHz with low losses (Reference 4). This capacitor type competes directly with tantalums in high frequency filtering applications, with the advantage of a broader range of values.

A more specialized high performance aluminum electrolytic capacitor type uses an organic semiconductor electrolyte (Reference 5). The *OS-CON* capacitors feature appreciably lower ESR and higher frequency range than do other electrolytic types, with an additional feature of low low-temperature ESR degradation.

Film capacitors are available in a very broad range of values and an array of dielectrics, including polyester, polycarbonate, polypropylene, and polystyrene. Because of the low dielectric constant of these films, their volumetric efficiency is quite low, and a 10 mF/50 V polyester capacitor (for example) is actually the size of one's hand. Metalized (as opposed to foil) electrodes do help to reduce size, but even the highest dielectric constant units among film types (polyester, polycarbonate) are still larger than any electrolytic, even using the thinnest films with the lowest voltage ratings

(50 V). Where film types excel is in their low dielectric losses, a factor that may not necessarily be a practical advantage for filtering switchers. For example, ESR in film capacitors can be as low as 10 mΩ or less, and the behavior of films generally is very high in terms of Q. In fact, this can cause problems of spurious resonance in filters, requiring damping components.

Film capacitors using a wound layer-type construction can be inductive. This can limit their effectiveness for high frequency filtering. Obviously, only noninductively made film caps are useful for switching regulator filters. One specific style that is noninductive is the *stacked film* type, where the capacitor plates are cut as small overlapping linear sheet sections from a much larger wound drum of dielectric/plate material. This technique offers the low inductance attractiveness of a plate-sheet style capacitor with conventional leads (see References 4, 5, 6). Obviously, minimal lead length should be used for best high frequency effectiveness. Very high current polycarbonate film types are also available, specifically designed for switching power supplies, with a variety of low inductance terminations to minimize ESL (Reference 7).

Depending upon their electrical and physical size, film capacitors can be useful at frequencies to well above 10 MHz. At the highest frequencies, only stacked film types should be considered. Some manufacturers are now supplying film types in leadless surface-mount packages, which eliminates the lead length inductance.

Ceramic is often the capacitor material of choice above a few MHz, due to its compact size, low loss, and availability up to several mF in the high-K dielectric formulations (X7R and Z5U), at voltage ratings up to 200 V (see ceramic families of Reference 3). NP0 (also called COG) types use a lower dielectric constant formulation, and have nominally zero TC, plus a low voltage coefficient (unlike the less stable high-K types). NP0 types are limited to values of 0.1 mF or less, with 0.01 mF representing a more practical upper limit.

Multilayer ceramic "chip caps" are very popular for bypassing/filtering at 10 MHz or higher, simply because their very low inductance design allows near optimum RF bypassing. For smaller values, ceramic chip caps have an operating frequency range to 1 GHz. For high frequency applications, a useful selection can be ensured by selecting a value with a self-resonant frequency *above* the highest frequency of interest.

All capacitors have some finite ESR. In some cases, the ESR may actually be helpful in reducing resonance peaks in filters, by supplying "free" damping. For example, in most electrolytic types, a nominally flat broad series resonance region can be noted by the impedance versus frequency plot. This occurs where |Z| falls to a minimum level, nominally equal to the capacitor's ESR at that frequency. This low Q resonance can generally cover a relatively wide frequency range of several octaves.

Hardware Design Techniques

Contrasted to the very high Q sharp resonances of film and ceramic caps, the low Q behavior of electrolytics can be useful in controlling resonant peaks.

In most electrolytic capacitors, ESR degrades noticeably at low temperature, by as much as a factor of 4 to 6 times at –55°C versus the room temperature value. For circuits where ESR is critical to performance, this can lead to problems. Some specific electrolytic types do address this problem; for example, within the HFQ switching types, the –10°C ESR at 100 kHz is no more than 2× that at room temperature. The OSCON electrolytics have an ESR versus temperature characteristic that is relatively flat.

As noted, all real capacitors have parasitic elements that limit their performance. The equivalent electrical network representing a real capacitor models both ESR and ESL as well as the basic capacitance, plus some shunt resistance (see Figure 10-33). In such a practical capacitor, at low frequencies the net impedance is almost purely capacitive. At intermediate frequencies, the net impedance is determined by ESR, for example, about 0.12 Ω to 0.4 Ω at 125 kHz, for several types. Above about 1 MHz these capacitor types become inductive, with impedance dominated by the effect of ESL. All electrolytics will display impedance curves similar in general shape to that of Figure 10-34. The minimum impedance will vary with the ESR, and the inductive region will vary with ESL (which in turn is strongly affected by package style).

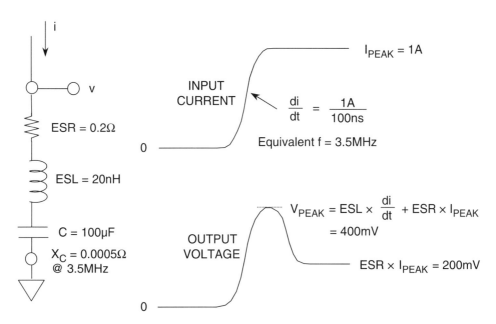

Figure 10-33: Capacitor Equivalent Circuit and Pulse Response

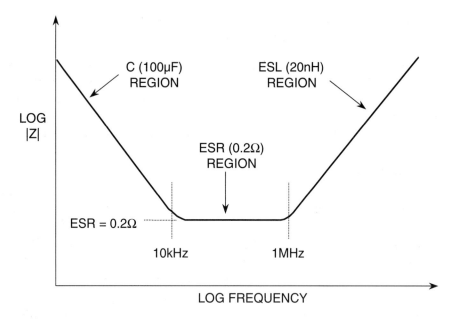

Figure 10-34: Electrolytic Capacitor Impedance vs. Frequency

Regarding inductors, *ferrites* (nonconductive ceramics manufactured from the oxides of nickel, zinc, manganese, or other compounds) are extremely useful in power supply filters (Reference 9). At low frequencies (<100 kHz), ferrites are inductive; thus they are useful in low-pass LC filters. Above 100 kHz, ferrites become resistive, an important characteristic in high frequency filter designs. Ferrite impedance is a function of material, operating frequency range, dc bias current, number of turns, size, shape, and temperature. Figure 10-35 summarizes a number of ferrite characteristics, and Figure 10-36 shows the impedance characteristic of several ferrite beads from Fair-Rite (http://www.fair-rite.com).

Several ferrite manufacturers offer a wide selection of ferrite materials from which to choose, as well as a variety of packaging styles (see References 10 and 11). A simple form is the *bead* of ferrite material, a cylinder of the ferrite which is simply slipped over the power supply lead to the decoupled stage. Alternately, the *leaded ferrite bead* is the same bead, premounted on a length of wire and used as a component (see Reference 11). More complex beads offer multiple holes through the cylinder for increased decoupling, plus other variations. Surface-mount beads are also available.

Hardware Design Techniques

- Ferrites Good for Frequencies Above 25 kHz
- Many Sizes and Shapes Available Including Leaded "Resistor Style"
- Ferrite Impedance at High Frequencies Primarily Resistive— Ideal for HF Filtering
- Low DC Loss: Resistance of Wire Passing Through Ferrite is Very Low
- High Saturation Current Versions Available
- Choice Depends Upon:
    - Source and Frequency of Interference
    - Impedance Required at Interference Frequency
    - Environmental: Temperature, AC and DC Field Strength, Size/Space Available
- Always Test the Design

Figure 10-35: Ferrites Suitable for High Frequency Filters

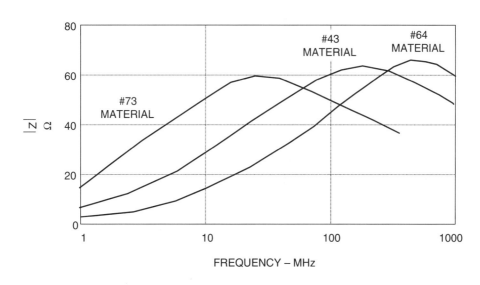

Courtesy: Fair-Rite Products Corp, Wallkill, NY
(http://www.fair-rite.com)

Figure 10-36: Impedance of Ferrite Beads

PSpice ferrite models for Fair-Rite materials are available, and allow ferrite impedance to be estimated (see Reference 12). These models have been designed to match measured impedances rather than theoretical impedances.

A ferrite's impedance is dependent upon a number of interdependent variables, and is difficult to quantify analytically; thus, selecting the proper ferrite is not straightforward. However, knowing the following system characteristics will make selection easier. First, determine the frequency range of the noise to be filtered. Second, the expected temperature range of the filter should be known, as ferrite impedance varies with temperature. Third, the peak dc current flowing through the ferrite must be known, to ensure that the ferrite does not saturate. Although models and other analytical tools may prove useful, the general guidelines given above, coupled with some experimentation with the actual filter connected to the supply output under system load conditions, should lead to a proper ferrite selection.

Using proper component selection, low and high frequency band filters can be designed to smooth a noisy switcher's dc output to produce an *analog ready* 5 V supply. It is most practical to do this over two (and sometimes more) stages, each stage optimized for a range of frequencies. A basic stage can be used to carry all of the dc load current, and filter noise by 60 dB or more up to a 1 MHz to 10 MHz range. This larger filter is used as a *card entry filter* providing broadband filtering for all power entering a PC card. Smaller, more simple local filter stages are also used to provide higher frequency decoupling right at the power pins of individual stages.

Switching Regulator Experiments

In order to better understand the challenge of filtering switching regulators, a series of experiments was conducted with a representative device, the ADP1148 synchronous buck regulator with a 9 V input and a 3.3 V/1 A output.

In addition to observing typical input and output waveforms, the objective of these experiments was to reduce the output ripple to less than 10 mV peak-to-peak, a value suitable for driving most analog circuits.

Measurements were made using a Tektronix wideband digitizing oscilloscope with the input bandwidth limited to 20 MHz so that the ripple generated by the switching regulators could be more readily observed. In a system, power supply ripple frequencies above 20 MHz are best filtered locally at each IC power pin with a low inductance ceramic capacitor and perhaps a series-connected ferrite bead.

Probing techniques are critical for accurate ripple measurements. A standard passive 10X probe was used with a "bayonet" probe tip adapter for making the ground connection as short as possible (see Figure 10-37). Use of the "ground clip lead" is not recommended in making this type of measurement because the lead length in the ground connection forms an unwanted inductive loop that picks up high frequency switching noise, thereby corrupting the signal being measured.

Hardware Design Techniques

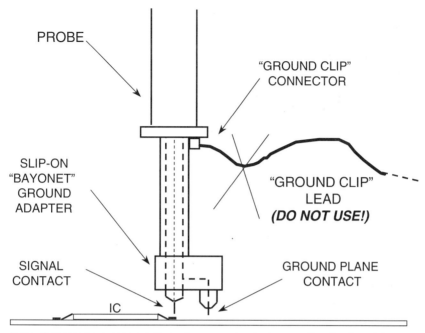

Note: Schematic representation of proper physical grounding is almost impossible. In all the following circuit schematics, the connections to ground are made to the ground plane using the shortest possible connecting path, regardless of how they are indicated in the actual circuit schematic diagram.

Figure 10-37: Proper Probing Techniques

The circuit for the ADP1148 9 V to 3.3 V/1 A buck regulator is shown in Figure 10-38.

The output waveform of the ADP1148 buck regulator is shown in Figure 10-39. The fundamental switching frequency is approximately 150 kHz, and the output ripple is approximately 40 mV.

Adding an output filter consisting of a 50 µH inductor and a 100 µF leaded tantalum capacitor reduced the ripple to approximately 3 mV as shown in Figure 10-40.

Linear regulators are often used following switching regulators for better regulation and lower noise. Low dropout (LDO) regulators such as the ADP3310 are desirable in these applications because they require only a small input-to-output series voltage to maintain regulation. This minimizes power dissipation in the pass device and may eliminate the need for a heat sink. Figure 10-41 shows the ADP1148 buck regulator configured for a 9 V input and a 3.75 V/1 A output. The output drives an ADP3310 linear LDO regulator configured for 3.75 V input and 3.3 V/1 A output. The input and output of the ADP3310 is shown in Figure 10-42. Notice that the regulator reduces the ripple from 40 mV to approximately 5 mV.

Section Ten

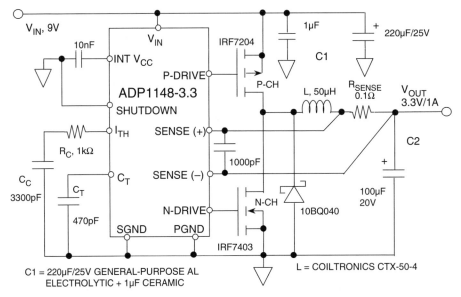

Figure 10-38: ADP1148 Buck Regulator Circuit

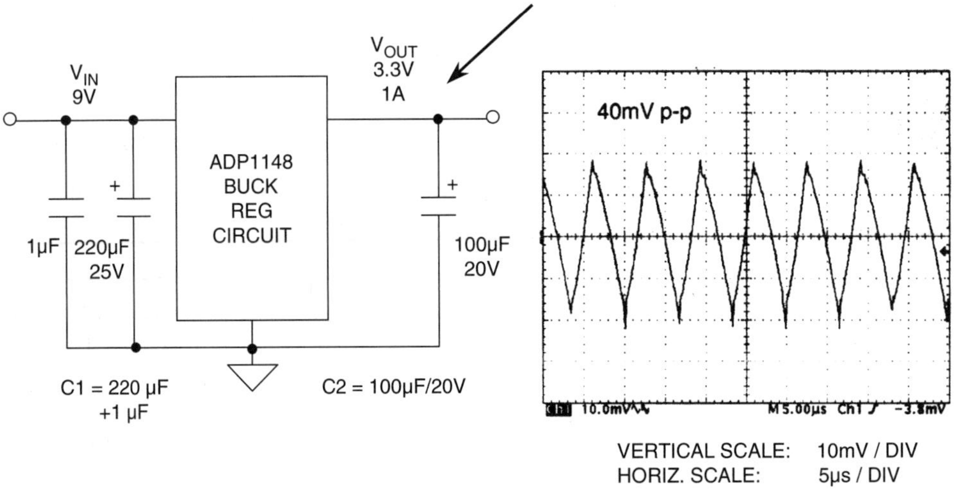

C1 = 1µF CERAMIC + 220µF/25V GENERAL-PURPOSE AL ELECTROLYTIC
C2 = 100µF/20V LEADED TANTALUM, KEMET T356-SERIES (ESR = 0.6Ω)

Figure 10-39: ADP1148 Buck Output Waveform

Hardware Design Techniques

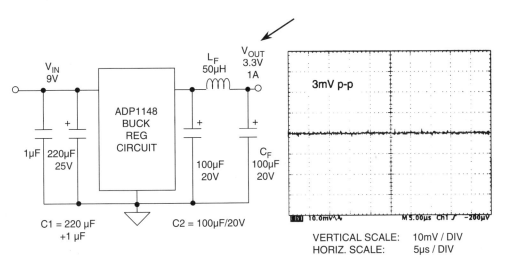

Figure 10-40: ADP1148 Buck Filtered Output

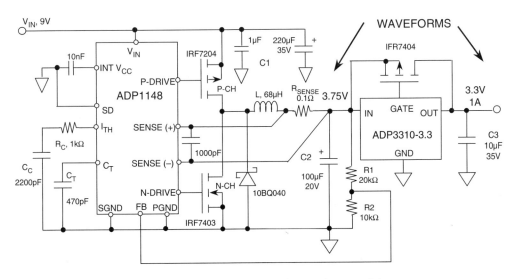

Figure 10-41: ADP1148 Buck Regulator Driving
ADP3310 Low Dropout Regulator

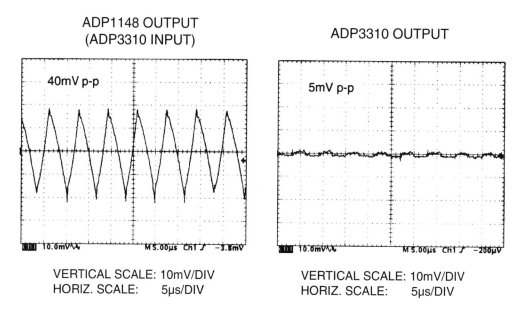

Figure 10-42: Waveforms for ADP1148 Buck Regulator Driving ADP3310 Low Dropout Regulator

There are many trade-offs in designing power supply filters. The success of any filter circuit is highly dependent upon a compact layout and the use of a large area ground plane. As has been stated earlier, all connections to the ground plane should be made as short as possible to minimize parasitic resistance and inductance.

Output ripple can be reduced by the addition of low ESL/ESR capacitors to the output. However, it may be more efficient to use an LC filter to accomplish the ripple reduction. In any case, proper component selection is critical. The inductor should not saturate under the maximum load current, and its dc resistance should be low enough as not to induce significant voltage drop. The capacitors should have low ESL and ESR and be rated to handle the required ripple current.

Low dropout linear postregulators provide both ripple reduction as well as better regulation, and can be effective, provided the sacrifice in efficiency is not excessive.

Finally, it is difficult to analytically predict the output ripple current, and there is no substitute for a prototype using the real-world components. Once the filter is proven to provide the desired ripple attenuation (with some added safety margin), care must be taken that parts substitutions or vendor changes are not made in the final production units without first testing them in the circuit for equivalent performance.

Hardware Design Techniques

- Proper Layout and Grounding (Using Ground Plane) Mandatory
- Low ESL/ESR Capacitors Give Best Results
- Parallel Capacitors Lower ESR/ESL and Increase Capacitance
- External LC Filters Very Effective in Reducing Ripple
- Linear Postregulation Effective for Noise Reduction and Best Regulation
- Completely Analytical Approach Difficult, Prototyping is Required for Optimum Results
- Once Design is Finalized, Do Not Switch Vendors or Use Parts Substitutions without First Verifying Their Performance in Circuit
- High Frequency Localized Decoupling at IC Power Pins is Still Required

Figure 10-43: Switching Supply Filter Summary

Localized High Frequency Power Supply Filtering

The LC filters described in the previous section are useful in filtering switching regulator outputs. It may be desirable, however, to place similar filters on the individual PC boards where the power first enters the board. Of course, if the switching regulator is placed on the PC board, the LC filter should be an integral part of the regulator design.

Localized high frequency filters may also be required at each IC power pin (see Figure 10-44). Surface-mount ceramic capacitors are ideal choices because of their low ESL. It is important to make the connections to the power pin and the ground plane as short as possible. In the case of the ground connection, a via directly to the ground plane is the shortest path. Routing the capacitor ground connection to another ground pin on the IC is not recommended due to the added inductance of the trace. In some cases, a ferrite bead in series with the power connection may also be desirable.

Section Ten

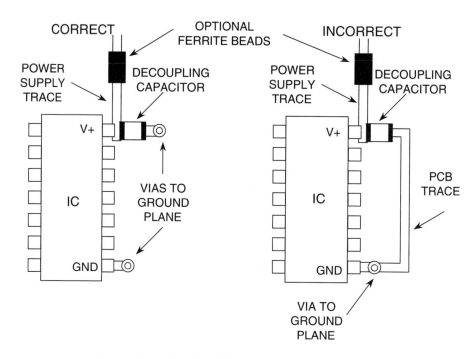

Figure 10-44: Localized Decoupling to Ground Plane Using Shortest Path

The following list summarizes the switching power supply filter layout/construction guidelines that will help ensure that the filter does the best possible job:

1. *Pick the highest electrical value and voltage rating for filter capacitors that is consistent with budget and space limits. This minimizes ESR, and maximizes filter performance. Pick chokes for low ΔL at the rated dc current, as well as low DCR.*

2. *Use short and wide PCB tracks to decrease voltage drops and minimize inductance. Make track widths at least 200 mils for every inch of track length for lowest DCR, and use 1 oz. or 2 oz. copper PCB traces to further reduce IR drops and inductance.*

3. *Use short leads or, better yet, leadless components, to minimize lead inductance. This minimizes the tendency to add excessive ESL and/or ESR. Surface-mount packages are preferred. Make all connections to the ground plane as short as possible.*

4. *Use a large area ground plane for minimum impedance.*

5. *Know what your components do over frequency, current, and temperature variations. Make use of vendor component models for the simulation of prototype designs, and make sure that lab measurements correspond reasonably with the simulation. While simulation is not* absolutely *necessary, it does instill confidence in a design when correlation is achieved* (see Reference 15).

Hardware Design Techniques

High Density DSP Localized Decoupling Considerations

High pin count DSP packages require special consideration with respect to localized decoupling due to their high digital transient currents. Typical decoupling arrangements are shown in Figure 10-45. The surface-mount capacitors are placed on the top side of the PC board in Figure 10-45A. For the SHARC family, eight 0.02 µF ceramic capacitors are recommended. They should be placed as close to the package as possible. The connections to the V_{DD} pins should be as short as possible using wide traces. The connections to ground should be made directly to the ground plane with vias. A less desirable method is shown in Figure 10-45B, where the capacitors are mounted on the back side of the PC board underneath the footprint of the package. If the ground plane underneath the package footprint is perforated with many signal vias, the capacitor return transient current must flow to the outside ground plane, which may be poorly connected to the inside ground plane due to the vias.

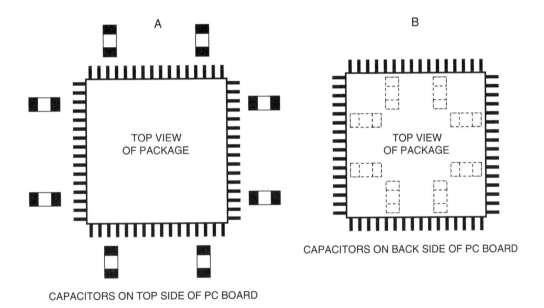

Figure 10-45: Decoupling High Pin Count DSPs in PQFP Packages

Section Ten

The PC board for a ball grid array (BGA) package is shown in Figure 10-46. Note that all connections to the balls must be made using vias to other layers of the board. The "dogbone" pattern shown is often used for the BGA packages. The shaded area indicates the location of the solder mask. As in the case of PQFP packages, the localized decoupling capacitors should be placed as close as possible to the package with short connections to the V_{DD} pins and direct connections to vias to the ground plane layer.

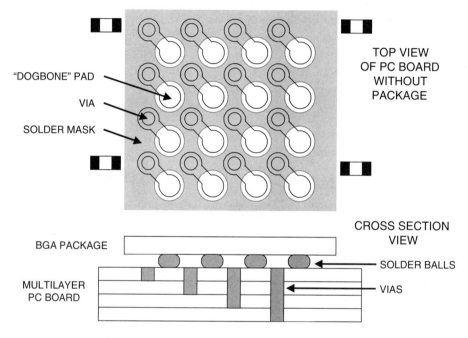

Figure 10-46: Decoupling High Pin Count DSPs in Ball Grid Array (BGA) Packages

The ADSP-21160 400-ball BGA 27 mm × 27 mm package approximate power and ground assignments are shown in Figure 10-47. The ball pitch is 1.27 mm. Approximately 84 balls are used in the center of the pattern for ground connections. The connections to the core voltage (40 balls) and the external voltage (46 balls) surround the ground balls. The remaining outer balls are used for the various signals.

The centrally located ground balls serve a dual function. Their primary function is to make a low impedance connection directly to the ground plane layer. Their secondary function is to conduct the package heat to the ground plane layer, which also acts as a heat sink, since the device must dissipate about 2.5 W under average operating conditions. The addition of an external heatsink as shown lowers the junction-to-ambient thermal resistance even further.

Hardware Design Techniques

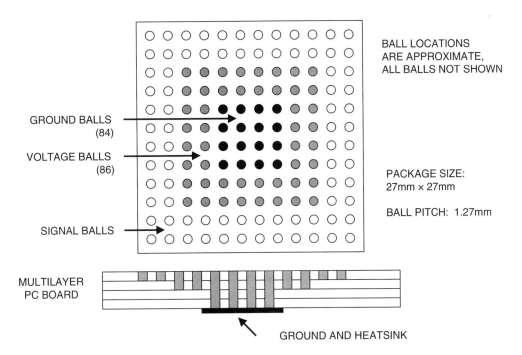

Figure 10-47: ADSP-21160 DSP 400-Lead PBGA Package Ball Locations

References on Noise Reduction and Filtering

1. **EMC Design Workshop Notes**, Kimmel-Gerke Associates, Ltd., St. Paul, MN. 55108, (612) 330-3728.

2. Walt Jung, Dick Marsh, *Picking Capacitors, Parts 1 & 2*, **Audio,** February, March, 1980.

3. Tantalum Electrolytic and Ceramic Capacitor Families, Kemet Electronics, Box 5928, Greenville, SC, 29606, (803) 963-6300.

4. Type HFQ Aluminum Electrolytic Capacitor and Type V Stacked Polyester Film Capacitor, Panasonic, 2 Panasonic Way, Secaucus, NJ, 07094, (201) 348-7000.

5. OS-CON Aluminum Electrolytic Capacitor 93/94 Technical Book, Sanyo, 3333 Sanyo Road, Forrest City, AK, 72335, (501) 633-6634.

6. Ian Clelland, *Metalized Polyester Film Capacitor Fills High Frequency Switcher Needs*, **PCIM**, June 1992.

7. Type 5MC Metallized Polycarbonate Capacitor, Electronic Concepts, Inc., Box 1278, Eatontown, NJ, 07724, (908) 542-7880.

8. Walt Jung, *Regulators for High-Performance Audio, Parts 1 and 2*, **The Audio Amateur,** issues 1 and 2, 1995.

9. Henry Ott, **Noise Reduction Techniques in Electronic Systems, 2d Ed.,** 1988, Wiley.

10. Fair-Rite Linear Ferrites Catalog, Fair-Rite Products, Box J, Wallkill, NY, 12886, (914) 895-2055, http://www.fair-rite.com.

11. Type EXCEL Leaded Ferrite Bead EMI Filter, and Type EXC L Leadless Ferrite Bead, Panasonic, 2 Panasonic Way, Secaucus, NJ, 07094, (201) 348-7000.

12. Steve Hageman, *Use Ferrite Bead Models to Analyze EMI Suppression*, **The Design Center Source,** MicroSim Newsletter**,** January, 1995.

13. Type 5250 and 6000-101K Chokes, J. W. Miller, 306 E. Alondra Blvd., Gardena, CA, 90247, (310) 515-1720.

14. DIGI-KEY, PO Box 677, Thief River Falls, MN, 56701-0677, (800) 344-4539.

15. Tantalum Electrolytic Capacitor SPICE Models, Kemet Electronics, Box 5928, Greenville, SC, 29606, (803) 963-6300.

16. Eichhoff Electronics, Inc., 205 Hallene Road, Warwick, RI., 02886, (401) 738-1440, http://www.eichhoff.com.

17. **Practical Design Techniques for Power and Thermal Management**, Analog Devices, 1998, Chapter 8.

Dealing with High Speed Logic

Much has been written about terminating printed circuit board traces in their characteristic impedance to avoid reflections. A good rule of thumb to determine when this is necessary is: *Terminate the line in its characteristic impedance when the one-way propagation delay of the PCB track is equal to or greater than one-half the applied signal rise/fall time (whichever edge is faster).* A conservative approach is to use a 2-inch (PCB track length)/nanosecond (rise/fall time) criterion. For example, PCB tracks for high speed logic with rise/fall time of 1 ns should be terminated in their characteristic impedance if the track length is equal to or greater than 2 inches (including any meanders). Figure 10-48 shows the typical rise/fall times of several logic families including the SHARC DSPs operating on 3.3 V supplies. As would be expected, the rise/fall times are a function of load capacitance.

- GaAs: 0.1 ns
- ECL: 0.75 ns
- ADI SHARC DSPs: 0.5 ns to 1 ns (Operating on 3.3 V Supply)

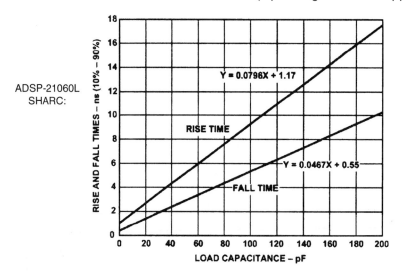

Figure 10-48: Typical DSP Output Rise Times and Fall Times

This same 2-inch/nanosecond rule of thumb should be used with analog circuits in determining the need for transmission line techniques. For instance, if an amplifier must output a maximum frequency of f_{max}, then the equivalent rise time, t_r, can be calculated using the equation $t_r = 0.35/f_{max}$. The maximum PCB track length is then calculated by multiplying the rise time by 2 inch/nanosecond. For example, a maximum output frequency of 100 MHz corresponds to a rise time of 3.5 ns, and a track carrying this signal greater than 7 inches should be treated as a transmission line.

Hardware Design Techniques

Equation 10.1 can be used to determine the characteristic impedance of a PCB track separated from a power/ground plane by the board's dielectric (microstrip transmission line):

$$Z_o(\Omega) = \frac{87}{\sqrt{\varepsilon_r + }}$$ Eq. 10.1

where ε_r = dielectric constant of printed circuit board material
 d = thickness of the board between metal layers, in mils
 w = width of metal trace, in mils
 t = thickness of metal trace, in mils

The one-way transit time for a single metal trace over a power/ground plane can be determined from Equation 10.2:

$$t_{pd}(ns/ft) = 1.0$$ Eq. 10.2

For example, a standard four-layer PCB board might use 8-mil-wide, 1 oz. (1.4 mils) copper traces separated by 0.021" FR-4 ($\varepsilon_r = 4.7$) dielectric material. The characteristic impedance and one-way transit time of such a signal trace would be 88 W and 1.7 ns/ft (7"/ ns), respectively.

The best ways to keep sensitive analog circuits from being affected by fast logic are to physically separate the two and use no faster logic family than dictated by system requirements. In some cases, this may require the use of several logic families in a system. An alternative is to use series resistance or ferrite beads to slow down the logic transitions where the speed is not required. Figure 10-49 shows two methods. In the first, the series resistance and the input capacitance of the gate form a low-pass filter. Typical CMOS input capacitance is 5 pF to 10 pF. Locate the series resistor close to the driving gate. The resistor minimizes transient currents and may eliminate the necessity of using transmission line techniques. The value of the resistor should be chosen such that the rise and fall times at the receiving gate are fast enough to meet system requirement, but no faster. Also, make sure that the resistor is not so large that the logic levels at the receiver are out of specification because of the voltage drop caused by the source and sink current that flow through the resistor. The second method is suitable for longer distances (>2 inches), where additional capacitance is added to slow down the edge speed. Notice that either one of these techniques increases delay and increases the rise/fall time of the original signal. This must be considered with respect to the overall timing budget, and the additional delay may not be acceptable.

Figure 10-50 shows a situation where several DSPs must connect to a single point, as would be the case when using read or write strobes bidirectionally connected from several DSPs. Small damping resistors shown in Figure 10-50A can minimize ringing provided the length of separation is less than about 2 inches. This method will also increase rise/fall times and propagation delay. If two groups of processors must be connected, a single resistor between the pairs of processors as shown in Figure 10-50B can serve to damp out ringing.

Section Ten

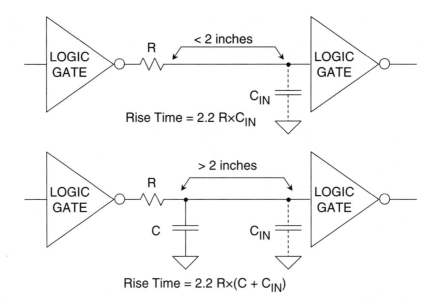

Figure 10-49: Damping Resistors Slow Down Fast Logic Edges to Minimize EMI/RFI Problems

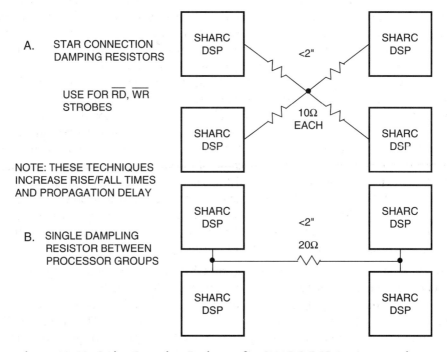

Figure 10-50: Series Damping Resistors for SHARC DSP Interconnections

Hardware Design Techniques

The only way to preserve 1 ns or less rise/fall times over distances greater than about 2 inches without ringing is to use transmission line techniques. Figure 10-51 shows two popular methods of termination: end termination, and source termination. The end termination method (Figure 10-51A) terminates the cable at its terminating point in the characteristic impedance of the microstrip transmission line. Although higher impedances can be used, 50 Ω is popular because it minimizes the effects of the termination impedance mismatch due to the input capacitance of the terminating gate (usually 5 pF to 10 pF). In Figure 10-51A, the cable is terminated in a Thevenin impedance of 50 Ω terminated to 1.4 V (the midpoint of the input logic threshold of 0.8 V and 2.0 V). This requires two resistors (90 Ω and 120 Ω), which adds about 50 mW to the total quiescent power dissipation to the circuit. Figure 10-51A also shows the resistor values for terminating with a 5 V supply (68 Ω and 180 Ω). Note that 3.3 V logic is much more desirable in line driver applications because of its symmetrical voltage swing, faster speed, and lower power. Drivers are available with less than 0.5 ns time skew, source and sink current capability greater than 25 mA, and rise/fall times of about 1 ns. Switching noise generated by 3.3 V logic is generally less than 5 V logic because of the reduced signal swings and lower transient currents.

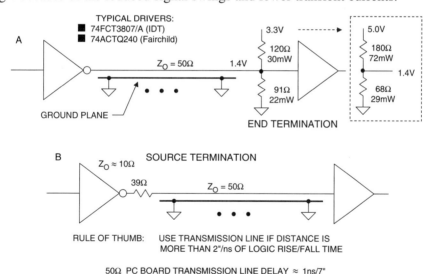

Figure 10-51: Termination Techniques for Controlled Impedance Microstrip Transmission Lines

The source termination method, shown in Figure 10-51B, absorbs the reflected waveform with an impedance equal to that of the transmission line. This requires about 39 Ω in series with the internal output impedance of the driver, which is generally about 10 Ω. This technique requires that the end of the transmission line be terminated in an open circuit; therefore no additional fanout is allowed.

Section Ten

The source termination method adds no additional quiescent power dissipation to the circuit.

Figure 10-52 shows a method for distributing a high speed clock to several devices. The problem with this approach is that there is a small amount of time skew between the clocks because of the propagation delay of the microstrip line (approximately 1 ns/7"). This time skew may be critical in some applications. It is important to keep the stub length to each device less than 0.5" in order to prevent mismatches along the transmission line.

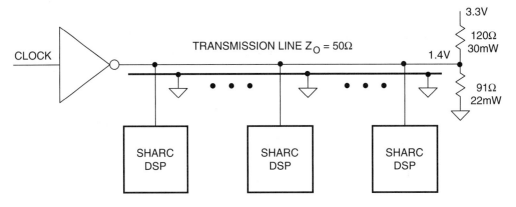

50Ω PC BOARD TRANSMISSION LINE DELAY ≈ 1ns/7"

NOTE: KEEP STUB LENGTH < 0.5"

Figure 10-52: Clock Distribution Using End-of-Line Termination

The clock distribution method shown in Figure 10-53 minimizes the clock skew to the receiving devices by using source terminations and making certain the length of each microstrip line is equal. There is no extra quiescent power dissipation, as would be the case using end termination resistors.

Figure 10-54 shows how source terminations can be used in bidirectional link port transmissions between SHARC DSPs. The output impedance of the SHARC driver is approximately 17 Ω, and therefore a 33 Ω series is required on each end of the transmission line for proper source termination.

The method shown in Figure 10-55 can be used for bidirectional transmission of signals from several sources over a relatively long transmission line. In this case, the line is terminated at both ends, resulting in a dc load impedance of 25 Ω. SHARC drivers are capable of driving this load to valid logic levels.

Hardware Design Techniques

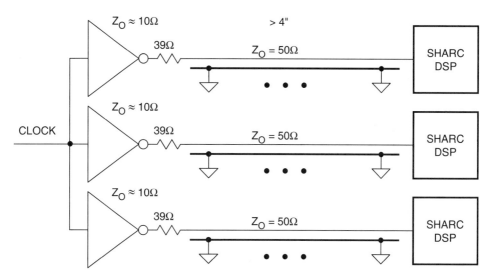

Figure 10-53: Preferred Method of Clock Distribution Using Source Terminated Transmission Lines

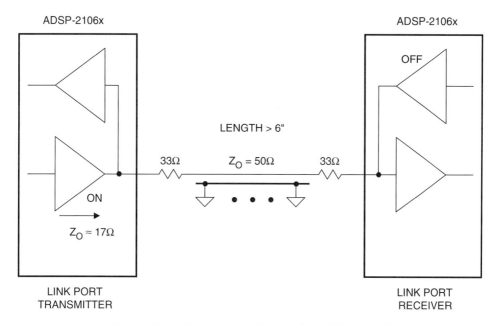

Figure 10-54: Source Termination for Bidirectional Transmission Between SHARC DSPs

Section Ten

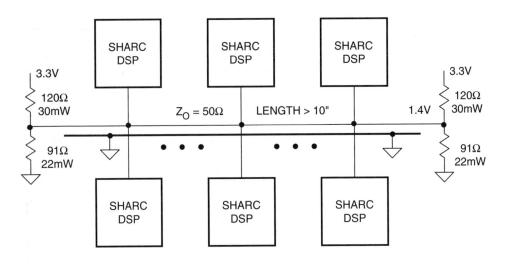

NOTE: KEEP STUB LENGTH < 0.5"

Figure 10-55: Single Transmission Line Terminated at Both Ends

References on Dealing with High Speed Logic

1. Howard W. Johnson and Martin Graham, **High-Speed Digital Design**, PTR Prentice Hall, 1993.

2. *EDN's Designer's Guide to Electromagnetic Compatibility*, **EDN**, January, 20, 1994, material reprinted by permission of Cahners Publishing Company, 1995.

3. *Designing for EMC (Workshop Notes)*, Kimmel Gerke Associates, Ltd., 1994.

4. Mark Montrose, **EMC and the Printed Circuit Board**, IEEE Press, 1999 (IEEE Order Number PC5756).

Index

- Subject Index
- Analog Devices Parts Index

Index

A

AA Alkaline Battery Discharge Characteristics, 334
Absolute value amplifier, 89
Active and Passive Electrical Wave Catalog, 56
AD185x, multibit sigma-delta DAC, 109
AD260/AD261:
   digital isolators
      key specifications, 358
      schematic, 358
AD820, op amp, 68-69
AD1819B SoundPort, codec, for audio/modem, 311
AD1836, mixed-signal IC, codec, 311-312
AD1852, sigma-delta audio DAC, 77
AD1853:
   dual 24-bit DAC, 110
   sigma-delta audio DAC, 77
AD1854, sigma-delta audio DAC, 77
AD1877:
   16-bit sigma-delta ADC
      characteristics, 76
      FIR filter characteristics, 77
AD189x, sample rate converters, 181
AD77xx, 24-bit sigma-delta ADC, low frequency/high resolution, filters, 77
AD77XX-Series Data Sheets, 93
AD773x, 24-bit ADC, with PGA, 7
AD977x, 14-bit TxDAC, 104
AD983x, 10-bit DDS system, 114
AD985x, 14-bit TxDAC, 104
AD5322:
   12-bit dual DAC
      block diagram, 264
      power-down feature, 263
      serial interface, 264-265
AD5340:
   12-bit parallel input DAC
      block diagram, 256
      interface diagram, 258
   wait states, 257

AD6521, voiceband/baseband mixed-signal codec, 295-296
AD6522, DSP-based baseband processor, 295-296
AD6523, Zero-IF Transceiver, 297
AD6524, Multi-Band Synthesizer, 297
AD6600, diversity receiver ADC, 304
AD6622, quad digital TSP, 304-305
AD6624, quad digital RSP, 304-305
AD6640:
   12-bit ADC
      with MagAmps, 84-85
      multitone testing, diagram, 44
AD6644, 14-bit ADC, 304-305
AD7722, 16-bit ADC, 335
AD7725:
   16-bit sigma-delta ADC
      with on-chip PulseDSP filter, 148
      programmable digital filter, 313
         block diagram, 314
      Systolix PulseDSP, 314
   sigma-delta ADC, with Systolix PulseDSP processor, 79-80
AD7730, sigma-delta ADC, 335
AD7731, sigma-delta ADC, 335
AD7853/7853L:
   12-bit ADC, block diagram, 262
   serial ADC interface, 263
AD7853L:
   serial clock, 261-17
      output timing, 262
AD7854/AD7854L:
   12-bit ADC
      block diagram, 251
      interface diagram, 252-253
      key interface timing comparisons, 251-252
AD7858/59:
   12-bit ADC, 68-69
   SAR ADC, circuit, 68-69
AD7892, 12-bit SAR ADC, 335

Index

AD9042:
    12-bit ADC, NPR, diagram, 46
    12-bit wideband ADC, high SFDR, 41-42
AD9201, 10-bit, dual-channel ADC, 268-269
AD9220:
    12-bit ADC, SINAD and ENOB, diagram, 39
    12-bit CMOS ADC, latency or pipeline delay, 87
AD9221, 12-bit CMOS ADC, latency or pipeline delay, 87
AD9223, 12-bit CMOS ADC, latency or pipeline delay, 87
AD9288-100, 8-bit dual ADC, functional diagram, 92
AD9410, 10-bit flash ADC, interpolation, diagram, 83
AD9761, 10-bit, dual channel DAC, 268-269
AD9772:
    14-bit interpolating DAC, 305
    14-bit oversampling interpolating TxDAC, 107-109
        block diagram, 108
    14-bit TxDAC, 104
        diagram, 104
        SFDR, 53-54
    TxDAC 14-bit DAC, 304-305
AD9850, DDS/DAC synthesizer system, diagram, 114
AD73322:
    block diagram, 265-266
    simplified interface timing, 267
AD73422, dspConverter, specifications, 268
AD20msp430:
    baseband processing chipset, 295-297
    components, 296-297
    construction, 295
Adams, R.W., 93, 94
Adaptive filter, 181-185
    basic concept, 181-182
    speech compression and synthesis, 182
ADC, 4
    1-bit, comparator, 71
    3-bit unipolar, transfer characteristics, 22-23
    10-bit, theoretical NPR, 46
    11-bit, theoretical NPR, 46
    12-bit
        pipelined architecture, diagram, 67
        theoretical NPR, 46
    24-bit, with PGAs, 7
    analog bandwidth, 39-40
    analog input, 24
    bit-per-stage, 61
    boot memory select, 248
    communications application, SFDR, 40
    conversion complete output, 247-248
    conversion process, 16
    convert start, 247-248
    data memory address, 250
    data memory select, 248
    digital output, grounding, 342-344
    DNL errors, 36
    DSP applications, 61-95
        high speed architectures, 62
        sigma-delta, 62
        successive approximation, 61-69
        types, 61
    dynamic performance, quantification, 35
    ENOB versus frequency, diagram, 40
    equivalent input referred noise, 34-35
    excess DNL, and missing codes, 26-27
    flash, 61
    high speed, pipelined, latency, 66, 68
    ideal 12-bit
        FFT, noise floor, 33
        SFDR sampling clock to input frequency ratio, 32
    ideal, distortion and noise, 29-33
    ideal N-bit
        dynamic performance analysis, diagram, 31
        errors, 29
        quantization noise, diagram, 30
    input/output memory select, 248
    low power/low voltage, design issues, 61
    memory address bus, 248
    memory read, 248
    memory select line, 248
    non-ideal 3-bit, transfer function, diagram, 27
    non-monotonic, 26
    normalized signal to reference, 24
    output, quantization, 21
    output enable/read, 248
    oversampling ratio, 71
    pipelined, 61
        latency or pipeline delay, 85
        timing, diagram, 67
    practical
        distortion and noise, 33-34
        noise and distortion sources, 34
    processor interrupt request line, 247-248
    program memory select, 248

Index

quantization error, 23-24
quantization uncertainty, 23-24
quantized output, 23, 24
ripple, 61
sampling clock jitter, 47
sampling frequency, versus antialiasing characteristics, 19
SAR, 63
    external high frequency clock, 66
    fundamental timing, 65
    resolutions, table, 65
    typical timing, 65
sigma-delta, 61-62, 69-81
    circuitry, 69-70
    as oversampling converter, 20
    programmable digital filter, 313-315
    VLSI technology, 69
    see also Sigma-delta ADC
sign-magnitude, use, 24
single-tone sinewave FFT testing, 32
SNR decrease with input frequency, 47
static transfer functions, DC errors, 21-28
subranging, 61
    pipelined converter, 66
successive approximation, 61-69
    basic, 63
    resolutions, table, 65
    see also SAR ADC
thermal noise, 34-35
Address bus, DAC, 255
ADI DSP collaborative, 243
ADI modified Harvard architecture, in microprocessor, 196-197
ADI SHARC floating point DSPs, 215-219
ADMC300, motor controller, 308
ADMC331, motor controller, 308
ADMC401, motor controller, 308
ADMCF326, motor controller, 308
ADMCF328:
    motor controller, 308
    block diagram, 309
ADP1148:
    synchronous buck regulator, 366-370
    circuit, 367-368
    driving ADP3308
        circuit diagram, 369
        waveforms, 370
    filtered output, 369
    output waveform, 367-368
ADP3310, linear LDO regulator, 367, 369-370

ADPCM, adaptive pulse code modulation, 4
ADSL, 285-290
    advantages, 286-287
    block diagram, 287
    data transmission capability, 289
    definition, 286
    installation advantages, 288
    modem, block diagram, 289
    modems, three-channel approach, 287-288
ADSP-21ESP202, codec, embedded speech processing, 310
ADSP-21mod870:
    digital modem processor, 284
    expanding central office capability, 283
    in voice-based RAS modem, 282-283
ADSP-21xx:
    16-bit fixed point DSP core, 196-212
    arithmetic, signed fractional format, 212-213
    assembly code for FIR filter, 10
    buses, 200-201, 201
    computational units, 201-203
    core architecture
        diagram, 199
        summary, 200
    data address generators and program sequencer, 203-204
    digital filter example, 197-200
    DSP optimized, 199
    FIR filter assembly code, 198
    fixed-point DSP, 197-198
        FIR filter, 157-158
        assembly code, 159
    internal buses, 200-201
    internal peripherals, powerdown, 208
    memory-write cycle timing diagram, 253-254
    on-chip peripherals, 204-212
        byte DMA port, 207
        internal DMA, 207
        SPORTs, 206
    read-cycle, timing diagram, 249
    serial ports
        block diagram, 259
        features, 259
        operation, 258-260
    SPORTs, 206
ADSP-218x:
    architecture, 205
    byte memory interface, 207
    implementation, multi-channel VOIP server, diagram, 285

Index

interface with codec, 266-268
internal direct-memory-access port, 206
memory-mapped peripherals, 206
modified Harvard architecture, 206
multiple core devices, 212
on-chip peripherals, memory interface, 205
roadmap, 212
VisualDSP software, 242

ADSP-219x:
   architecture, 209
   code compatibility with ADSP-218x, 209
   key specifications, 210
   roadmap, 212
   VisualDSP software, 242

ADSP-2100, core in ADSP-218x architecture, 205

ADSP-2100 EZ-KIT Lite Reference Manual, 244, 271

ADSP-2100 Family EZ Tools Manual, 244, 271

ADSP-2100 Family Users Manual, 3rd Edition, 244, 271

ADSP-2106x:
   characteristics, 217
   external ports, 218
   host interface, 218
   I/O processor, 218
   IEEE Standard P1149.1 Joint Test Action Group standard, 219
   instruction set, 217
   internal memory, 219
   multiprocessing systems, 218
   on-chip DMA, 219
   SHARC processors, 215-216
   Super Harvard Architecture, 215-217

ADSP-2106x SHARC EZ-KIT Lite Manual, 244, 271

ADSP-2106x SHARC User's Manual, 2nd Edition, July 1996, 244, 271

ADSP-2116x:
   32-bit DSP, second-generation, 220
   SIMD core architecture, 220-225
      diagram, 220
   SIMD DSP, 222
   SIMD features, 221

ADSP-2189M:
   69-tap FIR filter, design example, 168
   75 MIPS DSP
      filter subroutine, 158
      throughput time, 174
   75 MIPS processor, 166

EZ-KIT Lite, 236
fixed-point DSP, 147
key timing specifications, 255-256
parallel read timing, 250, 253
parallel write interface, timing specifications, 257
serial port, receive timing diagram, 260-261
system interface, full memory mode, diagram, 270

ADSP-21000 Family Application Handbook, 143, 186

ADSP-21060L SHARC, DSP output rise and fall times, 378

ADSP-21065L:
   connected to ADC and DAC, 268-269
   EZ-KIT Lite, 238
   SHARC, DSP benchmarks, 223

ADSP-21065L SHARC EZ-LAB User's Manual, 244, 271

ADSP-21065L SHARC User's Manual, Sept. 1, 1998, 244, 271

ADSP-21160:
   16-channel audio mixer, SIMD architecture, 313
   32-bit SHARC, key features, 221
   DSP, BGA package locations, 374-375
   integrated peripherals, 221
   SIMD/multiple channels, DSP benchmarks, 223
   SISD, DSP benchmarks, 223

ADSP-21160 SHARC DSP Hardware Reference, 244, 271

ADSP-TS001:
   features, 229
   TigerSHARC
      16-bit fixed-point DSP, 136-137
      architecture, 225-233
      benchmarks, 233
      diagram, 227

ADuM1100A, digital isolator, 356-357
Advanced mobile phone service *see* AMPS
Agilent HCPL-7720, 356
Aiken, Howard, 197
Alfke, P., 334
Aliasing, 16-3
   frequency domain, representation, 17
All-pole lattice filter, parameters, from speech samples, 184-185
Alternate framing mode, 260
Aluminum electrolytic capacitor, 360-361

Index

Amplifier Applications Guide (1992), 57, 95
Amplitude shift keying, POTS, 276
Analog bandwidth, ADC, 39-40
Analog cellular base station, 301-302
Analog Devices Inc., 16-bit DSP, roadmap, 212
Analog Devices' Motor Control Website, 317
Analog filter:
    frequency response, 150
    popular types, 170, 173-174
    requirements, 149
        for oversampling, 108
    versus digital, 149
Analog front end, 265-268
Analog receiver design, 300-301
Analog return current, 336-337
Analog signal:
    characteristics, 3-2
    discrete time sampling, 16-21
        diagram, 15
    normalized ratio, 23
    quantization, 16
        diagram, 15
    sampling, aliasing, 17
Analog superheterodyne receiver, 301
Andreas, D., 93
ANSI/IEEE Standard 754-1985, floating point arithmetic, 213-214
Anti-imaging filter, 148
Antialiasing filter, 148
    baseband sampling, oversampling, 19
    requirements, relaxing, 20
    specifications, 18-19
Aperture delay, 47-49
Aperture jitter, 47-49, 345
Apex-ICE, 238-239
    USB simulator, 239
Application of Digital Signal Processing in Motion Control Seminar (2000), 317
Architecture:
    computer
        ADI modified Harvard, 196
        Harvard, 196-197
        Von Neumann, 196
Arithmetic logic unit:
    in DSP, 194
    features, 201
Armstrong, Edwin H., Major, 301
The ARRL Handbook for Radio Amateurs, 115
Asymmetric digital subscriber line *see* ADSL
Audio system, synthesized, 5

Autocorrelation, 5
Automotive/home theater, using 32-bit SHARC, 312

B

Backplane ground plane, 338
Baines, Rupert, 316
Baker, Bonnie, 354
Ball grid array, 374
Band filter, 366
Bandpass filter:
    design, 169-170
    from low-pass and high-pass filters, 170
Bandpass sampling, 20-21
Bandpass sigma-delta ADC, undersampling, 78
Bandstop filter:
    design, 169-170
        equivalent impulse response, 169
    from low-pass and high-pass filters, 170
Baseband data signal, POTS, 276
Baseband sampling:
    antialiasing filter, 18-20
        oversampling, 19
    Nyquist zone, 20
Basestation, block diagram, 304
Basis function, 120
    correlation, DFT, 122
BDC binary coding, data converters, 24
Bennett, W.R., 56
Bessel filter, 170, 173-174
Best straight line, 25-26
Best, R.E., 115
Bilinear transformation, 174
Binary ADC:
    3-bit
        diagram, 89
        input and residue waveforms, 89
    single-stage, diagram, 88
Binary coding, data converters, 24
Binary DAC, 5-bit, architectures, diagram, 102
Bingham, John, 316
Bipolar converter, types, 24
Biquad, in IIR filter, 171
Bit reversal:
    for 8-point DFT, 132
    algorithm, 129-132
Bit-per-stage ADC, 87-92
    diagram, 87
Blackman window function, 140-141
Blackman, R.B., 57

Index

Block floating point, in FFT, 135
Boot memory select, ADC, 248
Bordeaux, Ethan, 244, 321
Boser, B., 93
Boyd, I., 316
Branch target buffer, 228, 230
Brannon, Brad, 316
Brokaw, Paul, 354
Bryant, James, 15, 61, 69, 99, 335, 354
Buck regulator, 366-370
Buffer register, 342-343
Bus:
    data memory address, 200
    data memory data, 200
    data transfer, 200-201
    internal result, 200-203
    program memory address, 200
    program memory data, 200
Buss wire, 336
Butterfly, 129-130
    DIF FFT, 133
    DIT FFT, 130
Butterworth filter, 107, 170, 173-174
    characteristics, 19
Byrne, Mike, 335

C

Cage jack, 336
Calhoun, George, 316
Capacitive coupling, doublet glitch, 51
Capacitor:
    equivalent circuit, pulse response, 363
    finite ESR, 362-363
    parasitic elements, 363
    types, 360-362
Card entry filter, 366
Carrier, 37
Cascaded biquads, 171
Cauer filter, 173-174
CCD image processing, 7
CDMA, digital telephone system, 291
Cellular phone:
    basic system, diagram, 290
    frequency reuse, diagram, 290
Ceramic, capacitor, 360-362
Charpentier, A., 93
Chebyshev filter, 8, 148-149, 170, 173-174
Chestnut, Bill, 359
Chip select, DAC, 255
Circular buffering:

DSP application, 195
DSP requirement, 195
in FIR filter, 156
FIR filter, 195
in FIR filter, output calculation, 157
FIR filter pseudocode, in DSP, 198
Clelland, Ian, 376
Clock distribution:
    end-of-line termination, diagram, 382
    source terminated transmission lines,
diagram, 383
Cluster multiprocessing, 223
    SHARC family, 225
CMOS IC, secondary I/O ring, diagram, 329
CMOS IC output driver, configuration, 323-324
Code division multiple access *see* CDMA
Code transition noise, and DNL, effects, 28
Codec, 7, 247
    interfacing, 265-268
    sampling rate, 266
    voiceband/audio applications, 310-313
codec and DSP, in voiceband and audio,
310-313
COder/DECoder, 247
Coding, types, 24
Coleman, Brendan, 57
Colotti, James J., 57
Comfort noise insertion, 294
Communications, external port, versus link
port, 224
Computational unit:
    arithmetic logic, 201
    multiplier-accumulator, 201-202
    shifter, 201-202
Computer, general purpose, Von Neumann
architecture, 196
Computing applications:
    CISC, 192
    data manipulation, 191
    mathematical calculation, 191
    RISC, 192-3
    tabular summary, 191
Concentrator, 281
Connelly, J.A., 354
Convert start, ADC, 247-248
Convolution, 5
    loop, 158
Convolving, filter responses, 169
Cooley, J.W., 128, 143
Cosier, G., 316

Coussens, P.J.M., 316
Crystal Oscillators, 354

D

DAC, 4
   3-bit switched capacitor, diagram, 64
   3-bit unipolar, transfer characteristics, 22-23
   address bus, 255
   binary weighted, 99
   in CD player, use of interpolation, 179
   chip select, 255
   conversion process, 16
   current output, diagram, 100
   in direct digital synthesis, 107
   distortion, specification, 52
   DNL errors, 36
   double-buffered, latches, 105
   DSP applications, 99-115
   dynamic performance, 49-55
   high-speed, 107
   interpolating, 107-108
   interrupt request, 255
   ladder networks, 99
   logic, 105-106
   low distortion, architectures, 101-105
   memory select, 255
   midscale glitch, diagram, 51
   monotonicity, 26
   non-ideal 3-bit, transfer function, diagram, 27
   non-monotonic, and DNL, 26
   output, quantization, 21
   reconstruction, output, 55
   segmented voltage, diagram, 101
   settling time, applications, 52
   sigma-delta, 109-110
     1-bit, in CD players, 107
   SIN (x)/x frequency roll-off, 55
   static transfer functions, DC errors, 21-28
   structures, 99-101
   transitions, with glitch, 50
   voltage output, Kelvin divider, 99
   write, 255
DAG, 203
Damping resistor:
   EMI/RFI minimization, 380
   series, for SHARC DSP interconnections, 380
DashDSP, motor controllers, 308-309
Data address generator:
   features, 203
   mode status register, 203
Data converter:
   analog bandwidth specification, 40
   applications, 22
   DC errors, 24, 24-25
   DC performance, 29
   gain error, diagram, 25
   integral and differential non-linearity
distortion effects, 36-37
   offset error, diagram, 25
   sampling and reconstruction systems, 22
Data manipulation, by computer, 191
Data memory address, ADC, 250
Data memory address bus, 200
Data memory buffer, circular, 195
Data memory data bus, 200
Data memory select, ADC, 248
Data reduction algorithms, in signal processing, 4
Data sampling, 15-58
   block diagram, 15
   distortion and noise, 29-33
   FFT, 15-16
   quantization error, 70
   quantization noise, 70
   real-time system, 15
Data scrambling, in sigma-delta DAC, 109
Data-flow multiprocessing, 223
Dattorro, J., 93
DC error, types, 24
DDS, 110-114
   basic architecture, 111
   code-dependent glitches, 52
   DAC, 101
     distortion, contributors, 53
   problems, 111
   system, flexible, diagram, 112
   tuning equation, 113
Dead time, DAC settling time, 49
Decimation, 70, 71
   multirate filter, diagram, 177
Decimation-in-frequency, DIF, 131-132
Decimation-in-time, DIT, 129
Decoupling, localized, diagram, 372
Decoupling points, diagram, 344
Deglitching, 85
Del Signore, B.P., 93
Delta phase register, 112
Designing for EMC (Workshop Notes), 385
DFT, 119-126
   8-point, 127, 130

Index

using DIT, 130-131
applications, 120
basis functions, 120
butterfly operation, 129-130
characteristics, 122
complex, 123
   equations, 124
   from real, 125
   input/output, 124
      real/imaginary values, 125
equations, real versus imaginary, 126
expansion, 127-128
FIR filter, 161
fundamental analysis equation, 120
inverse, 123
output spectrum, 120
real, 123
   input/output, 124
real versus imaginary
   equations, 125-126
   output conversion, 126
   relationship, 125-126
sampled time domain signal, 120
versus FFT, 129
DIF:
   butterfly, 131-132
   decimation-in-frequency, 131-132
Differential non-linearity:
   DNL, 26
   from encoding process, 36
DIGI-KEY, 376
Digital cellular base station, 302-305
Digital cellular base stations, advantages, 302
Digital cellular telephone, 290-294
   GSM system, 292-294
Digital communications services *see* DCS
Digital correction, for subranging ADC, 84
Digital filter, 147-185
   coefficient values, 147
   design procedure, 147
   diagram, 148
   filter coefficient modification, 181
   frequency response, 150
   general equation, 171-172
   lattice, 151
   moving average, 151-156
   programming ease in ADSP-21xx, 197-198
   real-time, 147, 149
      processing requirements, 150
   requirements, 149

types, 151
versus analog, 149
Digital filtering, 70
Digital FIR filter, 8
Digital isolation:
   by LED/photodiode optocouplers, 356
   by LED/phototransistor optocouplers, 355
Digital isolation technique, 355-358
Digital isolator, 356-358
Digital mobile radio, standards, 292
Digital receiver, characteristics, 301-302
Digital return current, 336-337
Digital signal, characteristics, 3-4
Digital Signal Processing Applications Using the ADSP-2100 Family, 143, 186, 316
Digital signal processor *see* DSP
Digital system, with ADC, quantization noise, 45
Digital telephone system, CDMA and TDMA, 291
Digital transmission, using adaptive equalization, 183
Direct data scrambling, in sigma-delta DAC, 109-110
Direct digital synthesis *see* DDS
Direct form 1 biquad filter, diagram, 172
Direct IF to digital conversion, 20-21
Direct memory access, DSP controller, 268
Direct-memory-access *see* DMA
Discontinuous transmission, 292-293, 294
Discrete Fourier transform *see* DFT
Discrete time Fourier series, 119
Discrete time sampling, 16
   analog signal, 16-21
Distortion, DAC performance, 49-55
Distortion and noise, practical ADCs, 33-34
DIT:
   decimation-in-time, 129
   FFT, algorithm, 129, 131
Dither signal, 32
DMA, internal, in ADSP-21xx, 207
DNL:
   ADC and DAc errors, 36
   converter error, definition, 26
   differential non-linearity, 26
Doernberg, Joey, 57
Double precision, 64-bit, floating point arithmetic, 214
Double-buffered DAC:
   advantages, 105-106

diagram, 106
latches, 105
Doublet glitch, 51
DSP, 4, 191-193
   16-bit, fixed-point family, history, 210-212
   32-bit, second-generation, 220
   ADSP-21xx, characteristics, 8-10
   analog versus digital, 6-7
      options, diagram, 7
   applications, 275-317
      summary, 315
   arithmetic
      fixed-point versus floating point, 212-215
      fixed-point versus floating-point, 212-215
   characteristics, 192
   code, compilation, 234
   and computer applications, 191
   core voltage, 322
   development tools, 236
   dot-product, 192
   efficiency, 208
   evaluation and development tools, 234-243
      VisualDSP and VisualDSP++, 241-243
   fundamental mathematical operation, 193
   grounding, 349-350
   hardware, 191-244
   high density, localized decoupling, 373-375
   interfacing, 247-271
   interfacing I/O ports, analog front ends, and codecs, 19-22
   internal phase-locked loops, grounding, 349-350
   kernel, 191-192
   optimization, 196
   output rise and fall times, diagram, 378
   parallel interfacing, 247-258
      block diagram, 254
      reading data from memory-mapped peripheral ADCs, 247-253
         to external ADC, block diagram, 248
      writing data to memory-mapped DACs, 253-258
   practical example, 7-10
   requirements, 193-196
      circular buffering, 195
      dual operand fetch, 194-195
      extended precision, 194
      fast arithmetic, 194
      summary, 196
      zero-overhead looping, 195-196

   sampled data, block diagram, 15
   serial interfacing, 258-265
      serial ADC, 260-265
      serial DAC, 263-265
   system interface, 269-270
   using VLSI, 6
   voiceband/audio applications, 310-313
DSP Designer's Reference (DSP Solutions), 244, 271
DSP Navigators: Interactive Tutorials about Analog Devices' DSP Architectures, 244
DSP Navigators: Interactive Tutorials about Analog Device's DSP Architectures, 271
Dual operand fetch:
   DSP application, 194-195
   DSP requirement, 194-195
Duracell MN1500 "AA" alkaline battery, discharge characteristics, 322
Dynamic performance analysis, ideal N-bit ADC, 31
Dynamic range compression, 6

E

Echo, voiceband telephone connection, 276
Eckbauer, F., 93
EDN's Designer's Guide to Electromagnetic Compatibility, 385
Edson, J.O., 56, 94
Effective aperture delay time, 48
   diagram, 49
Effective number of bits see ENOB
Eichhoff Electronics, Inc., 377
Electrolytic capacitor:
   characteristics, 360-361
   impedance versus frequency, diagram, 364
Elliptic filter, 170, 173-174
   characteristics, 19
Embedded speech processing, 310
EMC Design Workshop Notes, 376
End point, 25-26
Engelhardt, E., 93
ENOB, 35, 38-39, 70
   ADC, effective resolution, 70
   definition, 70
Equiripple FIR filter design, program inputs, 165-166
Equivalent input referred noise, thermal noise, 34-35

Index

Error voltage, 337
Euler's equation, 120
Extended precision:
   DSP application, 194
   DSP requirement, 194
   floating point arithmetic, 214
External clock jitter, 345
External port, versus link port, in communications, 224
EZ-ICE, in-circuit simulator, 238
EZ-KIT Lite, 234-235, 243
EZ-LAB, evaluation board, 234

F

Fair-Rite, 364-366
Fair10.Rite Linear Ferrites Catalog, 376
Fairchild 74VCX164245, 16-bit low voltage dual logic
  translators/transceivers, 330-335
Faraday shield, 342
Fast arithmetic:
   DSP application, arithmetic logic unit, 194
   DSP requirement, 194
Fast Fourier transform *see* FFT
Fast logic, and analog circuits, 379
FDMA:
   frequency division multiple access, 4
   telephone system, 4
FDMA communications link, NPR testing, 44
Ferguson, P. Jr., 94
Ferguson, P.F. Jr., 93
Ferrite, 364
   beads, 336, 340, 349, 350
      impedance, 365-366
   suitable for high-frequency filters, 365
FFT, 15, 119-143, 127-135, 192
   8-point DIF, 133
   8-point DIT algorithm, 131
   and DFT, 128
   DIF, 131-132
   FIR filter, 161
   first Nyquist zone aliases, 18
   floating point DSP, 136
   frame-based systems, 136
   hardware implementation and benchmarks, 135-136
   noise floor determination, 137
   processing gain, 32-33, 138
   processor, 135
   radix-2, 134

   radix-2 complex, hardware benchmark comparisons, 136
   radix-4, DIT butterfly, 134-135
   real-time
      considerations, 138
      DSP requirements, 136-138
      processing, 137
   sinewave
      integral number of cycles, 139
      nonintegral number of cycles, 140
   spectral leakage and windowing, 139-142
   twiddle factors, 127-10
   versus DFT, 129
Film capacitor, 360-362
Filter:
   analog, for oversampling, 108
   analog versus digital, 148-150
      frequency response comparison, 9
   antialiasing, specifications, 18-5
   band, 366
   bandpass, design, 169-170
   bandstop, design, 169-170
   baseband antialiasing, 18-20
   Butterworth, characteristics, 19
   capacitor, types, 360
   Chebyshev, 8
   corner frequency, 18
   design, and FFT, 119
   digital, diagram, 8
   digital FIR, 8
   elliptic, characteristics, 19
   FIR versus IIR, comparisons, 176
   high frequency, 371
   high-frequency, ferrites, 365
   high-pass, design, 169-170
   impulse convolving, 169
   low-pass, analog versus digital, in sampled data system, 7-6
   power supply, 359-377
   roll-off, sharpness, 155
   switching supply, summary, 371
   transfer function, 73
Filtering, 5, 6
FilterWizard, compiler, 314
Finite amplitude resolution due to quantization, 16
Finite impulse response *see* FIR
FIR filter, 151-176
   4-tap, output calculation, via circular buffer, 157

arbitrary frequency response, design, 163-164
CAD design programs, 164-165
characteristics, 160
circular buffering, 156-159
in circular buffering, 195
circular buffering, fixed boundary RAM, 156
coefficients, 161
compared with IIR filter, 176
computational efficiency increase, 178
design
    Fourier series method, with windowing, 163
    frequency sampling method, 163-18
    fundamental concepts, 159
    Parks-McClellan program, 164-168
    programming ease in ADSP-21xx, 197-198
    windowed-sinc method, 162
designing, 159-170
implementation, circular buffering, 156-159
impulse response, and coefficients, 160
Momentum Data Systems design
    frequency response, 167
    impulse response, 168
    step response, 167
N-tap, general form, 155-156
output decimation, 178
program outputs, 166
pseudocode, 158, 197-198
simplified diagram, 156
transfer function
    frequency domain, 161
    time domain, 161
versus IIR filter, 176
First-order sigma-delta ADC, 72
Fisher, J., 93
Fixed-point:
    16-bit, fractional format, 213
    DSP arithmetic, 212-215
    versus floating point, DSP arithmetic, 212-215
Flash ADC:
    10-bit, diagram, 83
    limited to 8-bits, 82
    parallel ADCs, 81
        diagram, 82
    problems, 81
Flash converter, 81-83
Floating-point:
    DSP arithmetic, 212-215
    versus fixed point arithmetic, 215
Folding converter, 89

Fourier series, 119
    FIR filter design, windowing, 163
Fourier transform, 119, 161
    and time domain signal, 120
Fourier, Jean Baptiste Joseph, 119
FPBW, full power bandwidth, 40
Framing mode, 260
Freeman, D.K., 316
Frequency division multiple access, FDMA, 4
Frequency domain, versus time domain, 159, 161
Frequency sampling method, FIR filter design, 163-164
Frequency synthesis, using PLLs and oscillators, diagram, 110
Fu, Dennis, 316
Full memory mode, 269
Full power bandwidth, FPBW, 40
Full-duplex hands-free car kit, diagram, 310
Fully decoded DAC:
    5-bit, diagram, 102
    diagram, 100

G

Gain error, 24
Galand, C., 316
Ganesan, A., 93, 94
Gardner, F.M., 115
Gaussian noise, 34-35, 44
Gaussian-filtered minimum-shift keying, 300
Geerling, Greg, 191
General DSP Training and Workshops, 244, 271
General purpose aluminum electrolytic capacitor, 360-261
Gerber files, 353
Ghausi, M.S., 58
Gibbs effect, 119
Glitch:
    code-dependent, effects, 52
    DAC performance, 49-55
    definition, 50
    energy, 51
    and harmonic distortion and SFDR, prediction, 53
    impulse area, 51
Global System for Mobile Communication *see* GSM
Glue logic, 247
Gold, B., 143, 186
Gold, Bernard, 58
Gosser, Roy, 56, 94

Index

Graham, M., 334
Graham, Martin, 354, 385
Grame, Jerald, 354
Gray bit, 90
Gray code binary coding, data converters, 24
Gray, A.H., Jr., 186
Gray, G.A., 56
Ground loop, 340, 355
Ground plane, 350
   decoupling high frequency current, 335
   impedance, 340
   low-impedance return path, 335-336
   printed circuit board, 337-338
Ground screen, 339
Grounded-input histogram, 34-35
Grounding:
   DSP, internal phase-locked loops, diagram, 350
   mixed signal devices, in multicard system, 347-349
   mixed signal systems, 335-354
   mixed-signal ICs, 335
   multiple ground pins, 352
   philosophy, summary, 351
   points, diagram, 344
   separate analog and digital grounds, 339-340
   single-card versus multicard, concepts, 346-347
Groupe Speciale Mobile *see* GSM
GSM, block diagram, 292-293
GSM handset:
   components, 295-300
   using SoftFone baseband processor, and Othello radio, 295-300

H

Hageman, Steve, 376
Half-flash ADC, 83
Hamming window function, 140-141
Hanning window function, 140-141
Hard limiter, 43
Hardware design, techniques, 321-385
Harmonic distortion, 37-38
   DAC, 52
   definition, 37
   products, location, diagram, 37
Harmonic sampling, 20-21
Harmonic undersampling, 21
Harrington, M.B., 317
Harris, Fredrick J., 57, 143, 186

Harris, Steven, 94
Harvard architecture:
   in DSP, 195
   in microprocessor, 196-197
Hauser, Max W., 94
Haykin, S., 186
HDTV, high definition television, 4
Heise, B., 93
Hellwig, K., 316
Henning, H.H., 56, 94
Higgins, Richard J., 11, 58, 143, 186, 244, 271
High definition television, HDTV, 4
High density DSP:
   localized decoupling diagram, 373
   using BGA packages, 374
High Speed Design Techniques (1996), 95
High speed logic, 378-385
High-level language support, architectural features, 231
High-speed interfacing, 268-269
High-pass filter:
   design, 169-170
   using low-pass FIR, 169
Hilton, Howard E., 58
Hodges, David A., 57
Hofmann, R., 316
Honig, Michael L., 186
Horvath, Johannes, 321
Host memory mode, 269
HP Journal, 57-58
HP Product Note, 57

I

IC, low-voltage mixed signal, 322
IEEE-754 standard, 213-214
IEEE Trial-Use Standard for Digitizing Waveform Recorders, 58
IF sampling, 20-21
IIR biquad filter:
   basic, 171
   simplified notations, 173
IIR elliptic filter, 176
IIR filter, 170-173
   analog counterparts, 170
   CAD design, using Fletcher-Powell algorithm, 174
   characteristics, 171
   compared with FIR filter, 176
   design techniques, 173-175, 175

direct form implementation, 172
feedback, 151
IIR filter, 170-173
implementation, 171
throughput considerations, 175
versus FIR filter, 176
IMD:
    measurement, 42
    second- and third-order, diagram, 43
    two tone intermodulation distortion, 42-44
Impulse invariant transformation, 174
IMT-2000 protocol, 225
In-circuit simulator, 238
Indirect field-oriented control, 306
Induction motor:
    control, 306-309
        block diagram, 307
Infinite impulse response *see* IIR
Information, in signal, 4
Input filtering, 7
Input noise rejection, 7
Input/output memory select, ADC, 248
Instruction register, data address generator, 204
Integral linearity error, measurement, 25-26
Integral sample-and-hold, 33
Interface:
    2.5V/63V, diagram, 335
    63V/2.5V, 330-335
        diagram, 331-332
Interfacing, high-speed, 268-269
Interference, components, 359
Intermodulation distortion, DAC, 52
Internal aperture jitter, 345
Internal result bus, 200-203
Interpolation:
    frequency domain effects, 179-180
    implementation example, 179-180
    multirate filter, 178-179
Interrupt request, DAC, 255
Intranet, 281
I.Q. convention, 213
IRQ, 247

J
Jantzi, S.A., 93
JEDEC, 334
    specification, for packaging, 210
    standards bureau, 325
Jitter:
    aperture, 345

effects, 47
external clock, 345
internal apeture, 345
sampling clock, 345
Johnson noise, 22
Johnson, H., 334
Johnson, Howard W., 354, 385
Joint Electron Device Engineering Council *see* JEDEC
JTAG, 238
Jung, Walt, 359, 376

K
Kelvin divider, 99
    disadvantages, 99-100
    ladder network, 101
Kelvin-Varley divider, string DAC, 101
Kerr, Richard J., 115
Kester, Walt, 3, 15, 57, 61, 95, 99, 119, 147, 191, 247, 275, 321, 335, 354-355, 359
Kettle, P., 317
King, Dan, 191, 247
Koch, R., 93

L
Lagrange, Joseph Louis, 119
Laker, K.R., 58
Lane, Chuck, 56, 94
Laplace transform, 161
    conversion to z-transform, 174
Laplace, Pierre Simon de, 119
Latency, 85
Lattice filter, 151
Leaded ferrite bead, 364
Least significant bit:
    definition, 21
    quantization, chart, 22
    size, 22
Least-mean-square algorithm, for filter coefficients, 182
Lee, Hae-Seung, 57
Lee, Wai Laing, 93
Lee, W.L., 93
Levine, Noam, 191
Linear Design Seminar (1995), 56, 93, 94
Linear integrity error, 25
Linear predictive coding *see* LPC
Linear settling time, DAC settling time, 49
Linearity errors, 24

Index

Link descriptor file *see* LDF
Link port, versus external port, in communications, 224
Link port multiprocessing, 223
LMS, least-mean-square, 182
Logarithmic pulse code modulation, 293
Logic:
    CMOS IC output driver, configuration, 323-324
    high speed, 378-385
Logic translating transceiver:
    diagram, 331
    voltage compliance, 331
Long term prediction *see* LTP
Low voltage interface, 321-324
Low Voltage Logic Alliance, standards, 325
Low voltage logic level, standards, 324-325
Low voltage mixed-signal IC, 322
Low-pass filter, LPF, 71
Low-pass filter, design of other filters, 169
LPC, 182-183
    in all-pole lattice filter, 184-185
    model of speech production, 183
    speech companding system, 184
    speech processing system, 293
    speech system, digital filters, 184
LPF:
    analog, in sigma-delta ADC, 73
    low-pass filter, 71
LQFP, new package designation, 210
Lucey, D.J., 317
LVTTL, logic level, 324
Lyne, Niall, 317

M

MAC, 194, 201-202
McClellan, J.H., 186
MagAmp, 89
    3-bit folding ADC
        block diagram, 91
        functional equivalent circuit, 90
        input and residue waveforms, 91
Magnitude amplifier, 89
Magnitude-amplifier architecture, MagAmp, 84
Mahoney, Matthew, 58
Main lobe spreading, 140
Manolakis, Dimitris G., 143, 186
Markel, J.D., 186
Marsh, Dick, 376
Matched z-transform, 174
MathCad 4.0 software package, 58

Mathematical calculation, by computer, 191
Matsuya, Y., 93, 94
Maximally flat filter, 173
Mayo, J.S., 56, 94
Meehan, Pat, 57
Memory address bus, ADC, 248
Memory read, ADC, 248
Memory select:
    ADC, 248
    DAC, 255
Messerschmitt, David G., 186
Microcontroller, 191-193
    characteristics, 192
MicroConverter, precision analog circuitry, 5, 192
Microprocessor, 191-193
    architectures, comparison, 196
    characteristics, 192
Million operations per second *see* MOPS
Millions of instructions per second *see* MIPS
miniBGA package, 210-211
Minimum 4-term Blackman-Harris window function, 140-141
MIPS, 192, 208
Missing codes:
    in ADC, 26-27
    defining, 28
Mitola, Joe, 316
Mixed Signal Design Seminar (1991), 93
Mixed signal processing, 6
Mixed-signal device:
    grounding, 346-347
        diagram, 347
Mixed-signal IC:
    grounding, multiple printed circuit boards, 348-349
    high digital currents, grounding, 348-349
    low digital current, grounding and decoupling, 341-342
    low digital currents, grounding, 347-348
Mobile telephone service, superheterodyne receiver, diagram, 301
Mode status register, data address generator, 203
Modem:
    full-duplex, 275
    half-duplex, 275
    high performance, telephone service, 275-282
    high speed, 4
    RAS, 281-284

standards, chart, 277
V.90 analog
   block diagram, 278
   details, 280
V.90 analog versus V.34, diagram, 281
Momentum Data Systems, QED1000 program, for FIR filter design, 164
Momentum Data Systems, Inc., 186
Montrose, Mark, 354, 385
MOPS, 208
Moreland, Carl, 56, 94
Morley, Nick, 316
Morris, Jesse, 191
Morrison, Ralph, 354
Motchenbacher, C.D., 354
Motherboard, grounding, 338-339
Motor control, 306-309
   fully integrated, DashDSP, 309
Motor Control Products, Application Notes, and Tools, 317
Moving average filter, 151-156
   calculating output, 153
   diagram, 152
   frequency response, 154
   noise, 152-154
   step function response, 152-153
MSP, 6
Multichannel high frequency communication system, NPR, 45
Multichannel VOIP server, 284-285
Multitone SFDR, measurement, 43
Multicard mixed-signal system, grounding, 338-339
Multilayer ceramic chip caps, 362
Multiplier-accumulator unit, features, 194, 202
Multiplying DACs, 101
Multipoint ground, diagram, 339
Multiprocessing, using SHARCs, 223-225
Multirate filter, 177-181
   decimation, 177
   interpolation, 177
Murden, Frank, 56, 94
Murray, Aengus, 317

N

Narrowband, 301
Network:
   LAN, 4
   local area, 4
Nicholas, Henry T. III, 115

NMOS FET, bus switches, interfacing, 326-328
Noise:
   and grounding, 342
   power supply, 359-377
Noise power ratio see NPR
Noise shaping, using analog filter, 73
Non-ideal ADC, 3-bit, transfer function, diagram, 27
Non-ideal DAC, 3-bit, transfer function, diagram, 27
Normal framing mode, 260
NPR, 44-46
   measurement, diagram, 45
   NPR, 44-46
   theoretical, for various ADCs, 46
Numerically controlled oscillator, 111
   block diagram, 112
Nyquist band, 70
Nyquist bandwidth, 29, 31
   aliasing, 18
   definition, 18
Nyquist criteria, 16
Nyquist frequency, 55, 279
Nyquist zone, 18, 40-41
   baseband sampling, 20
   spurious frequency component, 18

O

Offset binary coding, data converters, 24
Offset error, 24
one's complement binary coding, data converters, 24
Oppenheim, A.V., 143, 186
Optocoupler, 355
Optoisolator, 355
OS-CON Aluminum Electrolytic Capacitor 93/94 Technical Book, 376
OS-CON electrolytic capacitor, 360-361, 363
Othello:
   radio, chipset, 295
   radio receiver
      advantages, 298
      block diagram, 298-299
      compactness, advantages, 300
      superhomodyne architecture, 297-298
Ott, Henry, 354, 376
Output enable/read, ADC, 248
Output ripple, reduction, 370
Oversampling, 70
   and baseband antialiasing filter, 19

Index

definition, 30-31
ratio, 71

P

Parallel ADC, diagram, 82
Parallel peripheral device:
   read interface, key requirements, 249
   write interface, key requirements, 255
Parasitic capacitance, 363
Parasitic inductance, 342
Parks-McClellan program:
   equiripple FIR filter design, program inputs, 165-166
      FIR filter, 174
         design, 164-168
Parks, T.W., 186
Parzefall, F., 93
Passband ripple, 148
   in filter, 8
Pattavina, Jeffrey S., 354
PCI emulator, 240
Peak glitch area, 51
Peak spurious spectral content, ADC, 40
Pentium-Series, 192
Pericom Semiconductor Corporation, 334
Permanent magnet synchronous machine, 307
PGA, programmable gain amplifier, 7
Phase accumulator, 112
Phase jitter, 47
Phase-locked loop, PLL, 110
Phase-Locked Loop Design Fundamentals, 115
PIC, 192
Picocell, 304
Pin socket, 336
Pipeline delay, 85
Pipelined ADC, 83-87
   12-bit CMOS ADC, block diagram, 86
   error correction, 85-86
Pipelined subranging ADC, 12-bit, digital error correction, block diagram, 85
Plain Old Telephone Service see POTS
PMOS transistor current switch, diagram, 105
Point of presence see POP
Polyester, capacitor, 360-361
POP, 282
POTS:
   block diagram, 276
   hybrid curcuit, 275
Power inverter, 306

Power plane, 335-337
Power supply:
   filtering, 359-377
   localized high frequency filter, 371-373
      construction guidelines, 372-373
   noise reduction, 359-377
   pins, decoupling, 336
   separate for analog and digital circuits, 343
PowerPC, 192
Practical Analog Design Techniques (1995), 56, 94
Practical Design Techniques for Power and Thermal Management, 377
Practical Design Techniques for Sensor Signal Conditioning, 11
Printed circuit board:
   double-sided versus multilayer, 337-338
   ground plane, 337
   impedance, calculation, 379
   mixed-signal system, layout guidelines, 351-353
   "motherboard", grounding, 338-339
   trace termination, 378
Proakis, John G., 143, 186
Processing gain:
   definition, 30-31
   FFT, 32-33
Processor interrupt request line, ADC, 247-248
Program memory address bus, 200
Program memory data bus, 200
Program memory select, ADC, 248
Program sequencer:
   data address generator, 204
   features, 204
Programmable gain amplifier, PGA, 7
Prom splitter, 234
Pseudocode, FIR filter, using DSP with circular buffering, 198
Pulse code modulation, 277

Q

QEDesign, 314
   filter package, 80
QS3384 Data Sheet, 334
QS3384 QuickSwitch, 327-328
   transient response, diagram, 328
Quadrature amplitude modulation, 300
   diagram, 279
   POTS, 277
Quantization:
   effects on ideal ADC, 29

size of LSB, chart, 22
Quantization error, 23-24
  signal, 29
Quantization noise:
  in digital system with ADC, 45
  for ideal N-bit ADC, diagram, 30
  shaping, 70
    in sigma-delta ADC, 74
  spectrum, diagram, 30
Quantization uncertainty, 23-24
QuickSwitch, in bidirectional interface, 326-328

R
Rabiner, Lawrence, 58
Rabiner, L.R., 143, 186, 187
RAMDAC, 5
Ramierez, Robert W., 57
Ramirez, R.W., 143
RAS:
  equipment, 284
  modem, 281-284
    as Internet gateway, diagram, 282
  on-switch based, 283
RAS/VOIP servers, 284
Receive data register, 260
Receive frame sync, 260
Receive shift register, 260
Reconstruction, definition, 28
Reconstruction filter, 148
Recovery time, DAC settling time, 49
Recursive filter, 170
Recursive-least-squares algorithm, for filter coefficients, 182
Reference frame theory, 307
Regular pulse excitation see RPE
Reidy, John, 57
Remez exchange algorithm, 164-165
Remote access server see RAS
Remote network access, 281
Rempfer, William C., 354
Ripple, passband, in filter, 8
Ripple ADC, 87-92
  diagram, 87
RISC, and DSP, 193
RLS, recursive-least-squares, 182
Roche, P.J., 317
Rolloff:
  sharpness, 155
  sidelobe, 140
Root-sum-square, RSS, 38

Rorabaugh, C. Britton, 143, 186, 244, 271
Rosso, M., 316
Ruscak, Steve, 56

S
Sample rate converter, using interpolator and decimator, 181
Sample-and-hold see SHA
Sampled data system, definition, 28
Sampling:
  above first Nyquist zone, 21
  bandpass, 20-21
  harmonic, 20-21
  IF, 20-21
  rate versus bandwidth, 21
Sampling clock, 344-346
  ground planes, diagram, 346
  jitter, effect on SNR, 345
Sampling rate, increased, 7
Samueli, Henry, 115
SAR, 62
SAR ADC, 63
  external high frequency clock, 66
  fundamental timing, 65
  resolutions, table, 65
  switched capacitor, 68
  with switched capacitor DAC, 63
  typical timing, 66
Sauerwald, Mark, 354
Scannell, J.R., 317
Schafer, R.W., 143, 186, 187
Schmid, Hermann, 56, 94
Schmitt trigger, 358
Schottky diode, 327, 340, 349, 350
Segmentation, in thermometer DAC, 103
Segmented DAC, 100-101
  10-bit, diagram, 103
Segmented voltage DAC, segmented voltage, 101
Sensor:
  as analog device, 6
  in industrial data acquisition and control systems, 5
  uses, 3-4
Serial ADC, 87-92
  diagram, 87
  to DSP interface, 260-263
Serial clock, 260
Serial DAC, to DSP interface, 263-265
Serial ports, in ADSP-21xx, 206
Serial-Gray, ADC architecture, 89

Index

Settling time:
   DAC performance, 49-55
      diagram, 50
   periods, 49
SFDR, 31, 35, 40-42, 49-55
   ADC, definition, 40
   DAC
      distortion, 52
      performance, 49-55
      test setup, 54
   diagram, 41
   sampling clock to input frequency ratio, for ideal 12-bit ADC, 32
SHA, 33-34
   and aperture jitter, 47
SHARC:
   32-bit, key features, 221
   architecture, 216
   coding, 222
   decoupling, 373
   DSP
      evaluation device, 238-239
      floating and fixed point arithmetic, 214
   DSP benchmarks, 223
   family roadmap, 222
   FFT butterfly processing, 215
   floating point DSP, 215-219
   floating-point DSP, 215-219
   key features, 216
   multi-function instruction, 222
   multiprocessing, 223-225
   multiprocessor communication, examples, 224
   program sequencer, 215
   VisualDSP++ software, 242
SHARC DSP:
   32-bit floating point, 311-312
   bidirectional transmission, source termination, diagram, 383
Sheingold, Dan, 58
Sheingold, Daniel H., 11, 94
Shifter unit, features, 202
Shunt resistance, 363
Sidelobe, 140
Sigma-delta ADC:
   band-pass filters, 78-79
   characteristics, table, 70
   first-order, diagram, 72
   fixed internal digital filter, 79
   and missing codes, 28
   multi-bit

      diagram, 78
      flash ADC with DAC, 77-78
   noise shaping, 72-73
   order versus oversampling, 74-75
   oversampling, 80
      converter, 20
   programmable digital filter, 313-315
   quantization noise shaping, 74
   second-order, block diagram, 75
   settling time, and digital filter, 76
   SNR, 80
   summary, 81
Sigma-delta audio DAC, 77
Sigma-delta DAC, 109-110
   architecture, diagrams, 109
   high resolution, oversampling, 109
Sign-magnitude bipolar converter, 24
Signal:
   amplitude, 4
   analog, 3
   bandwidth
      aliasing, 16
      Nyquist criteria, 16
   characteristics, table, 3
   conditioning, 3
      analog signal processor, 6
      sensors, 3-4
      successive approximation ADCs, 62
   continuous aperiodic, 119
   continuous periodic, 119
      sinusoidal waves, 119
   definition, 3, 4
   digital, 3-4
   frequency, 4
   processing
      frequency compression, 4
      information extraction, 4
      methods, 6-7
      mixed, 6
      real-time, comparison chart, 10
      reasons, 5
   real-world
      generation, 5
      origins, 3-4
      processing, 4-5
      units of measurement, 3-4
   recovery, 5
   sampled aperiodic, 119
   sampled periodic, 119
   spectral content, 4

timing, 4
units of measurement, table, 3
Signal-to-noise ratio *see* SNR
Signal-to-noise-and-distortion ratio, SINAD, 35, 38-39
Signed fractional format, DSP arithmetic, 212
Signed integer format, DSP arithmetic, 212
Silence descriptor, 294
SIMD architecture, 220
SINAD, 35, 38-39
   conversion to ENOB, equation, 39
Singer, Larry, 56
Single precision, floating point arithmetic, standard, 214
single-instruction, multiple data *see* SIMD
Single-instruction, single-data *see* SISD
Sinusoidal wave, continuous periodic signal, 119
SISD architecture, 220
Sluyter, R.J., 316
Small signal bandwidth, 39
Smith, Steven W., 143, 186, 244, 271
Snelgrove, M., 93
SNR, 29, 35, 38-39
   decrease with input frequency in ADC, 47
   definition, 29
   degradation, from external clock jitter, 345
   due to aperture and sampling clock jitter, diagram, 48
SNR-without-harmonics, calculation, 39
Sodini, C.G., 93
SoftCell:
   chipset
      block diagram, 304
      for digital phone receivers, 303
SoftFone, baseband processor, chipset, 295
Software development environment, 241
Software simulator, for debugging, 234
Southcott, C.B., 316
Spectral inversion, in filter design, 169
Spectral leakage:
   FFT, 139
   windowing, 141
Spectral reversal, in filter design, 169
Speech:
   compression, adaptive filter use, 182
   encoder/decoder, 292-293
   processing, standards, 292-293
   synthesis, adaptive filter use, 182
   synthesized, 5

SPORT:
   in ADSP-21xx, 206
   transfer rate, 266
Spurious free dynamic range *see* SFDR
SSBW, small signal bandwidth, 39
Stacked-film capacitor, 360, 362
Star ground, 336, 339, 346
Static branch prediction logic, 228
Static superscalar, 228
Stearns, S.D., 186
Stopband ripple, 148
String DAC:
   diagram, 100
   disadvantages, 99-100
   Kelvin divider, 99
Subranging ADC, 83-87
   8-bit, diagram, 84
   12-bit, digitally corrected, block diagram, 85
   digital correction, 84
Successive approximation ADC, 61-69
   basic, 63
Successive approximation register *see* SAR
SUMMIT-ICE, PCI emulator, 240
Super Harvard Architecture, for 32-bit DSP, 216-217
Superheterodyne architecture, 297-298
Superhomodyne architecture:
   diagram, 298-299
   operation, 299-300
Supply voltage reduction, 321
Surface mount ferrite bead, 364
Swanson, E.J., 93
Switching capacitor, characteristics, 360-261
Switching regulator, 366-371
   filtering, experiment, 366-370
   noise, reduction tools, 360
Switching supply, filter, summary, 371
Switching time, DAC settling time, 49
System Applications Guide (1993), 57, 93, 95
Systolix FilterExpress, 314
Systolix PulseDSP, 314
   filter core, 148
   processor, 79-80

T

T-Carrier system, 4
Tantalum electrolytic capacitor, 360-361
Tantalum Electrolytic Capacitor SPICE Models, 377
Tantalum Electrolytic and Ceramic Capacitor

Index

Families, 376
Tant, M.J., 56
TDMA, 4
Telephone, FDMA, 4
Tesla, Nikola, 306
THD+N, 37-38
  definition, 38
THD, 35, 37-38
  definition, 38
Thermal noise, equivalent input referred noise, 34-35
Thermometer code, 81
Thermometer DAC:
  5-bit, diagram, 102
  diagram, 100
  disadvantage, 103
  segmentation, 103
Thevenin impedance, 381
Thompson filter, 173-174
TigerSHARC:
  ADSP-TS001 static superscalar DSP, 225-233
  architecture, 227
    branch target buffer, 228, 230
    features, 228
    flexibility, 225
    key features, 226
    multiprocessor key elements, 225-226
    static branch prediction logic, 228
    static superscalar, 228
  development tools, 243
  DSP, 304-305
  key features, 229
  multiprocessing implementation, sample configuration, 231-232
  peak computation rates, 230
  roadmap, 232
Time division multiple access *see* TDMA
Time domain, versus frequency domain, 159, 161
Time sampling, analog signal, 16-21
Total harmonic distortion *see* THD
Total harmonic distortion plus noise *see* THD+N
TQFP package designation, new designations, 210
Transmission line:
  controlled impedance microstrip, termination, 381
  termination at both ends, 384
Transmit data register, 260
Transmit shift register, 260
TREK-ICE, Ethernet emulator, 240
TTL, logic level, 324
Tukey, J.W., 57, 128, 143
Twiddle factors, 127-128
Two tone intermodulation distortion, IMD, 42-44
two's complement binary coding, data converters, 24
Type 1 Chebyshev filter, 173-174
Type 2 Chebyshev filter, 173
Type 5MC Metallized Polycarbonate Capacitor, 376
Type 5250 and 6000-101K chokes, 376
Type EXCEL leaded ferrite bead EMI filter, and type EXC L leadless
  ferrite ead, 376
Type HFQ Aluminum Electrolytic Capacitor and type V Stacked Polyester
  Film Capacitor, 376

U

Undersampling, 20-21
  diagram, 20
  harmonic sampling, 21
  Nyquist zone, 20
Unipolar converter, 23-24
Unsigned fractional format, DSP arithmetic, 212
Unsigned integer format, DSP arithmetic, 212
Using the ADSP-2100 Family, 244, 271

V

Vary, P., 316
VCX device, 329
74VCX164245 Data Sheet, 334
Vector control, 306-307
Very Large Scale Integration, VLSI, 6
Very long instruction word, in SHARC architecture, 226
Video raster scan display system, 5
Virtual-IF transmitter, in AD6523, 297
VisualDSP++, DSP development software, 241
VisualDSP:
  DSP development software, 241
  test drive, 243
Viterbi algorithm, 280
Viterbi decoder, 279
Voice activity detector, 294
Voice over the Internet protocol, 231
Voice-over-Internet-provider *see* VOIP
Voltage:

identification pin, 321
low, interfaces, 321-334
supply, reduction, 321
Voltage compliance, 325-326
definition, 326
internally created, 328-330
considerations, 329-330
Voltage tolerance, 325-326
definition, 325-326
internally created, 328-330
Voltage-controlled-oscillator, 110
Von Neumann architecture, 196

W

Waldhauer, F.D., 56, 89, 94
Waurin, Ken, 191
Weaver, Lindsay A., 115
Weeks, Pat, 57
Welland, D.R., 93
Wepman, Jeffery, 316
Wideband, 301
Widrow, B., 186
Window function, 140, 141, 162
characteristics, 142
frequency responses, 142
Windowed-sinc method:
FIR filter design, 162
responses, diagrams, 162
Witte, Robert A., 57
Wooley, Bruce, 93
Worst harmonic, 37-38
Write, DAC, 255

X

xDSL protocol, 225

Analog Devices Parts Index

AD260/AD261, 356-358
AD820, 68-69
AD974, 65
AD2S80A, 307-308
AD2S82A, 307-308
AD2S83A, 307-308
AD2S90A, 307-308
AD185x, 109
AD189x, 181
AD1819B, 7, 311
AD1836, 311-312
AD1852, 77
AD1853, 77, 110
AD1854, 77
AD1877, 76-77
AD1890, 181
AD1891, 181
AD1892, 181
AD1893, 181
AD1896, 181
AD5322, 263-265
AD5340, 256-258
AD6521, 295-296
AD6522, 295-296
AD6523, 296-297, 299
AD6524, 296-297, 299
AD6600, 304
AD6622, 304-305
AD6624, 304-305
AD6640, 43-44, 84-85
AD6644, 304-305
AD7472, 65
AD77xx, 77
AD773x, 7
AD977x, 104
AD983x, 114
AD984x, 7
AD985x, 104
AD7660, 65
AD7664, 65
AD7722, 335
AD7725, 79, 148, 313-315
AD7730, 335
AD7731, 335
AD7853/7853L, 261-264
AD7853L, 261-263
AD7854/AD7854L, 250-253
AD7856/67, 65
AD7858/59, 65, 68-69
AD7861, 307
AD7862, 307
AD7863, 307
AD7864, 307
AD7865, 307

Index

AD7887/88, 65
AD7891, 65
AD7892, 335
AD9042, 41, 45-46
AD9201, 268-269
AD9220, 38-39, 86
AD9221, 86
AD9223, 86
AD9288-100, 92
AD9410, 83
AD9761, 268-269
AD9772, 53, 104, 107-108, 304-305
AD9814, 7
AD9815, 7
AD9850, 113-114
AD73311, 247
AD73322, 7, 247, 265-268, 310
AD73422, 267-268
AD74222-80, 267-268
AD20msp430, 295-296
AD20msp910, 286, 289
AD20msp918, 289
ADMC200/ADMC201, 307
ADMC300, 308
ADMC326, 308
ADMC328, 308
ADMC331, 308
ADMC401, 308
ADMCF5xx, 309
ADMCF326, 308
ADMCF328, 308
ADP1148, 366
ADP3310, 367, 369-370
ADP33xx, 296
ADP34xx, 296
ADSP-21ESP202, 310
ADSP-21mod870, 282-284
ADSP-21mod970, 284
ADSP-21mod980, 284
ADSP-21modxxx, 283
ADSP-21msp5x, 211
ADSP-21xx, 8-10, 157-159, 174, 192, 194, 196-213, 247-249, 253-255, 258-
  260, 279, 282, 307, 329

ADSP-210x, 211
ADSP-216x, 211
ADSP-217x, 208, 308-309
ADSP-218x, 197, 201, 205, 208, 212, 239, 241-243, 266-268, 279,
  282, 284-285, 296, 310-311
ADSP-218xL/M, 321
ADSP-219x, 197, 201, 209-211, 212, 241-243, 309
ADSP-2100, 199, 205
ADSP-2105, 206
ADSP-2106x, 215-219
ADSP-2111, 211
ADSP-2116x, 220-225
ADSP-2181/3, 211
ADSP-2183, 289
ADSP-2184/L, 211
ADSP-2185/L/M, 211
ADSP-2185L/86L, 267-268
ADSP-2186/L, 211
ADSP-2187L/M, 211
ADSP-2188M, 210, 211, 284-285
ADSP-2189M, 136, 147, 158, 166, 168, 174, 211, 236, 248-253, 255-
  258, 260-261, 269-270
ADSP-21000, 215-216
ADSP-21060, 218-219, 222
ADSP-21060L, 378
ADSP-21061, 218-219, 222
ADSP-21062, 218-219, 222
ADSP-21065, 218-219, 222
ADSP-21065L, 216, 223, 238, 268-269, 311-312
ADSP-21160, 136, 216, 220-221, 223, 225, 312-313, 349-351, 374-375
ADSP-21160M, 222
ADSP-21161N, 222
ADSP-TS001 TigerSHARC, 136-137, 225-233, 304-305
ADuM1100A, 356-357
ADuM1100B, 356-357
EZ-ICE, 238
EZ-KIT Lite, 234-238
EZ-LAB, 234